Genetic Algorithms and Genetic Programming

Modern Concepts and Practical Applications

Numerical Insights

The Numerical Insights series aims to show how numerical simulations provide valuable insights into the mechanisms and processes involved in a wide range of disciplines. Such simulations provide a way of assessing theories by comparing simulations with observations. These models are also powerful tools which serve to indicate where both theory and experiment can be improved.

In most cases the books will be accompanied by software on disk demonstrating working examples of the simulations described in the text.

The editors will welcome proposals using modelling, simulation and systems analysis techniques in the following disciplines: physical sciences; engineering; environment; ecology; biosciences; economics.

Volume 1
Numerical Insights into Dynamic Systems: Interactive Dynamic System Simulation with Microsoft® Windows™ and NT™
Granino A. Korn

Volume 2
Modelling, Simulation and Control of Non-Linear Dynamical Systems: An Intelligent Approach using Soft Computing and Fractal Theory
Patricia Melin and Oscar Castillo

Volume 3
Principles of Mathematical Modeling: Ideas, Methods, Examples
A.A. Samarskii and A. P. Mikhailov

Volume 4
Practical Fourier Analysis for Multigrid Methods
Roman Wienands and Wolfgang Joppich

Volume 5
Effective Computational Methods for Wave Propagation
Nikolaos A. Kampanis, Vassilios A. Dougalis, and John A. Ekaterinaris

Volume 6
Genetic Algorithms and Genetic Programming: Modern Concepts and Practical Applications
Michael Affenzeller, Stephan Winkler, Stefan Wagner, and Andreas Beham

Genetic Algorithms and Genetic Programming

Modern Concepts and Practical Applications

Michael Affenzeller, Stephan Winkler,
Stefan Wagner, and Andreas Beham

CRC Press
Taylor & Francis Group
Boca Raton London New York

CRC Press is an imprint of the
Taylor & Francis Group, an **informa** business

A CHAPMAN & HALL BOOK

CRC Press
Taylor & Francis Group
6000 Broken Sound Parkway NW, Suite 300
Boca Raton, FL 33487-2742

First issued in paperback 2017

© 2009 by Taylor & Francis Group, LLC
CRC Press is an imprint of Taylor & Francis Group, an Informa business

ISBN-13: 978-1-58488-629-7 (hbk)
ISBN-13: 978-1-138-11427-2 (pbk)

DOI: 10.1201/9781420011326

Library of Congress Cataloging-in-Publication Data

Genetic algorithms and genetic programming : modern concepts and practical
   applications / Michael Affenzeller ... [et al.].
      p. cm. -- (Numerical insights ; v. 6)
   Includes bibliographical references and index.
   ISBN 978-1-58488-629-7 (hardcover : alk. paper)
      1. Algorithms. 2. Combinatorial optimization. 3. Programming (Mathematics)
   4. Evolutionary computation. I. Affenzeller, Michael. II. Title. III. Series.

QA9.58.G46 2009
006.3'1--dc22   2009003656

Visit the Taylor & Francis Web site at
http://www.taylorandfrancis.com

and the CRC Press Web site at
http://www.crcpress.com

Contents

List of Tables   xi

List of Figures   xv

List of Algorithms   xxiii

Introduction   xxv

1 Simulating Evolution: Basics about Genetic Algorithms   1
   1.1 The Evolution of Evolutionary Computation   1
   1.2 The Basics of Genetic Algorithms   2
   1.3 Biological Terminology   3
   1.4 Genetic Operators 6
        1.4.1 Models for Parent Selection 6
        1.4.2 Recombination (Crossover)   7
        1.4.3 Mutation   9
        1.4.4 Replacement Schemes 9
   1.5 Problem Representation 10
        1.5.1 Binary Representation 11
        1.5.2 Adjacency Representation 12
        1.5.3 Path Representation 13
        1.5.4 Other Representations for Combinatorial Optimization
               Problems 13
        1.5.5 Problem Representations for Real-Valued Encoding . . 14
   1.6 GA Theory: Schemata and Building Blocks 14
   1.7 Parallel Genetic Algorithms 17
        1.7.1 Global Parallelization 18
        1.7.2 Coarse-Grained Parallel GAs 19
        1.7.3 Fine-Grained Parallel GAs 20
        1.7.4 Migration 21
   1.8 The Interplay of Genetic Operators 22
   1.9 Bibliographic Remarks 23

2 Evolving Programs: Genetic Programming   25
   2.1 Introduction: Main Ideas and Historical Background 26
   2.2 Chromosome Representation 28
        2.2.1 Hierarchical Labeled Structure Trees 28

2.2.2 Automatically Defined Functions and Modular Genetic Programming 35
2.2.3 Other Representations 36
2.3 Basic Steps of the GP-Based Problem Solving Process 37
2.3.1 Preparatory Steps 37
2.3.2 Initialization 39
2.3.3 Breeding Populations of Programs 39
2.3.4 Process Termination and Results Designation 41
2.4 Typical Applications of Genetic Programming 43
2.4.1 Automated Learning of Multiplexer Functions 43
2.4.2 The Artificial Ant 44
2.4.3 Symbolic Regression 46
2.4.4 Other GP Applications 49
2.5 GP Schema Theories 50
2.5.1 Program Component GP Schemata 51
2.5.2 Rooted Tree GP Schema Theories 52
2.5.3 Exact GP Schema Theory 54
2.5.4 Summary 59
2.6 Current GP Challenges and Research Areas 59
2.7 Conclusion 62
2.8 Bibliographic Remarks 62

3 Problems and Success Factors **65**
3.1 What Makes GAs and GP Unique among Intelligent Optimization Methods? 65
3.2 Stagnation and Premature Convergence 66

4 Preservation of Relevant Building Blocks **69**
4.1 What Can Extended Selection Concepts Do to Avoid Premature Convergence? 69
4.2 Offspring Selection (OS) 70
4.3 The Relevant Alleles Preserving Genetic Algorithm (RAPGA) 73
4.4 Consequences Arising out of Offspring Selection and RAPGA 76

5 SASEGASA – More than the Sum of All Parts **79**
5.1 The Interplay of Distributed Search and Systematic Recovery of Essential Genetic Information 80
5.2 Migration Revisited 81
5.3 SASEGASA: A Novel and Self-Adaptive Parallel Genetic Algorithm 82
5.3.1 The Core Algorithm 83
5.4 Interactions among Genetic Drift, Migration, and Self-Adaptive Selection Pressure 86

6 Analysis of Population Dynamics   **89**

6.1   Parent Analysis 89

6.2   Genetic Diversity 90

     6.2.1   In Single-Population GAs 90

     6.2.2   In Multi-Population GAs 91

     6.2.3   Application Examples 92

7 Characteristics of Offspring Selection and the RAPGA   **97**

7.1   Introduction 97

7.2   Building Block Analysis for Standard GAs 98

7.3   Building Block Analysis for GAs Using Offspring Selection . 103

7.4   Building Block Analysis for the Relevant Alleles Preserving GA (RAPGA) 113

8 Combinatorial Optimization: Route Planning   **121**

8.1   The Traveling Salesman Problem 121

     8.1.1   Problem Statement and Solution Methodology 122

     8.1.2   Review of Approximation Algorithms and Heuristics . 125

     8.1.3   Multiple Traveling Salesman Problems 130

     8.1.4   Genetic Algorithm Approaches 130

8.2   The Capacitated Vehicle Routing Problem 139

     8.2.1   Problem Statement and Solution Methodology 140

     8.2.2   Genetic Algorithm Approaches 147

9 Evolutionary System Identification   **157**

9.1   Data-Based Modeling and System Identification 157

     9.1.1   Basics 157

     9.1.2   An Example 159

     9.1.3   The Basic Steps in System Identification 166

     9.1.4   Data-Based Modeling Using Genetic Programming . . 169

9.2   GP-Based System Identification in HeuristicLab 170

     9.2.1   Introduction 170

     9.2.2   Problem Representation 171

     9.2.3   The Functions and Terminals Basis 173

     9.2.4   Solution Representation 178

     9.2.5   Solution Evaluation 182

9.3   Local Adaption Embedded in Global Optimization 188

     9.3.1   Parameter Optimization 189

     9.3.2   Pruning 192

9.4   Similarity Measures for Solution Candidates 197

     9.4.1   Evaluation-Based Similarity Measures 199

     9.4.2   Structural Similarity Measures 201

10 Applications of Genetic Algorithms: Combinatorial Optimization **207**
  10.1 The Traveling Salesman Problem 208
    10.1.1 Performance Increase of Results of Different Crossover Operators by Means of Offspring Selection 208
    10.1.2 Scalability of Global Solution Quality by SASEGASA 210
    10.1.3 Comparison of the SASEGASA to the Island-Model Coarse-Grained Parallel GA 214
    10.1.4 Genetic Diversity Analysis for the Different GA Types 217
  10.2 Capacitated Vehicle Routing 221
    10.2.1 Results Achieved Using Standard Genetic Algorithms 222
    10.2.2 Results Achieved Using Genetic Algorithms with Offspring Selection 226

11 Data-Based Modeling with Genetic Programming **235**
  11.1 Time Series Analysis 235
    11.1.1 Time Series Specific Evaluation 236
    11.1.2 Application Example: Design of Virtual Sensors for Emissions of Diesel Engines 237
  11.2 Classification 251
    11.2.1 Introduction 251
    11.2.2 Real-Valued Classification with Genetic Programming 251
    11.2.3 Analyzing Classifiers 252
    11.2.4 Classification Specific Evaluation in GP 258
    11.2.5 Application Example: Medical Data Analysis 263
  11.3 Genetic Propagation 285
    11.3.1 Test Setup 285
    11.3.2 Test Results 286
    11.3.3 Summary 288
    11.3.4 Additional Tests Using Random Parent Selection . . . 289
  11.4 Single Population Diversity Analysis 292
    11.4.1 GP Test Strategies 292
    11.4.2 Test Results 293
    11.4.3 Conclusion 297
  11.5 Multi-Population Diversity Analysis 300
    11.5.1 GP Test Strategies 300
    11.5.2 Test Results 301
    11.5.3 Discussion 303
  11.6 Code Bloat, Pruning, and Population Diversity 306
    11.6.1 Introduction 306
    11.6.2 Test Strategies 307
    11.6.3 Test Results 309
    11.6.4 Conclusion 318

Conclusion and Outlook **321**

Symbols and Abbreviations **325**

References **327**

Index **359**

List of Tables

7.1   Parameters for test runs using a conventional GA.   99
7.2   Parameters for test runs using a GA with offspring selection. 104
7.3   Parameters for test runs using the relevant alleles preserving
      genetic algorithm.   113

8.1   Exemplary edge map of the parent tours for an ERX operator. 138

9.1   Data-based modeling example: Training data.   160
9.2   Data-based modeling example: Test data.   164

10.1   Overview of algorithm parameters.   209
10.2   Experimental results achieved using a standard GA.   209
10.3   Experimental results achieved using a GA with offspring se-
       lection.   209
10.4   Parameter values used in the test runs of the SASEGASA
       algorithms with single crossover operators as well as with a
       combination of the operators.   211
10.5   Results showing the scaling properties of SASEGASA with
       one crossover operator (OX), with and without mutation. .   211
10.6   Results showing the scaling properties of SASEGASA with
       one crossover operator (ERX), with and without mutation.   212
10.7   Results showing the scaling properties of SASEGASA with
       one crossover operator (MPX), with and without mutation.   212
10.8   Results showing the scaling properties of SASEGASA with a
       combination of crossover operators (OX, ERX, MPX), with
       and without mutation.   213
10.9   Parameter values used in the test runs of a island model GA
       with various operators and various numbers of demes. . . .   215
10.10 Results showing the scaling properties of an island GA with
       one crossover operator (OX) using roulette-wheel selection,
       with and without mutation.   215
10.11 Results showing the scaling properties of an island GA with
       one crossover operator (ERX) using roulette-wheel selection,
       with and without mutation.   216
10.12 Results showing the scaling properties of an island GA with
       one crossover operator (MPX) using roulette-wheel selection,
       with and without mutation.   216

10.13 Parameter values used in the CVRP test runs applying a standard GA.   223

10.14 Results of a GA using roulette-wheel selection, 3-tournament selection and various mutation operators.   226

10.15 Parameter values used in CVRP test runs applying a GA with OS.   228

10.16 Results of a GA with offspring selection and population sizes of 200 and 400 and various mutation operators. The configuration is listed in Table 10.15.   232

10.17 Showing results of a GA with offspring and a population size of 500 and various mutation operators. The configuration is listed in Table 10.15.   234

11.1   Linear correlation of input variables and the target values (NO_x) in the NO_x data set I.   240

11.2   Mean squared errors on training data for the NO_x data set I.   241

11.3   Statistic features of the identification relevant variables in the NO_x data set II.   246

11.4   Linear correlation coefficients of the variables relevant in the NO_x data set II.   248

11.5   Statistic features of the variables in the NOx data set III. .   250

11.6   Linear correlation coefficients of the variables relevant in the NO_x data set III.   250

11.7   Exemplary confusion matrix with three classes   253

11.8   Exemplary confusion matrix with two classes   254

11.9   Set of function and terminal definitions for enhanced GP-based classification.   264

11.10 Experimental results for the *Thyroid* data set.   270

11.11 Summary of the best GP parameter settings for solving classification problems.   271

11.12 Summary of training and test results for the *Wisconsin* data set: Correct classification rates (average values and standard deviation values) for 10-fold CV partitions, produced by GP with offspring selection.   279

11.13 Comparison of machine learning methods: Average test accuracy of classifiers for the *Wisconsin* data set.   280

11.14 Confusion matrices for average classification results produced by GP with OS for the *Melanoma* data set.   280

11.15 Comparison of machine learning methods: Average test accuracy of classifiers for the *Melanoma* data set.   281

11.16 Summary of training and test results for the *Thyroid* data set: Correct classification rates (average values and standard deviation values) for 10-fold CV partitions, produced by GP with offspring selection.   282

11.17 Comparison of machine learning methods: Average test accuracy of classifiers for the *Thyroid* data set. 283

11.18 GP test strategies. 285

11.19 Test results. 286

11.20 Average overall genetic propagation of population partitions. 287

11.21 Additional test strategies for genetic propagation tests. . . . 289

11.22 Test results in additional genetic propagation tests (using random parent selection). 290

11.23 Average overall genetic propagation of population partitions for random parent selection tests. 290

11.24 GP test strategies. 293

11.25 Test results: Solution qualities. 294

11.26 Test results: Population diversity (average similarity values; avg., std.). 295

11.27 Test results: Population diversity (maximum similarity values; avg., std.). 296

11.28 GP test strategies. 302

11.29 Multi-population diversity test results of the GP test runs using the *Thyroid* data set. 303

11.30 Multi-population diversity test results of the GP test runs using the NO_x data set III. 304

11.31 GP parameters used for code growth and bloat prevention tests. 307

11.32 Summary of the code growth prevention strategies applied in these test series. 308

11.33 Performance of systematic and ES-based pruning strategies. 310

11.34 Formula size progress in test series (d). 311

11.35 Quality of results produced in test series (d). 311

11.36 Formula size and population diversity progress in test series (e). 312

11.37 Formula size and population diversity progress in test series (f). 313

11.38 Quality of results produced in test series (f). 313

11.39 Formula size and population diversity progress in test series (g). 314

11.40 Quality of results produced in test series (g). 314

11.41 Formula size and population diversity progress in test series (h). 315

11.42 Quality of results produced in test series (h). 316

11.43 Comparison of best models on training and validation data (b_t and b_v, respectively). 317

11.44 Formula size and population diversity progress in test series (i). 320

11.45 Quality of results produced in test series (i). 320

List of Figures

1.1   The canonical genetic algorithm with binary solution encod-
      ing.   4
1.2   Schematic display of a single point crossover.   8
1.3   Global parallelization concepts: A panmictic population struc-
      ture (shown in left picture) and the corresponding master–
      slave model (right picture).   18
1.4   Population structure of a coarse-grained parallel GA.   19
1.5   Population structure of a fine-grained parallel GA; the special
      case of a cellular model is shown here.   20

2.1   Exemplary programs given as rooted, labeled structure trees.   30
2.2   Exemplary evaluation of program (a).   31
2.3   Exemplary evaluation of program (b).   32
2.4   Exemplary crossover of programs (1) and (2) labeled as *par-
      ent1* and *parent2*, respectively. *Child1* and *child2* are possible
      new offspring programs formed out of the genetic material of
      their parents.   34
2.5   Exemplary mutation of a program: The programs *mutant1*,
      mutant2, and *mutant3* are possible mutants of *parent*. . . .   35
2.6   Intron-augmented representation of an exemplary program in
      PDGP [Pol99b].   38
2.7   Major preparatory steps of the basic GP process.   38
2.8   The genetic programming cycle [LP02].   40
2.9   The GP-based problem solving process.   41
2.10 GA and GP flowcharts: The conventional genetic algorithm
      and genetic programming.   42
2.11 The Boolean multiplexer with three address bits; (a) general
      black box model, (b) addressing data bit d_5.   44
2.12 A correct solution to the 3-address Boolean multiplexer prob-
      lem [Koz92b].   44
2.13 The Santa Fe trail.   45
2.14 A Santa Fe trail solution. The black points represent nodes
      referencing to the `Prog3` function.   46
2.15 A symbolic regression example.   48
2.16 Exemplary formulas.   49
2.17 Programs matching Koza's schema `H=[(+ x 3), y]`.   51

2.18   The rooted tree GP schema $*(=,= (x,=))$ and three exemplary programs of the schema's semantics.   53

2.19   The GP schema H $= +(*(=,x),=)$ and exemplary u and l schemata. Cross bars indicate crossover points; shaded regions show the parts of H that are replaced by "don't care" symbols.   56

2.20   The GP hyperschema $*(\#,= (x,=))$ and three exemplary programs that are a part of the schema's semantics.   56

2.21   The GP schema $H = +(*(=,x),=)$ and exemplary U and L hyperschema building blocks. Cross bars indicate crossover points; shaded regions show the parts of H that are modified.   57

2.22   Relation between approximate and exact schema theorems for different representations and different forms of crossover (in the absence of mutation).   58

2.23   Examples for bloat.   60

4.1   Flowchart of the embedding of offspring selection into a genetic algorithm.   71

4.2   Graphical representation of the gene pool available at a certain generation. Each bar represents a chromosome with its alleles representing the assignment of the genes at the certain loci.   74

4.3   The left part of the figure represents the gene pool at generation i and the right part indicates the possible size of generation $i + 1$ which must not go below a minimum size and also not exceed an upper limit. These parameters have to be defined by the user.   74

4.4   Typical development of actual population size between the two borders (lower and upper limit of population size) displaying also the identical chromosomes that occur especially in the last iterations.   76

5.1   Flowchart of the reunification of subpopulations of a SASEGASA (light shaded subpopulations are still evolving, whereas dark shaded ones have already converged prematurely).   84

5.2   Quality progress of a typical run of the SASEGASA algorithm.   85

5.3   Selection pressure curves for a typical run of the SASEGASA algorithm.   86

5.4   Flowchart showing the main steps of the SASEGASA.   87

6.1   Similarity of solutions in the population of a standard GA after 20 and 200 iterations, shown in the left and the right charts, respectively.   93

6.2 Histograms of the similarities of solutions in the population of a standard GA after 20 and 200 iterations, shown in the left and the right charts, respectively. 94

6.3 Average similarities of solutions in the population of a standard GA over for the first 2,000 and 10,000 iterations, shown in the upper and lower charts, respectively. 95

6.4 Multi-population specific similarities of the solutions of a parallel GA's populations after 5,000 generations. 96

6.5 Progress of the average multi-population specific similarity values of a parallel GA's solutions, shown for 10,000 generations. 96

7.1 Quality progress for a standard GA with OX crossover for mutation rates of 0%, 5%, and 10%. 99

7.2 Quality progress for a standard GA with ERX crossover for mutation rates of 0%, 5%, and 10%. 101

7.3 Quality progress for a standard GA with MPX crossover for mutation rates of 0%, 5%, and 10%. 102

7.4 Distribution of the alleles of the global optimal solution over the run of a standard GA using OX crossover and a mutation rate of 5% (remaining parameters are set according to Table 7.1). 103

7.5 Quality progress for a GA with offspring selection, OX, and a mutation rate of 5%. 105

7.6 Quality progress for a GA with offspring selection, MPX, and a mutation rate of 5%. 106

7.7 Quality progress for a GA with offspring selection, ERX, and a mutation rate of 5%. 107

7.8 Quality progress for a GA with offspring selection, ERX, and no mutation. 108

7.9 Quality progress for a GA with offspring selection using a combination of OX, ERX, and MPX, and a mutation rate of 5%. 109

7.10 Success progress of the different crossover operators OX, ERX, and MPX, and a mutation rate of 5%. The plotted graphs represent the ratio of successfully produced children to the population size over the generations. 110

7.11 Distribution of the alleles of the global optimal solution over the run of an offspring selection GA using ERX crossover and a mutation rate of 5% (remaining parameters are set according to Table 7.2). 111

7.12 Distribution of the alleles of the global optimal solution over the run of an offspring selection GA using ERX crossover and no mutation (remaining parameters are set according to Table 7.2). 112

7.13   Quality progress for a relevant alleles preserving GA with OX and a mutation rate of 5%. 114

7.14   Quality progress for a relevant alleles preserving GA with MPX and a mutation rate of 5%. 115

7.15   Quality progress for a relevant alleles preserving GA with ERX and a mutation rate of 5%. 115

7.16   Quality progress for a relevant alleles preserving GA using a combination of OX, ERX, and MPX, and a mutation rate of 5%. 116

7.17   Quality progress for a relevant alleles preserving GA using a combination of OX, ERX, and MPX, and mutation switched off. 116

7.18   Distribution of the alleles of the global optimal solution over the run of a relevant alleles preserving GA using a combination of OX, ERX, and MPX, and a mutation rate of 5% (remaining parameters are set according to Table 7.3). . . . 118

7.19   Distribution of the alleles of the global optimal solution over the run of a relevant alleles preserving GA using a combination of OX, ERX, and MPX without mutation (remaining are set parameters according to Table 7.3). 119

8.1   Exemplary nearest neighbor solution for a 51-city TSP instance ([CE69]). 126

8.2   Example of a 2-change for a TSP instance with 7 cities. . . 128

8.3   Example of a 3-change for a TSP instance with 11 cities. . . 129

8.4   Example for a partially matched crossover. 134

8.5   Example for an order crossover. 135

8.6   Example for a cyclic crossover. 136

8.7   Exemplary result of the sweep heuristic for a small CVRP. . 144

8.8   Exemplary sequence-based crossover. 149

8.9   Exemplary route-based crossover. 151

8.10   Exemplary relocate mutation. 152

8.11   Exemplary exchange mutation. 152

8.12   Example for a 2-opt mutation for the VRP. 153

8.13   Example for a 2-opt* mutation for the VRP. 153

8.14   Example for an or-opt mutation for the VRP. 154

9.1   Data-based modeling example: Training data. 160

9.2   Data-based modeling example: Evaluation of an optimally fit linear model. 161

9.3   Data-based modeling example: Evaluation of an optimally fit cubic model. 162

9.4   Data-based modeling example: Evaluation of an optimally fit polynomial model ($n = 10$). 162

9.5   Data-based modeling example: Evaluation of an optimally fit polynomial model ($n = 20$). 163

9.6   Data-based modeling example: Evaluation of an optimally fit linear model (evaluated on training and test data). 163

9.7   Data-based modeling example: Evaluation of an optimally fit cubic model (evaluated on training and test data). 164

9.8   Data-based modeling example: Evaluation of an optimally fit polynomial model ($n = 10$) (evaluated on training and test data). 165

9.9   Data-based modeling example: Summary of training and test errors for varying numbers of parameters n. 165

9.10 The basic steps of system identification. 167

9.11 The basic steps of GP-based system identification. 170

9.12 Structure tree representation of a formula. 179

9.13 Structure tree crossover and the functions basis. 181

9.14 Simple examples for pruning in GP. 195

9.15 Simple formula structure and all included pairs of ancestors and descendants (genetic information items). 202

10.1 Quality improvement using offspring selection and various crossover operators. 210

10.2 Degree of similarity/distance for all pairs of solutions in a SGA's population of 120 solution candidates after 10 generations. 218

10.3 Genetic diversity in the population of a conventional GA over time. 219

10.4 Genetic diversity of the population of a GA with offspring selection over time. 219

10.5 Genetic diversity of the entire population over time for a SASEGASA with 5 subpopulations. 220

10.6 Quality progress of a standard GA using roulette wheel selection on the left and 3-tournament selection the right side, applied to instances of the Taillard CVRP benchmark: tai75a (top) and tai75b (bottom). 223

10.7 Genetic diversity in the population of a GA with roulette wheel selection (shown on the left side) and 3-tournament selection (shown on the right side). 225

10.8 Box plots of the qualities produced by a GA with roulette and 3-tournament selection, applied to the problem instances tai75a (top) and tai75b (bottom). 227

10.9 Quality progress of the offspring selection GA for the instances (from top to bottom) tai75a and tai75b. The left column shows the progress with a population size of 200, while in the right column the GA with offspring selection uses a population size of 400. 229

10.10 Influence of the crossover operators SBX and RBX on each generation of an offspring selection algorithm. The lighter line represents the RBX; the darker line represents the SBX.   230

10.11 Genetic diversity in the population of an GA with offspring selection and a population size of 200 on the left and 400 on the right for the problem instances tai75a and tai75b (from top to bottom).   231

10.12 Box plots of the offspring selection GA with a population size of 200 and 400 for the instances tai75a and tai75b.   233

10.13 Box plots of the GA with 3-tournament selection against the offspring selection GA for the instances tai75a (shown in the upper part) and tai75b (shown in the lower part).   233

11.1 Dynamic diesel engine test bench at the Institute for Design and Control of Mechatronical Systems, JKU Linz.   238

11.2 Evaluation of the best model produced by GP for test strategy (1).   241

11.3 Evaluation of the best model produced by GP for test strategy (2).   242

11.4 Evaluation of models for particulate matter emissions of a diesel engine (snapshot showing the evaluation of the model on validation / test samples).   244

11.5 Errors distribution of models for particulate matter emissions. 244

11.6 Cumulative errors of models for particulate matter emissions. 245

11.7 Target NO_x values of NO_x data set II, recorded over approximately 30 minutes at 20Hz recording frequency yielding $\sim$36,000 samples.   247

11.8 Target $HoribaNOx$ values of NO_x data set III.   248

11.9 Target $HoribaNOx$ values of NO_x data set III, samples 6000 – 7000.   249

11.10 Two exemplary ROC curves and their area under the ROC curve (AUC).   255

11.11 An exemplary graphical display of a multi-class ROC (MROC) matrix.   257

11.12 Classification example: Several samples with original class values C_1, C_2, and C_3 are shown; the class ranges result from the estimated values for each class and are indicated as cr_1, cr_2, and cr_3.   261

11.13 An exemplary hybrid structure tree of a combined formula including arithmetic as well as logical functions.   265

11.14 Graphical representation of the best result we obtained for the *Thyroid* data set, CV-partition 9: Comparison of original and estimated class values.   272

11.15 ROC curves and their area under the curve (AUC) values for classification models generated for *Thyroid* data, CV-set 9.   273

11.16 MROC charts and their maximum and average area under the curve (AUC) values for classification models generated for *Thyroid* data, CV-set 9. 274

11.17 Graphical representation of a classification model (formula), produced for 10-fold cross validation partition 3 of the *Thyroid* data set. 275

11.18 pc_{total} values for an exemplary run of series I. 287

11.19 pc_{total} values for an exemplary run of series II. 287

11.20 pc_{total} values for an exemplary run of series III. 288

11.21 Selection pressure progress in two exemplary runs of test series III and V (extended GP with gender specific parent selection and strict offspring selection). 291

11.22 Distribution of similarity values in an exemplary run of NO_x test series A, generation 200. 297

11.23 Distribution of similarity values in an exemplary run of NO_x test series A, generation 4000. 298

11.24 Distribution of similarity values in an exemplary run of NO_x test series (D), generation 20. 298

11.25 Distribution of similarity values in an exemplary run of NO_x test series (D), generation 95. 299

11.26 Population diversity progress in exemplary *Thyroid* test runs of series (A) and (D) (shown in the upper and lower graph, respectively). 299

11.27 Exemplary multi-population diversity of a test run of *Thyroid* series F at iteration 50, grayscale representation. 305

11.28 Code growth in GP without applying size limits or complexity punishment strategies (left: standard GP, right: extended GP). 310

11.29 Progress of formula complexity in one of the test runs of series (1g), shown for the first ~400 iterations. 315

11.30 Progress of formula complexity in one of the test runs of series (1h) (shown left) and one of series (2h) (shown right). ... 316

11.31 Model with best fit on training data: Model structure and full evaluation. 318

11.32 Model with best fit on validation data: Model structure and full evaluation. 318

11.33 Errors distributions of best models: Charts I, II, and III show the errors distributions of the model with best fit on training data evaluated on training, validation, and test data, respectively; charts IV, V, and VI show the errors distributions of the model with best fit on validation data evaluated on training, validation, and test data, respectively. 319

11.34 A simple workbench in HeuristicLab 2.0. 323

List of Algorithms

1.1   Basic workflow of a genetic algorithm.   3
4.1   Definition of a genetic algorithm with offspring selection. . . .   72
9.1   Exhaustive pruning of a model m using the parameters h_1, h_2, $minimizeModel$, cp_{max}, and det_{max}.   196
9.2   Evolution strategy inspired pruning of a model m using the parameters λ, $maxUnsuccRounds$, h_1, h_2, $minimizeModel$, cp_{max}, and det_{max}.   198
9.3   Calculation of the evaluation-based similarity of two models m_1 and m_2 with respect to data base $data$ 200
9.4   Calculation of the structural similarity of two models m_1 and m_2   205

Introduction

DOI: 10.1201/9781420011326-1

Essentially, this book is about algorithmic developments in the context of genetic algorithms (GAs) and genetic programming (GP); we also describe their applications to significant combinatorial optimization problems as well as structure identification using HeuristicLab as a platform for algorithm development. The main issue of the theoretical considerations is to obtain a better understanding of the basic workflow of GAs and GP, in order to establish new bionic, problem independent theoretical concepts and to substantially increase the achievable solution quality.

The book is structured into a theoretical and an empirical part. The aim of the theoretical part is to describe the important and characteristic properties of the basic genetic algorithm as well as the main characteristics of the algorithmic extensions introduced here. The empirical part of the book elaborates two case studies: On the one hand, the traveling salesman problem (TSP) and the capacitated vehicle routing problem (CVRP) are used as representatives for GAs applied to combinatorial optimization problems. On the other hand, GP-based nonlinear structure identification applied to time series and classification problems is analyzed to highlight the properties of the algorithmic measures in the field of genetic programming. The borderlines between theory and practice become indistinct in some parts as it is also necessary to describe theoretical properties on the basis of practical examples in the first part of the book. For this purpose we go back to some small-dimensional TSP instances that are perfectly suited for theoretical GA considerations.

Research concerning the self-adaptive interplay between selection and the applied solution manipulation operators (crossover and mutation) is the basis for the algorithmic developments presented in this book. The ultimate goal in this context is to avoid the disappearance of relevant building blocks and to support the combination of those alleles from the gene pool that carry solution properties of highly fit individuals. As we show in comparative test series, in conventional GAs and GP this *relevant genetic information* is likely to get lost quite early in the standard variants of these algorithms and can only be reintroduced into the population's gene pool by mutation. This dependence on mutation can be drastically reduced by new generic selection principles based upon either self-adaptive selection pressure steering (*offspring selection, OS*) or self-adaptive population size adjustment as proposed in the *relevant alleles preserving genetic algorithm (RAPGA)*. Both algorithmic extensions certify the survival of essential genetic information by supporting the *survival of rel-*

evant alleles rather than the survival of above average chromosomes. This is achieved by defining the survival probability of a new child chromosome depending on the child's fitness in comparison to the fitness values of its own parents. With these measures it becomes possible to channel the relevant alleles, which are initially scattered in the entire population, to single chromosomes at the end of the genetic search process.

The SASEGASA algorithm is a special coarse-grained parallel GA; the acronym "SASEGASA" hereby stands for *S*elf-*A*daptive *Se*gregative *G*enetic *A*lgorithm including aspects of *S*imulated *A*nnealing. SASEGASA combines offspring selection with enhanced parallelization concepts in order to avoid premature convergence, one of the major problems with GAs. As we will show for the TSP, it becomes possible to scale the robustness and particularly the achievable solution quality by the number of subpopulations.

Due to the high focus on sexual recombination, evolution strategies (ES) are not considered explicitly in this book. Nevertheless, many of the theoretical considerations are heavily inspired by evolution strategies, especially the aspect of selection after reproduction and (self-)adaptive selection pressure steering. Aside from other variants of evolutionary computation, further inspirations are borrowed from fields, as for example, population genetics. The implementation of bionic ideas for algorithmic developments is quite pragmatic and ignores debates on principles that are discussed in natural sciences. Of course, we are always aware of the fact that artificial evolution as performed in an evolutionary algorithm is situated on a high level of abstraction compared to the biological role model in any case.

The problem-oriented part of the book is dedicated to the application of the algorithmic concepts described in this book to benchmark as well as real world problems. Concretely, we examine the traveling salesman problem and the capacitated vehicle routing problem (which is thematically related to the TSP), but more in step with actual practice, as representatives of combinatorial optimization problems.

Time series and classification analysis are used as application areas of data-based structure identification with genetic programming working with formula trees representing mathematical models. As a matter of principle, we use standard problem representations and the appropriate problem-specific genetic operators known from GA and GP theory for the experiments shown in these chapters. The focus is set on the comparison of results achievable with standard GA and GP implementations to the results achieved using the extended algorithmic concepts described in this book. These enhanced concepts do not depend on a concrete problem representation and its operators; their influences on population dynamics in GA and GP populations are analyzed, too.

Additional material related to the research described in this book is provided on the book's homepage at `http://gagp2009.heuristiclab.com`. Among other information this website provides some of the software used as well as dynamical presentations of representative test runs as additional material.

Chapter 1

Simulating Evolution: Basics about Genetic Algorithms

DOI: 10.1201/9781420011326-2

1.1 The Evolution of Evolutionary Computation

Work on what is nowadays called evolutionary computation started in the sixties of the 20th century in the United States and Germany. There have been two basic approaches in computer science that copy evolutionary mechanisms: evolution strategies (ES) and genetic algorithms (GA). Genetic algorithms go back to Holland [Hol75], an American computer scientist and psychologist who developed his theory not only under the aspect of solving optimization problems but also to study self-adaptiveness in biological processes. Essentially, this is the reason why genetic algorithms are much closer to the biological model than evolution strategies. The theoretical foundations of evolution strategies were formed by Rechenberg and Schwefel (see for example [Rec73] or [Sch94]), whose primary goal was optimization. Although these two concepts have many aspects in common, they developed almost independently from each other in the USA (where GAs were developed) and Germany (where research was done on ES).

Both attempts work with a population model whereby the genetic information of each individual of a population is in general different. Among other things this genotype includes a parameter vector which contains all necessary information about the properties of a certain individual. Before the intrinsic evolutionary process takes place, the population is initialized arbitrarily; evolution, i.e., replacement of the old generation by a new generation, proceeds until a certain termination criterion is fulfilled.

The major difference between evolution strategies and genetic algorithms lies in the representation of the genotype and in the way the operators are used (which are mutation, selection, and eventually recombination). In contrast to GAs, where the main role of the mutation operator is simply to avoid stagnation, mutation is the primary operator of evolution strategies.

Genetic programming (GP), an extension of the genetic algorithm, is a domain-independent, biologically inspired method that is able to create computer programs from a high-level problem statement. In fact, virtually all problems in artificial intelligence, machine learning, adaptive systems, and automated learning can be recast as a search for a computer program; genetic programming provides a way to search for a computer program in the space of computer programs (as formulated by Koza in [Koz92a]). Similar to GAs, GP works by imitating aspects of natural evolution, but whereas GAs are intended to find arrays of characters or numbers, the goal of a GP process is to search for computer programs (or, for example, formulas) solving the optimization problem at hand. As in every evolutionary process, new individuals (in GP's case, new programs) are created. They are tested, and the fitter ones in the population succeed in creating children of their own whereas unfit ones tend to disappear from the population.

In the following sections we give a detailed description of the basics of genetic algorithms in Section 1.2, take a look at the corresponding biological terminology in Section 1.3, and characterize the operators used in GAs in Section 1.4. Then, in Section 1.5 we discuss problem representation issues, and in Section 1.6 we summarize the schema theory, an essentially important concept for understanding not only how, but also why GAs work. Parallel GA concepts are given in Section 1.7, and finally we discuss the interplay of genetic operators in Section 1.8.

1.2 The Basics of Genetic Algorithms

Concerning its internal functioning, a genetic algorithm is an iterative procedure which usually operates on a population of constant size and is basically executed in the following way:

An initial population of individuals (also called "solution candidates" or "chromosomes") is generated randomly or heuristically. During each iteration step, also called a "generation," the individuals of the current population are evaluated and assigned a certain fitness value. In order to form a new population, individuals are first selected (usually with a probability proportional to their relative fitness values), and then produce offspring candidates which in turn form the next generation of parents. This ensures that the expected number of times an individual is chosen is approximately proportional to its relative performance in the population. For producing new solution candidates genetic algorithms use two operators, namely crossover and mutation:

- Crossover is the primary genetic operator: It takes two individuals, called parents, and produces one or two new individuals, called offspring, by combining parts of the parents. In its simplest form, the operator works by swapping (exchanging) substrings before and after a randomly selected crossover point.

- The second genetic operator, mutation, is essentially an arbitrary modification which helps to prevent premature convergence by randomly sampling new points in the search space. In the case of bit strings, mutation is applied by simply flipping bits randomly in a string with a certain probability called mutation rate.

Genetic algorithms are stochastic iterative algorithms, which cannot guarantee convergence; termination is hereby commonly triggered by reaching a maximum number of generations or by finding an acceptable solution or more sophisticated termination criteria indicating premature convergence. We will discuss this issue in further detail within Chapter 3.

The so-called standard genetic algorithm (SGA), which represents the basis of almost all variants of genetic algorithms, is given in Algorithm 1.1 (which is formulated as in [Tom95], for example).

Algorithm 1.1 Basic workflow of a genetic algorithm.

Produce an initial population of individuals
Evaluate the fitness of all individuals
while termination condition not met **do**
    Select fitter individuals for reproduction and produce new individuals (crossover and mutation)
    Evaluate fitness of new individuals
    Generate a new population by inserting some new "good" individuals and by erasing some old "bad" individuals
end while

A special and quite restricted GA variant, that has represented the basis for theoretical considerations for a long period of time, is given in Figure 1.1. This chart sketches a GA with binary representation operating with generational replacement, a population of constant size, and the following genetic operators: roulette wheel selection, single point crossover, and bit flip mutation. This special type of genetic algorithms, which is the basis for theoretical GA research such as the well known schema theorem and accordingly the building block hypothesis, is also called the canonical genetic algorithm (CGA).

1.3   Biological Terminology

The approximative way of solving optimization problems by genetic algorithms holds a strong analogy to the basic principles of biological evolution. The fundamentals of the natural evolution theory, as it is considered nowadays, mainly refer to the theories of Charles Darwin, which were published

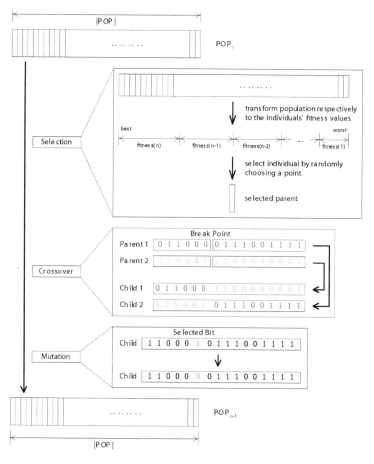

FIGURE 1.1: The canonical genetic algorithm with binary solution encoding.

in 1859 in his well-known work "The Origin of Species By Means of Natural Selection or the Preservation of Favoured Races in the Struggle for Life" (revised edition: [Dar98]). In this work Darwin states the following five major ideas:

- Evolution, change in lineages, occurs and occurred over time.

- All creatures have common descent.

- Natural selection determines changes in nature.

- Gradual change, i.e., nature changes somehow successively.

- Speciation, i.e., Darwin claimed that the process of natural selection results in populations diverging enough to become separate species.

Although some of Darwin's proposals were not new, his ideas (particularly those on common descent and natural selection) provided the first solid foundation upon which evolutionary biology has been built.

At this point it may be useful to formally introduce some essential parts of the biological terminology which are used in the context of genetic algorithms:

- All living organisms consist of cells containing the same set of one or more chromosomes, i.e., strings of DNA. A gene can be understood as an "encoder" of a characteristic, such as eye color. The different possibilities for a characteristic (e.g., brown, green, blue, gray) are called alleles. Each gene is located at a particular position (locus) on the chromosome.

- Most organisms have multiple chromosomes in each cell. The sum of all chromosomes, i.e., the complete collection of genetic material, is called the genome of the organism and the term genotype refers to the particular set of genes contained in a genome. Therefore, if two individuals have identical genomes, they are said to have the same genotype.

- Organisms whose chromosomes are arranged in pairs are called diploid, whereas organisms with unpaired chromosomes are called haploid. In nature, most sexually reproducing species are diploid. Humans for instance have 23 pairs of chromosomes in each somatic cell in their body. Recombination (crossover) occurs during sexual reproduction in the following way:

- For producing a new child, the genes of the parents are combined to eventually form a new diploid set of chromosomes. Offspring are subject to mutation where elementary parts of the DNA (nucleotides) are changed. The fitness of an organism (individual) is typically defined as its probability to reproduce, or as a function of the number of offspring the organism has produced.

For the sake of simplification, in genetic algorithms the term chromosome refers to a solution candidate (in the first GAs encoded as a bit). The genes are either single bits or small blocks of neighboring bits that encode a particular element of the solution. Even if an allele usually is either 0 or 1, for larger alphabets more alleles are possible at each locus.

As a further simplification to the biological role model, crossover typically operates by exchanging genetic material between two haploid parents whereas mutation is implemented by simply flipping the bit at a randomly chosen locus.

Finally it is remarkable that most applications of genetic algorithms employ haploid single-chromosome individuals, although the evolution of mankind has inspired the GA-community at most. This is most probably due to the easier and more effective representation and implementation of single-chromosome individuals.

1.4   Genetic Operators

In the following, the main genetic operators, namely parent selection, crossover, mutation, and replacement are to be described. The focus hereby lies on a functional description of the principles rather than to give a complete overview of operator concepts; for more details about genetic operators the interested reader is referred to textbooks as for example [DLJD00].

1.4.1   Models for Parent Selection

In genetic algorithms a fitness function assigns a score to each individual in a population; this fitness value indicates the quality of the solution represented by the individual. The fitness function is often given as part of the problem description or based on the objective function; developing an appropriate fitness function may also involve the use of simulation, heuristic techniques, or the knowledge of an expert. Evaluating the fitness function for each individual should be relatively fast due to the number of times it will be invoked. If the evaluation is likely to be slow, then concepts of parallel and distributed computing, an approximate function evaluation technique, or a technique, that only considers elements that have changed, may be employed.

Once a population has been generated and its fitness has been measured, the set of solutions, that are selected to be "mated" in a given generation, is produced. In the standard genetic algorithm (SGA) the probability, that a chromosome of the current population is selected for reproduction, is proportional to its fitness.

In fact, there are many ways of accomplishing this selection. These include:

- Proportional selection (roulette wheel selection):
  The classical SGA utilizes this selection method which has been proposed in the context of Holland's schema theorem (which will be explained in detail in Section 1.6). Here the expected number of descendants for an individual i is given as $p_i = \frac{f_i}{\bar{f}}$ with $f : \mathcal{S} \to \mathbb{R}^+$ denoting the fitness function and $\bar{f}$ representing the average fitness of all individuals. Therefore, each individual of the population is represented by a space proportional to its fitness. By repeatedly spinning the wheel, individuals are chosen using random sampling with replacement. In order to make proportional selection independent from the dimension of the fitness values, so-called windowing techniques are usually employed.

Further variants of proportional selection aim to reduce the dominance of a single or a group of highly fit individuals ("super individuals") by stochastic sampling techniques (as for example explained in [DLJD00]).

- Linear-rank selection:
  In the context of linear-rank selection the individuals of the population are ordered according to their fitness and copies are assigned in such a way that the best individual receives a pre-determined multiple of the number of copies the worst one receives [GB89]. On the one hand rank selection implicitly reduces the dominating effects of "super individuals" in populations (i.e., individuals that are assigned a significantly better fitness value than all other individuals), but on the other hand it warps the difference between close fitness values, thus increasing the selection pressure in stagnant populations. Even if linear-rank selection has been used with some success, it ignores the information about fitness differences of different individuals and violates the schema theorem.

- Tournament selection:
  There are a number of variants on this theme. The most common one is k-tournament selection where k individuals are selected from a population and the fittest individual of the k selected ones is considered for reproduction. In this variant selection pressure can be scaled quite easily by choosing an appropriate number for k.

1.4.2 Recombination (Crossover)

In its easiest formulation, which is suggested in the canonical GA for binary encoding, crossover takes two individuals and cuts their chromosome strings at some randomly chosen position. The produced substrings are then swapped to produce two new full length chromosomes.
Conventional crossover techniques for binary representation include:

- Single point crossover:
  A single random cut is made, producing two head sections and two tail sections. The two tail sections are then swapped to produce two new individuals (chromosomes); Figure 1.2 schematically sketches this crossover method which is also called one point crossover.

- Multiple point crossover:
  One natural extension of the single point crossover is the multiple point crossover: In a n-point crossover there are n crossover points and substrings are swapped between the n points. According to some researchers, multiple-point crossover is more suitable to combine good features present in strings because it samples uniformly along the full length of a chromosome [Ree95]. At the same time, multiple-point crossover becomes more and more disruptive with an increasing number of crossover

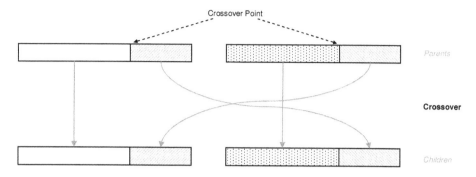

FIGURE 1.2: Schematic display of a single point crossover.

points, i.e., the evolvement of longer building blocks becomes more and more difficult. Decreasing the number of crossover points during the run of the GA may be a good compromise.

- Uniform crossover:
  Given two parents, each gene in the offspring is created by copying the corresponding gene from one of the parents. The selection of the corresponding parent is undertaken via a randomly generated crossover mask: At each index, the offspring gene is taken from the first parent if there is a 1 in the mask at this index, and otherwise (if there is a 0 in the mask at this index) the gene is taken from the second parent. Due to this construction principle uniform crossover does not support the evolvement of higher order building blocks.

The choice of an appropriate crossover operator depends very much on the representation of the search space (see also Section 1.5). Sequencing problems as routing problems for example often require operators different from the ones described above as almost all generated children may be situated outside of the space of valid solutions.

In higher order representations, a variety of real-number combination operators can be employed, such as the average and geometric mean. Domain knowledge can be used to design local improvement operators which sometimes allow more efficient exploration of the search space around good solutions. For instance, knowledge could be used to determine the appropriate locations for crossover points.

As the number of proposed problem-specific crossover-techniques has been growing that much over the years, it would go beyond the scope of the present book even to discuss the more important ones. For a good discussion of crossover-related issues and further references the reader is referred to [Mic92] and [DLJD00].

1.4.3   Mutation

Mutations allow undirected jumps to slightly different areas of the search space. The basic mutation operator for binary coded problems is bitwise mutation. Mutation occurs randomly and very rarely with a probability p_m; typically, this mutation rate is less than ten percent. In some cases mutation is interpreted as generating a new bit and in others it is interpreted as flipping the bit.

In higher order alphabets, such as integer numbering formulations, mutation takes the form of replacing an allele with a randomly chosen value in the appropriate range with probability p_m. However, for combinatorial optimization problems, such mutation schemes can cause difficulties with chromosome legality; for example, multiple copies of a given value can occur which might be illegal for some problems (including routing). Alternatives suggested in literature include pairwise swap and shift operations as for instance described in [Car94].

In addition, adaptive mutation schemes similar to mutation in the context of evolution strategies are worth mentioning. Adaptive mutation schemes vary either the rate, or the form of mutation, or both during a GA run. For instance, mutation is sometimes defined in such a way that the search space is explored uniformly at first and more locally towards the end, in order to do a kind of local improvement of candidate solutions [Mic92].

1.4.4   Replacement Schemes

After having generated a new generation of descendants (offspring) by crossover and mutation, the question arises which of the new candidates should become members of the next generation. In the context of evolution strategies this fact determines the life span of the individuals and substantially influences the convergence behavior of the algorithm. A further strategy influencing replacement quite drastically is offspring selection which will be discussed separately in Chapter 4. The following schemes are possible replacement mechanisms for genetic algorithms:

- Generational Replacement:
  The entire population is replaced by its descendants. Similar to the (μ, λ) evolution strategy it might therefore happen that the fitness of the best individual decreases at some stage of evolution. Additionally, this strategy puts into perspective the dominance of a few individuals which might help to avoid premature convergence [SHF94].

- Elitism:
  The best individual (or the n best individuals, respectively) of the previous generation are retained for the next generation which theoretically allows immortality similar to the $(\mu + \lambda)$ evolution strategy and might

be critical with respect to premature convergence. The special and commonly applied strategy of just retaining one (the best) individual of the last generation is also called the "golden cage model," which is a special case of n-elitism with $n = 1$. If mutation is applied to the elite in order to prevent premature convergence, the replacement mechanism is called "weak elitism."

- Delete-n-last:
  The n weakest individuals are replaced by n descendants. If $n \ll |POP|$ we speak of a steady-state replacement scheme; for $n = 1$ the changes between the old and the new generation are certainly very small and $n = |POP|$ gives the already introduced generational replacement strategy.

- Delete-n:
  In contrast to the delete-n-last replacement strategy, here not the n weakest but rather n arbitrarily chosen individuals of the old generation are replaced, which on the one hand reduces the convergence speed of the algorithm but on the other hand also helps to avoid premature convergence (compare elitism versus weak elitism).

- Tournament Replacement:
  Competitions are run between sets of individuals from the last and the actual generation, with the winners becoming part of the new population.

A detailed description of replacement schemes and their effects can be found for example in [SHF94], [Mic92], [DLJD00], and [Mit96].

1.5 Problem Representation

As already stated before, the first genetic algorithm presented in literature [Hol75] used binary vectors for the representation of solution candidates (chromosomes). Consequently, the first solution manipulation operators (single point crossover, bit mutation) have been developed for binary representation. Furthermore, this very simple GA, also commonly known as the canonical genetic algorithm (CGA), represents the basis for extensive theoretical inspections, resulting in the well known schema theorem and the building block hypothesis ([Hol75], [Gol89]). This background theory will be examined separately in Section 1.6, as it defines the scope of almost any GA as it should ideally be and distinguishes GAs from almost any other heuristic optimization technique.

The unique selling point of GAs is to compile so-called building blocks, i.e., somehow linked parts of the chromosome which become larger as the algorithm proceeds, advantageously with respect to the given fitness function. In other words, one could define the claim of a GA as to be an algorithm which is able to assemble the basic modules of highly fit or even globally optimal solutions (which the algorithm of course does not know about). These basic modules are with some probability already available in the initial population, but widespread over many individuals; the algorithm therefore has to compile these modules in such a clever way that continuously growing sequences of highly qualified alleles, the so-called building blocks, are formed.

Compared to heuristic optimization techniques based on neighborhood search (as tabu search [Glo86] or simulated annealing [KGV83], for example), the methodology of GAs to combine partial solutions (by crossover) is potentially much more robust with respect to getting stuck in local but not global optimal solutions; this tendency of neighborhood-based searches denotes a major drawback of these heuristics. Still, when applying GAs the user has to draw much more attention on the problem representation in order to help the algorithm to fulfill the claim stated above. In that sense the problem representation must allow the solution manipulation operators, especially crossover, to combine alleles of different parent individuals. This is because crossover is responsible for combining the properties of two solution candidates which may be located in very different regions of the search space so that valid new solution candidates are built. This is why the problem representation has to be designed in a way that crossover operators are able to build valid new children (solution candidates) with a genetic make up that consists of the union set of its parent alleles.

Furthermore, as a tribute to the general functioning of GAs, the crossover operators also have to support the potential development of higher-order building blocks (longer allele sequences). Only if the genetic operators for a certain problem representation show these necessary solution manipulator properties, the corresponding GA can be expected to work as it should, i.e., in the sense of a generalized interpretation of the building block hypothesis.

Unfortunately, a lot of more or less established problem representations are not able to fulfill these requirements, as they do not support the design of potentially suited crossover operators. Some problem representations will be considered exemplarily in the following attracting notice to their ability to allow meaningful crossover procedures. Even if mutation, the second solution manipulation concept of GAs, is also of essential importance, the design of meaningful mutation operators is much less challenging as it is a lot easier to fulfill the requirements of a suited mutation operator (which in fact is to introduce a small amount of new genetic information).

1.5.1   Binary Representation

In the early years of GA research there was a strong focus on binary encoding of solution candidates. To some extent, an outgrowth of these ambitions is certainly the binary representation for the TSP. There have been different ways how to use binary representation for the TSP, the most straightforward one being to encode each city as a string of $log_2 n$ bits and a solution candidate as a string of $n(log_2 n)$ bits. Crossover is then simply performed by applying single-point crossover as proposed by Holland [Hol75]. Further attempts using binary encoding have been proposed using binary matrix representation ([FM91], [HGL93]). In [HGL93], Homaifar and Guan for example defined a matrix element in the i-th row and the j-th column to be 1 if and only if in the tour city j is visited after city i; they also applied one- or two- point crossover on the parent matrices, which for one-point crossover means that the child tour is created by just taking the column vectors left of the crossover point from one parent, and the column vectors right of the crossover point from the other parent.

Obviously, these strategies lead to highly illegal tours which are then repaired by additional repair strategies [HGL93], which is exactly the point where a GA can no longer act as it is supposed to. As the repair strategies have to introduce a high amount of genetic information which is neither from the one nor from the other parent, child solutions emerge whose genetic make-up has only little in common with its own parents; this counteracts the general functioning of GAs as given in a more general interpretation of the schema theorem and the according building block hypothesis.

1.5.2   Adjacency Representation

Using the adjacency representation for the TSP (as described in [LKM+99], e.g.), a city j is listed in position i if and only if the tour leads from city i to city j. Based on the adjacency representation, the so-called alternating edges crossover has been proposed for example which basically works as follows: First it chooses an edge from one parent and continues with the position of this edge in the other parent representing the next edge, etc. The partial tour is built up by choosing edges from the two parents alternatingly. In case this strategy would produce a cycle, the edge is not added, but instead the operator randomly selects an edge from the edges which do not produce a cycle and continues in the way described above.

Compared to the crossover operators based on binary encoding, this strategy has the obvious advantage that a new child is built up from edges of its own parents. However, also this strategy is not very well suited as a further claim to crossover is not fulfilled at all: The alternating edges crossover cannot inherit longer tour segments and therefore longer building blocks cannot establish. As a further development to the alternating edges crossover, the so-called sub-tour chunks crossover aims to put things right by not alternating the edges but sub-tours of the two parental solutions. However, the capabilities of this strategy are also rather limited.

1.5.3   Path Representation

The most natural representation of a TSP tour is given by the path representation. Within this representation, the n cities of a tour are put in order according to a list of length n, so that the order of cities to be visited is given by the list entries with an imaginary edge from the last to the first list entry. A lot of crossover and mutation operators have been developed based upon this representation, and most of the nowadays used TSP solution methods using GAs are realized using path representation. Despite obvious disadvantages like the equivocality of this representation (the same tour can be described in $2n$ different ways for a symmetrical TSP and in n different ways for an asymmetrical TSP) this representation has allowed the design of quite powerful operators like the order crossover (OX) or the edge recombination crossover (ERX) which are able to inherit parent sub-tours to child solutions with only a rather small ratio of edges stemming from none of its own parents which is essential for GAs. A detailed description of these operators is given in Chapter 8.

1.5.4   Other Representations for Combinatorial Optimization Problems

Combinatorial optimization problems that are more in step with actual practice than the TSP require more complex problem representations, which makes it even more difficult for the designer of genetic solution manipulation operators to construct crossover operators that fulfill the essential requirements.

Challenging optimization tasks arise in the field of logistics and production planning optimization where the capacitated vehicle routing problem with (CVRPTW, [Tha95]) and without time windows (CVRP, [DR59]) as well as the job shop scheduling problem (JSSP [Tai93]) denote abstracted standard formulations which are used for the comparison of optimization techniques on the basis of widely available standardized benchmark problems. Tabu search [Glo86] and genetic algorithms are considered the most powerful optimization heuristics for these rather practical combinatorial optimization problems [BR03].

Cheng et al. as well as Yamada and Nakano give a comprehensive review of problem representations and corresponding operators for applying Genetic Algorithms to the JSSP in [CGT99] and [YN97], respectively.

For the CVRP, Bräysy and Gendreau give a detailed overview about the application of local search algorithms in [BG05a] and about the application of metaheuristics in [BG05b]; concrete problem representations and crossover operators for GAs are outlined in [PB96] and [Pri04]. Furthermore, the application of extended GA concepts to the CVRP will be covered in the practical part of this book within Chapter 10.

1.5.5   Problem Representations for Real-Valued Encoding

When using real-valued encoding, a solution candidate is represented as a real-valued vector in which the dimension of the chromosomes is constant and equal to the dimension of the solution vectors. Crossover concepts are distinguished into discrete and continuous recombination where the discrete variants copy the exact allele values of the parent chromosomes to the child chromosome whereas the continuous variants perform some kind of averaging.

Mutation operators for real-valued encoding either slightly modify all positions of the gene or introduce major changes to only some (often just one) position. Often a mixture of different crossover and mutation techniques leads to the best results for real-valued GAs. A comprehensive review of crossover and mutation techniques including also more sophisticated techniques like multi-parent recombination is given in [DLJD00].

Although real-valued encoding is a problem representation which is especially suited for evolution strategies or particle swarm optimization rather than for GAs, a lot of operators have been established also for GAs which are quite similar to modern implementations of ES that make use of recombination [Bey01]. Real-valued encoding for GAs distinguishes itself from typical discrete representations for combinatorial optimization problems in that point that the evolvement of longer and longer building block sequences in terms of adjacent alleles is of minor or no importance. Nevertheless, GA-based techniques like offspring selection have proven to be a very powerful optimization technique also for this kind of problem representation especially in case of highly multimodal fitness landscapes [AW05].

1.6   GA Theory: Schemata and Building Blocks

Researchers working in the field of GAs have put a lot of effort into the analysis of the genetic operators (crossover, mutation, selection). In order to achieve better analysis and understanding, Holland has introduced a construct called schema [Hol75]:

Under the assumption of a canonical GA with binary string representation of individuals, the symbol alphabet {0,1,#} is considered where {#}(don't care) is a special wild card symbol that matches both, 0 and 1.

A schema is a string with fixed and variable symbols. For example, the schema [0#11#01] is a template that matches the following four strings: [0011001], [0011101], [0111001], and [0111101]. The symbol # is never actually manipulated by the genetic algorithm; it is just a notational device that makes it easier to talk about families of strings.

Essentially, Holland's idea was that every evaluated string actually gives partial information about the fitness of the set of possible schemata of which the string is a member. Holland analyzed the influence of selection, crossover, and mutation on the expected number of schemata, when going from one generation to the next. A detailed discussion of related analysis can be found in [Gol89]; in the context of the present work we only outline the main results and their significance.

Assuming fitness proportional replication, the number m of individuals of the population belonging to a particular schema H at time $t + 1$ is related to the same number at the time t as

$$m(H, t + 1) = m(H, t)\frac{f_H(t)}{\overline{f}(t)} \tag{1.1}$$

where $f_H(t)$ is the average fitness value of the string representing schema H, while $\overline{f}(t)$ is the average fitness value over all strings within the population. Assuming that a particular schema remains above the average by a fixed amount $c\overline{f}(t)$ for a number t of generations, the solution of the equation given above can be formulated as the following exponential growth equation:

$$m(H, t) = m(H, 0)(1 + c)^t \tag{1.2}$$

where $m(H, 0)$ stands for the number of schemata H in the population at time 0, c denotes a positive integer constant, and $t \geq 0$.

The importance of this result is the exponentially increasing number of trials to above average schemata.

The effect of crossover which breaks strings apart (at least in the case of canonical genetic algorithms) is that they reduce the exponential increase by a quantity that is proportional to the crossover rate p_c and depends on the defining length δ of a schema on the string of length l:

$$p_c\frac{\delta(H)}{l - 1} \tag{1.3}$$

The defining length δ of a schema is the distance between the first and the last fixed string position. For example, for the schema $[\#\#\#0\#0101]$ $\delta = 9 - 4 = 5$. Obviously, short defining length schemata are less likely to be disrupted by a single point crossover operator. The main result is that above average schemata with short defining lengths will still be sampled at an exponential increasing rate. These schemata with above average fitness and short defining length are the so-called building blocks and play an important role in the theory of genetic algorithms.

The effects of mutation are described in a rather straightforward way: If the bit mutation probability is p_m, then the probability of survival of a single bit is $1 - p_m$; since single bit mutations are independent, the total survival probability is therefore $(1 - p_m)^l$ with l denoting the string length. But in the context of schemata only the fixed, i.e., non-wildcard, positions matter. This number is called the order $o(H)$ of schema H and equals to l minus the number of "don't care" symbols. Then the probability of surviving a mutation for a certain schema H is $(1 - p_m)^{o(H)}$ which can be approximated by $1 - o(H)p_m$ for $p_m \ll 1$.

Summarizing the described effects of mutation, crossover, and reproduction, we end up with Holland's well known schema theorem [Hol75]:

$$m(H, t+1) \geq m(H, t)\frac{f_H(t)}{\bar{f}(t)}[1 - p_c\frac{\delta(H)}{l-1} - o(H)p_m] \qquad (1.4)$$

The result essentially says that the number of short schemata with low order and above average quality grows exponentially in subsequent generations of a genetic algorithm.

Still, even if the schema theorem is a very important result in GA theory, it is obtained under idealized conditions that do not hold for most practical GA applications. Both the individual representation and the genetic operators are often different from those used by Holland. The building block hypothesis has been found reliable in many cases but it also depends on the representation and on the genetic operators. Therefore, it is easy to find or to construct problems for which it is not verified. These so-called deceptive problems are studied in order to find out the inherent limitations of GAs, and which representations and operators can make them more tractable. A more detailed description of the underlying theory can for instance be found in [Raw91] or [Whi93].

The major drawback of the building block theory is given by the fact that the underlying GA (binary encoding, proportional selection, single-point crossover, strong mutation) is applicable only to very few problems as it requires more sophisticated problem representations and corresponding operators to tackle challenging real-world problems. Therefore, a more general theory is an intense topic in GA research since its beginning. Some theoretically interesting approaches like the *forma theory* of Radcliffe and Surry [RS94], who consider a so-called *forma* as a more general schema for arbitrary representations, state requirements to the operators, which cannot be fulfilled for practical problems with their respective constraints.

By the end of the last millennium, Stephens and Waelbroeck ([SW97], [SW99]) developed an exact GA schema theory. The main idea is to describe the total transmission probability α of a schema H so that $\alpha(H, t)$ is the probability that at generation t the individuals of the GA's population will match H (for a GA working on fixed-length bit strings). Assuming a crossover probability p_{xo}, $\alpha(H, t)$ is calculated as[1]:

$$\alpha(H, t) = (1 - p_{xo})p(H, t) + \frac{p_{xo}}{N-1}\sum_{i=1}^{N-1} p(L(H, i), t)p(R(H, i), t) \qquad (1.5)$$

with $L(H, i)$ and $R(H, i)$ being the left and right parts of schema H, respectively, and $p(H, t)$ the probability of selecting an individual matching H to

---

[1]We here give the slightly modified version as stated in [LP02]; it is equivalent to the results in [SW97] and [SW99] assuming $p_m = 0$.

become a parent. The "left" part of a schema H is thereby produced by replacing all elements of H at the positions from the given index i to N with "don't care" symbols (with N being the length of the bit strings); the "right" part of a schema H is produced by replacing all elements of H from position 1 to i with "don't care." The summation is over all positions from 1 to $N-1$, i.e., over all possible crossover points.
Stephens later generalized this GA schema theory to variable-length GAs; see for example [SPWR02].

Keeping in mind that the ultimate goal of any heuristic optimization technique is to approximately and efficiently solve highly complex real-world problems rather than stating a mathematically provable theory that holds only under very restricted conditions, our intention for an extended building block theory is a not so strict formulation that in return can be interpreted for arbitrary GA applications. At the same time, the enhanced variants of genetic algorithms and genetic programming proposed in this book aim to support the algorithms in their intention to operate in the sense of an extended building block interpretation discussed in the following chapters.

1.7   Parallel Genetic Algorithms

The basic idea behind many parallel and distributed programs is to divide a task into partitions and solve them simultaneously using multiple processors. This divide-and-conquer approach can be used in different ways, and leads to different methods to parallelize GAs where some of them change the behavior of the GA whereas others do not. Some methods (as for instance fine-grained parallel GAs) can exploit massively parallel computer architectures, while others (coarse-grained parallel GAs, e.g.) are better qualified for multi-computers with fewer and more powerful processing elements. Detailed descriptions and classifications of distributed GAs are given in [CP01], [CP97] or [AT99] and [Alb05]; the scalability of parallel GAs is discussed in [CPG99]. A further and newer variant of parallel GAs which is based on offspring selection (see Chapter 4) is the so-called SASEGASA algorithm which is discussed in Chapter 5.

In a rough classification, parallel GA concepts established in GA textbooks (as for example [DLJD00]) can be classified into global parallelization, coarse-grained parallel GAs, and fine-grained parallel GAs, where the most popular model for practical applications is the coarse-grained model, also very well known as the island model.

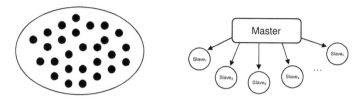

FIGURE 1.3: Global parallelization concepts: A panmictic population structure (shown in left picture) and the corresponding master–slave model (right picture).

1.7.1   Global Parallelization

Similar to the sequential GA, in the context of global parallelization there is only one single panmictic[2] population and selection considers all individuals, i.e., every individual has a chance to mate with any other. The behavior of the algorithm remains unchanged and the global GA has exactly the same qualitative properties as a sequential GA. The most common operation that is parallelized is the evaluation of the individuals as the calculation of the fitness of an individual is independent from the rest of the population. Because of this the only necessary communication during this phase is in the distribution and collection of the workload.

One master node executes the GA (selection, crossover, and mutation), and the evaluation of fitness is divided among several slave processors. Parts of the population are assigned to each of the available processors, in that they return the fitness values for the subset of individuals they have received. Due to their centered and hierarchical communication order, global parallel GAs are also known as single-population master–slave GAs.

Figure 1.3 shows the population structure of a master–slave parallel GA: This panmictic GA has all its individuals (indicated by the black spots) in the same population. The master stores the population, executes the GA operations, and distributes individuals to the slaves; the slaves compute the fitness of the individuals. As a consequence, global parallelization can be efficient only if the bottleneck in terms of runtime consumption is the evaluation of the fitness function.

Globally parallel GAs are quite easy to implement, and they can be a quite efficient method of parallelization if the evaluation requires considerable computational effort compared to the effort required for the operations carried out by the master node. However, they do not influence the qualitative properties of the corresponding sequential GA.

---

[2]In general, a population is called panmictic when all individuals are possible mating partners.

1.7.2 Coarse-Grained Parallel GAs

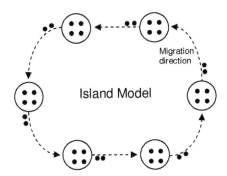

FIGURE 1.4: Population structure of a coarse-grained parallel GA.

In the case of a coarse-grained parallel GA, the population is divided into multiple subpopulations (also called islands or demes) that evolve mostly isolated from each other and only occasionally exchange individuals during phases called migration. This process is controlled by several parameters which will be explained later in Section 1.7.4. In contrast to the global parallelization model, coarse-grained parallel GAs introduce fundamental changes in the structure of the GA and have a different behavior than a sequential GA. Coarse-grained parallel GAs are also known as distributed GAs because they are usually implemented on computers with distributed memories. Literature also frequently uses the notation "island parallel GAs" because there is a model in population genetics called the island model that considers relatively isolated demes.

Figure 1.4 schematically shows the design of a coarse-grained parallel GA: Each circle represents a simple GA, and there is (infrequent) communication between the populations. The qualitative performance of a coarse-grained parallel GA is influenced by the number and size of its demes and also by the information exchange between them (migration). The main idea of this type of parallel GAs is that relatively isolated demes will converge to different regions of the solution-space, and that migration and recombination will combine the relevant solution parts [SWM91]. However, at present there is only one model in the theory of coarse-grained parallel GAs that considers the concept of selection pressure for recombining the favorable attributes of solutions evolved in the different demes, namely the SASEGASA algorithm (which will be described later in Chapter 5). Coarse-grained parallel GAs are the most frequently used parallel GA concept, as they are quite easy to implement and are a natural extension to the general concept of sequential GAs making use of commonly available cluster computing facilities.

1.7.3   Fine-Grained Parallel GAs

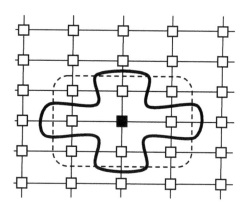

FIGURE 1.5: Population structure of a fine-grained parallel GA; the special case of a cellular model is shown here.

Fine-grained models consider a large number of very small demes; Figure 1.5 sketches a fine-grained parallel GA. This class of parallel GAs has one spatially distributed population; it is suited for massively parallel computers, but it can also be implemented on other supercomputing architectures. A typical example is the diffusion model [Müh89] which represents an intrinsic parallel GA-model.

The basic idea behind this model is that the individuals are spread throughout the global population like molecules in a diffusion process. Diffusion models are also called cellular models. In the diffusion model a processor is assigned to each individual and recombination is restricted to the local neighborhood of each individual.

A recent research topic in the area of parallel evolutionary computation is the combination of certain aspects of the different population models resulting in so-called hybrid parallel GAs. Most of the hybrid parallel GAs are coarse-grained at the upper level and fine-grained at the lower levels. Another way to hybridize parallel GAs is to use coarse-grained GAs at the high as well as at the low levels in order to force stronger mixing at the low levels using high migration rates and a low migration rate at the high level [CP01]. Using this strategy, computer cluster environments at different locations can collectively work on a common problem with only little communication overhead (due to the low migration rates at the high level).

1.7.4 Migration

Especially for coarse-grained parallel GAs the concept of migration is considered to be the main success criterion in terms of achievable solution quality. The most important parameters for migration are:

- The communication topology which defines the interconnections between the subpopulations (demes)

- The migration scheme which controls which individuals (best, random) migrate from one deme to another and which individuals should be replaced (worst, random, doubles)

- The migration rate which determines how many individuals migrate

- The migration interval or migration gap that determines the frequency of migrations

The most essential question concerning migration is when and to which extent migration should take place. Much theoretical work considering this has already been done; for a survey of these efforts see [CP97] or [Alb05]. It is very usual for parallel GAs that migration occurs synchronously meaning that it occurs at predetermined constant intervals. However, synchronous migration is known to be slow and inefficient in some cases [AT99]. Asynchronous migration schemes perform communication between demes only after specific events. The migration rate which determines how many individuals undergo migration at every exchange can be expressed as a percentage of the population size or as an absolute value. The majority of articles in this field suggest migration rates between 5% and 20% of the population size. However, the choice of this parameter is considered to be very problem dependent [AT99]. A recent overview of various migration techniques is given in [CP01].

Recent theory of self-adaptive selection pressure steering (see Chapters 4 and 5) plays a major role in defying the conventions of recent parallel GA-theory. Within these models it becomes possible to detect local premature convergence, i.e., premature convergence in a certain deme. Thus, local premature convergence can be detected independently in all demes, which should give a high potential in terms of efficiency especially for parallel implementations. Furthermore, the fact that selection pressure is adjusted self-adaptively with respect to the potential of genetic information stored in the certain demes makes the concept of a parallel GA much more independent in terms of migration parameters (see [Aff05] and Chapter 5).

1.8 The Interplay of Genetic Operators

In order to allow an efficient performance of a genetic algorithm, a beneficial interplay of exploration and exploitation should be possible. Critical factors for this interplay are the genetic operators selection, crossover, and mutation.

The job of crossover is to advantageously combine alleles of selected (above average) chromosomes which may stem from different regions of the search space. Therefore, crossover is considered to rather support the aspect of breadth search. Mutation slightly modifies certain chromosomes at times and thus brings new alleles into the gene pool of a population in order to avoid stagnation. As mutation modifies the genetic make-up of certain chromosomes only slightly it is primarily considered as a depth search operator. However, via mutation newly introduced genetic information does also heavily support the aspect of breadth search if crossover is able to "transport" this new genetic information to other chromosomes in other search space regions. As we will show later in this book, this aspect of mutation is of prime importance for an efficient functioning of a GA.

The aspect of migration in coarse-grained parallel GAs should also be mentioned in our considerations about the interplay of operators. In this kind of parallel GAs, migration functions somehow like a meta-model of mutation introducing new genetic information into certain demes at the chromosome-level whereas mutation introduces new genetic information at the allele level. Concerning migration, a well-adjusted interplay between breadth and depth search is aimed to function in the way that breadth search is supported in the intra-migration phases by allowing the certain demes to drift to different regions of the search space until a certain stage of stagnation is reached; the demes have expanded over the search space. Then migration comes into play by introducing new chromosomes stemming from other search space regions in order to avoid stagnation in the certain demes; this then causes the demes to contract again slightly which from a global point of view tends to support the aspect of depth search in the migration phases. The reason for this is that migration causes an increase of genetic diversity in the specific demes on the one hand, but on the other hand it decreases the diversity over all islands. This global loss of genetic diversity can be interpreted as an exploitation of the search space.

This overall strategy is especially beneficial in case of highly multimodal search spaces as it is the case for complex combinatorial optimization problems.

1.9 Bibliographic Remarks

There are numerous books, journals, and articles available that survey the field of genetic algorithms. In this section we summarize some of the most important ones. Representatively, the following books are widely considered very important sources of information about GAs (in chronological order):

- J. H. Holland: *Adaptation in Natural and Artificial Systems* [Hol75]

- D. E. Goldberg: *Genetic Algorithms in Search, Optimization and Machine Learning* [Gol89]

- Z. Michalewicz: *Genetic Algorithms + Data Structures = Evolution Programs* [Mic92]

- D. Dumitrescu et al.: *Evolutionary Computation* [DLJD00]

The following journals are dedicated to either theory and applications of genetic algorithms or evolutionary computation in general:

- *IEEE Transactions on Evolutionary Computation* (IEEE)

- *Evolutionary Computation* (MIT Press)

- *Journal of Heuristics* (Springer)

Moreover, several conference and workshop proceedings include papers related to genetic and evolutionary algorithms and heuristic optimization. Some examples are the following ones:

- *Genetic and Evolutionary Computation Conference (GECCO)*, a recombination of the *International Conference on Genetic Algorithms* and the *Genetic Programming Conference*

- *Congress on Evolutionary Computation (CEC)*

- *Parallel Problem Solving from Nature (PPSN)*

Of course there is a lot of GA-related information available on the internet including theoretical background and practical applications, course slides, and source code. Publications of the *Heuristic and Evolutionary Algorithms Laboratory (HEAL)* (including several articles on GAs and GP) are available at http://www.heuristiclab.com/publications/.

Chapter 2

Evolving Programs: Genetic Programming

DOI: 10.1201/9781420011326-3

In the previous chapter we have summarized and discussed genetic algorithms; it has been illustrated how this kind of algorithms is able to produce high quality results for a variety of problem classes.

Still, a GA is by itself not able to handle one of the most challenging tasks in computer science, namely getting a computer to solve problems without programming it explicitly. As Arthur Samuel stated in 1959 [Sam59], this central task can be formulated in the following way:

> *How can computers be made to do what needs to be done,*
> *without being told exactly how to do it?*

In this chapter we give a compact description and discussion of an extension of the genetic algorithm called genetic programming (GP). Similar to GAs, genetic programming works on populations of solution candidates for a given problem and is based on Darwinian principles of survival of the fittest (selection), recombination (crossover), and mutation; it is a domain-independent, biologically inspired method that is able to create computer programs from a high-level problem statement.[1]

Research activities in the field of genetic programming started in the 1980s; still, it took some time until GP was widely received by the computer science community. Since the beginning of the 1990s GP has been established as a human-competitive problem solving method. The main factors for its widely accepted success in the academic world as well as in industries can be summarized in the following way [Koz92b]:

- Virtually all problems in artificial intelligence, machine learning, adaptive systems, and automated learning can be recast as a search for computer programs, and

- genetic programming provides a way to successfully conduct the search in the space of computer programs.

---

[1]Please note that we here in general see computer programs as entities that receive inputs, perform computations, and produce output.

In the following we

- give an overview of the main ideas and foundations of genetic programming in Sections 2.1 and 2.2,

- summarize basic steps of the GP-based problem solving process (Section 2.3),

- report on typical application scenarios (Section 2.4),

- explain theoretical foundations (GP schema theories, Section 2.5),

- discuss current GP challenges and research areas in Section 2.6,

- summarize this chapter on GP in Section 2.7, and finally

- refer to a range of outstanding literature in the field of theory and praxis of GP in Section 2.8.

2.1 Introduction: Main Ideas and Historical Background

As has already been mentioned, one of the central tasks in artificial intelligence is to make computers do what needs to be done without telling them exactly how to do it. This does not seem to be unnatural since it demands of computers to mimic the human reasoning process - humans are able to learn what needs to be done, and how to do it. In short, interactions of networks of neurons are nowadays believed to be the basis of human brain information processing; several of the earliest approaches in artificial intelligence aimed at imitating this structure using connectionist models and artificial neural networks (ANNs, [MP43]). Suitable network training algorithms enable ANNs to learn and generalize from given training examples; ANNs are in fact a very successful distributed computation paradigm and are frequently used in real-world applications where exact algorithmic approaches are too difficult to implement or even not known at all. Pattern recognition, classification, data-based modeling (regression) are some examples of AI areas in which ANNs have been applied in numerous ways. Unlike this network-based approach, genetic algorithms were developed using main principles of natural evolution. As has been explained in Chapter 1, GAs are population-based optimization algorithms that imitate natural evolution: Starting with a primordial ooze of thousands of randomly created solution candidates appropriate to the respective problem, populations of solutions are progressively evolved over many generations using the Darwinian principles.

Similar to the GA, GP is an evolutionary algorithm inspired by biological evolution to find computer programs that perform a user-defined computational task. It is therefore a machine learning technique used to optimize a population of computer programs according to a fitness landscape determined by a program's ability to perform the given task; it is a domain-independent, biologically inspired method that is able to create computer programs from a high-level problem statement (with computer programs being here defined as entities that receive inputs, perform computations, and produce output).

The first research activities in the context of GP have been reported in the early 1980s. For example, Smith reported on a learning system based on GAs [Smi80], and in [For81] Forsyth presented a computer package producing decision-rules (i.e., small computer programs) in forensic science for the UK police by induction from a database (where these rules are Boolean expressions represented by tree structures). In 1985, Cramer presented a representation for the adaptive generation of simple sequential programs [Cra85]; it is widely accepted that this article on genetic programming is the first paper to describe the tree-like representation and operators for manipulating programs by genetic algorithms.

Even though there was noticeable research activity in the field of GP going on by the middle of the 1980s, still it took some time until GP was widely received by the computer science community. GP is very intensive from a computational point of view and so it was mainly used to solve relatively simple problems until the 1990s. But thanks to the enormous growth in CPU power that has been going on since the 1980s, the field of applications for GP has been extended immensely yielding human competitive results in areas such as data-based modeling, electronic design, game playing, sorting, searching, and many more; examples (and respective references) are going to be given in the following sections.

One of the most important GP publications was "Genetic Programming: On the Programming of Computers by Means of Natural Selection" [Koz92b] by John R. Koza, professor for computer science and medical informatics at Stanford University who has since been one of the main proponents of the GP idea. Based on extensive theoretical background as well as test results in many different problem domains he demonstrated GP's ability to serve as an automated invention machine producing novel and outstanding results for various kinds of problems. By now there have been three more books on GP by Koza (and his team), but also several other very important publications (for example by Banzhaf, Langdon, Poli and many others); a short summary is given in Section 2.8.

Along with these ad hoc engineering approaches there was an increasing interest in how and why GP works. Even though GP was applied successfully for solving problems in various areas, the development of a GP theory was considered rather difficult even through the 1990s. Since the early 2000s it has finally been possible to establish a theory of GP showing a rapid development since then. A book that has to be mentioned in this context is clearly

"Foundations of Genetic Programming" [LP02] by Langdon and Poli since it presents exact GP schema analysis.

As we have now summarized the historical background of GP, it is now high time to describe how it really works and how typical applications are designed. This is exactly what the reader can find in the following sections.

2.2   Chromosome Representation

As in the context of any GA-based problem solving process, the representation of problem instances and solution candidates is a key issue also in genetic programming. On the one hand, the representation scheme should enable the algorithm to find suitable solutions for the given problem class, but on the other hand the algorithm should be able to directly manipulate the coded solution representation. The use of fixed-length strings (of bits, characters, or integers, e.g.) enables the conventional GA to solve a huge amount of problems and also allows the construction of a solid theoretical foundation, namely the schema theorem. Still, in the context of GP the most natural representation for a solution is a hierarchical computer program of variable size [Koz92b].

2.2.1   Hierarchical Labeled Structure Trees

2.2.1.1   Basics

So, how can hierarchical computer programs be represented? The representation that is most common in literature and is used by Koza ([Koz92b], [Koz94], [KIAK99], [KKS⁺03b]), Langdon and Poli ([LP02]), and many other authors is the point-labeled structure tree. Originally, these structure trees were for example seen as graphical representations of so-called S-expressions of the programming language LISP ([McC60], [Que03], [WH87]) which have for example been used by Koza in [Koz92b] and [Koz94].[2] Here we do not strictly stick to LISP-syntax for the examples given, but the main paradigms of S-expressions are used.

The following key facts are relevant in the context of structure tree based genetic programming:

- All tree-nodes are either *functions* or *terminals*.

---

[2]In fact, of course, any higher programming language is suitable for implementing a GP-framework and for representing hierarchical computer programs. Koza, for example, switched to the C programming language as described in [KIAK99], and the HeuristicLab framework and the GP-implementation, which is realized as plug-ins for it, are programmed in C# using the .NET framework - this is to be explained in further detail later.

- *Terminals* are evaluated directly, i.e., their return values can be calculated and returned immediately.

- All *functions* have child nodes which are evaluated before using the children's calculated return values as inputs for the parents' evaluation.

- The probably most convenient string representation is the *prefix* notation, also called *Polish* or *Lukasiewicz*[3] notation: Function nodes are given before the child nodes' representations (optionally using parentheses). Evaluation is executed recursively, depth-first way, starting from the left; operators are thus placed to the left of their operands.
  In case of fixed arities of the functions (i.e., if the numbers of function's inputs is fixed and known), no parentheses or brackets are needed.

In a more formal way this program representation structure schema can be summarized as follows [ES03]:

- Symbolic expressions can be defined using

  - a terminal set T, and
  - a function set F.

- The following general recursive definition is applied:

  - Every $t \in T$ is a correct expression,
  - $f(e_1, \ldots, e_n)$ is a correct expression if $f \in F$, $arity(f) = n$ and $e_1, \ldots, e_n$ are correct expressions, and
  - there are no other forms of correct expressions.

- In general, expressions in GP are not typed (closure property: any $f \in F$ can take any $g \in F$ as argument). Still, as we see in the discussion of genetic operators in Section 2.2.1.3, this might be not true in certain cases depending on the function and terminal sets chosen.

In the following we give exemplary simple programs. We thereby give conventional as well as prefix (not exactly following LISP notation) textual notations:

- (a) IF (Y>X OR Y<4) THEN i:=(i+1), ELSE i:=0.
  Prefix notation: IF(OR(>(Y,X),<(Y,4)),:=(i,+(i,1)),:=(i,0)).

- (b) $\frac{X+5}{2Y}$. Prefix notation: DIV(ADD(X,5),MULT(2,Y)).

Graphical representations of the programs (given as rooted, point-labeled structure trees) are given in Figure 2.1.

---

[3] Jan Łukasiewicz (1878–1956), a Polish mathematician, invented the prefix notation which is also the basis of the recursive stack ("last in, first out"; [Ham58], [Ham62]). In reference to his nationality the notation is also referred to as "Polish" notation.

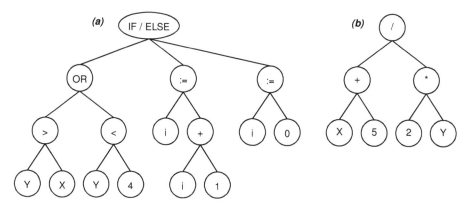

FIGURE 2.1: Exemplary programs given as rooted, labeled structure trees.

2.2.1.2   Evaluation

As already mentioned previously, the execution (evaluation) of GP chromosomes representing hierarchical computer programs as structure trees is done recursively, depth-first way, and starting from the left. In order to demonstrate this we here simulate the evaluation of the example programs given in Section 2.2.1.1; graphical representations are given in Figures 2.2 and 2.3.

- (a) Internal states before execution: $X = 7$, $Y = 3$, $i = 2$.
  Execution:
  ```
  IF(OR(>(Y,X),<(Y,4)),:=(i,+(i,1)),:=(i,0))
  ⇒ IF(OR(>(3,7),<(Y,4)),:=(i,+(i,1)),:=(i,0))
  ⇒ IF(OR(FALSE,<(Y,4)),:=(i,+(i,1)),:=(i,0))
  ⇒ IF(OR(FALSE,<(3,4)),:=(i,+(i,1)),:=(i,0))
  ⇒ IF(OR(FALSE,TRUE),:=(i,+(i,1)),:=(i,0))
  ⇒ IF(TRUE,:=(i,+(i,1)),:=(i,0))
  ⇒ :=(i,+(i,1))
  ⇒ :=(i,+(2,1))
  ⇒ :=(i,3).
  ```
  Internal states after execution: $X = 7$, $Y = 3$, $i = 3$.

- (b) Internal states before execution: $X = 7$, $Y = 3$.
  Execution:
  ```
  DIV(ADD(X,5),MULT(2,Y))
  ⇒ DIV(ADD(7,5),MULT(2,Y))
  ⇒ DIV(12,MULT(2,Y))
  ⇒ DIV(12,MULT(2,3))
  ⇒ DIV(12,6)
  ⇒ 2
  ```
  Return value: 2; internal states after execution: $X = 7$, $Y = 3$.

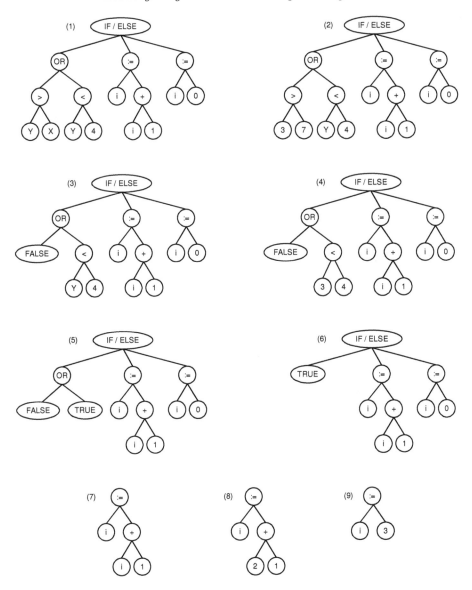

FIGURE 2.2: Exemplary evaluation of program (a).

2.2.1.3 Genetic Operations: Crossover and Mutation

As genetic programming is an extension to the genetic algorithm, GP also uses two main operators for producing new solution candidates in the search

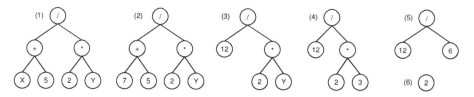

FIGURE 2.3: Exemplary evaluation of program (b).

space, namely crossover and mutation.

As we already know from Chapter 1, crossover, the most important reproduction operator, takes two parent individuals and produces new offspring by swapping parts of the parents. Here we immediately see one of the major advantages of hierarchical tree representations of computer programs: Singlepoint crossover can be simply performed by replacing a subtree of (a copy of) one of the parents by a subtree of the other parent; these subtrees are chosen at random. There are several different strategies for selecting these subtrees as it might be reasonable to choose either rather small, rather big, or completely randomly chosen parts.

Mutation can be seen as an arbitrary modification introduced to prevent premature convergence by randomly sampling new points in the search space. In the case of genetic programming, mutation is applied by modifying a randomly chosen node of the respective structure tree:

- A subtree could be deleted or replaced by a randomly re-initialized subtree.

- A function node could for example change its function type or turn into a terminal node.

Numerous other mutation variants are possible, many of them depending on the problem and chromosome representation chosen. In Chapter 11, for example, we describe mutation variants applicable for GP-based structure identification (related to symbolic regression, see Section 2.4.3).

Figure 2.4 illustrates examples for sexual reproduction using the exemplary programs (1) and (2) as parents, labeled as *parent1* and *parent2*, respectively. It thereby becomes obvious that in the context of GP there can be the chance of creating invalid chromosomes: The second offspring (*child2*) seems to be incorrect since it includes the comparison of a Boolean value (Y>X OR Y<4) and a number (2*Y). Thus, also in GP there are certain constraints that affect the crossover of solution candidates; these constraints have to be considered when it comes to designing and implementing a GP framework.

Of course, it again depends on the chosen implementation, if the evaluation of this syntactically dubious program can be executed or not. In case of real-valued representation of Boolean values (TRUE represented by 1.0, FALSE represented by 0.0, e.g.) this structure tree represents a valid program that can be calculated without any further problems.

Figure 2.5 illustrates exemplary results of applying mutation to program (1). In the first case, a Boolean function node ($<$) is turned into another type of Boolean function node ($>$) yielding *mutant1*; *mutant2* is produced by omitting a subtree, namely the second child of the *OR* function node. While these two first mutants are syntactically correct, *mutant3* is an example for an invalid mutation example: The first child of the conditional (*IF*) node has been deleted leaving the root node with only two children - the evaluation of this program is not possible.

Again, real-valued representation of Boolean values can help here. In this case the value calculated by the first child of such a conditional node would have to be interpreted as a Boolean value triggering the execution of the second child subtree, the *then*-branch. As there is no third child node there is also no *else*-branch, thus there is probably no action if the first (condition) node is evaluated (or at least interpreted) as *false*.

These two examples of syntactically incorrect programs demonstrate what was hinted in Section 2.2.1.1: Even though expressions are in general not typed in GP, there are cases in which this is not true - a fact which has to be considered during the design and implementation of a GP-based problem solving system.

2.2.1.4 Advantages

As we are going to see later, the hierarchical structure tree is not the only way how programs can be modeled and used in the GP process. Still, the cumulation of the following reasons strongly favors the choice of this program representation schema[4]:

- Even though structure trees show an (at least for many people) rather unusual appearance and syntax, most programming language compilers internally convert given programs into parse trees representing the underlying programs (i.e., their compositions of functions and terminals). In most programming languages, these parse trees are not (conveniently) accessible to the programmer; here we present the programs directly as parse trees as we need to genetically manipulate parts of the programs (sub-trees).

- As evaluation is executed recursively starting from the root node, a newly generated or manipulated program can be (re-)evaluated immediately without any intermediate transformation step.

---

[4]In fact, these reasons partially correlate to Koza's reasons for choosing LISP for his GP implementation reported on in [Koz92b] and [Koz94], for example.

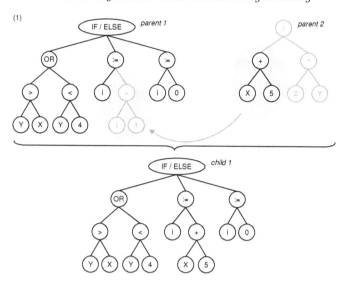

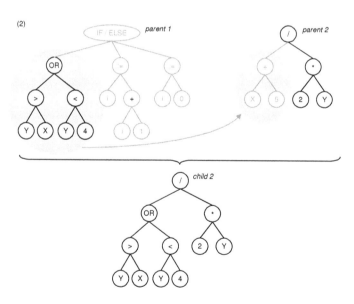

FIGURE 2.4: Exemplary crossover of programs (1) and (2) labeled as *parent1* and *parent2*, respectively. *Child1* and *child2* are possible new offspring programs formed out of the genetic material of their parents.

- Structure trees allow the representation of programs whose size and shape change dynamically.

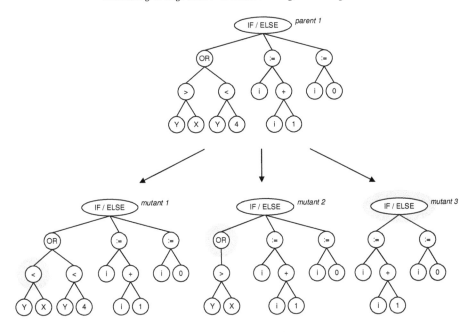

FIGURE 2.5: Exemplary mutation of a program: The programs *mutant1*, *mutant2*, and *mutant3* are possible mutants of *parent*.

2.2.2 Automatically Defined Functions and Modular Genetic Programming

Numerous variations and extensions to Koza's structure tree based genetic programming have been proposed since its publication at the beginning of the 1990s. The probably best known and most frequently used one is the concept of *automatically defined functions* (ADFs) proposed in "Genetic Programming II: Automatic Discovery of Reusable Programs" [Koz94].

The main idea of ADFs is that program code (which has been evolved during the GP process) is organized into useful groups (subroutines); this enables the parameterized reuse and hierarchical invocation of evolved code as functions that have not been taken from the original functions set F but are rather defined automatically. The (re-)use of subroutines (subprograms, procedures) is enabled in this way. In the meantime the idea of ADFs has been extended; automatically defined iterations, loops, macros, recursions, and stores have since then been proposed and their use demonstrated for example in [Koz94], [KIAK99], and [KKS+03b].

With ADFs a GP chromosome program is split into a main program tree (which is called and executed from outside) and arbitrarily many separate trees representing ADFs. These separate functions can take arguments as well as be called by the main program or another ADF.

Different approaches realizing modular genetic programming which have gained popularity and are well known in the GP community are the genetic library (presented by Angeline in [Ang93] and [Ang94], e.g.) and the adaptive representation through learning (ARL) algorithm (proposed by Rosca, see for example [Ros95a] or [RB96]). In both approaches, some parts of the evolved code are automatically extracted from programs (usually of those that show rather good fitness values). These extracted code fragments are then fixed and kept in the GP library, thus they are available for the evolving programs in the GP population.

Other advanced GP concepts that extend the tree concept are discussed in [Gru94], [KBAK99], [Jac99], and [WC99]; basic features of these modular GP approaches can be combined with multi-tree (multi-agent) systems which shall be described a bit later.

2.2.3   Other Representations

We are not going to say much about GP systems that are not based on trees in the context of this book; still, the reader could be prone to suspect that there might be computer program representations other than the tree-based approach. In fact, there are two other forms of GP that shall be mentioned here whose program encoding differs significantly from the approach described before: *Linear* and *graphical genetic programming*.

2.2.3.1   Linear Genetic Programming

The main difference between linear GP and tree-based GP is that in linear GP individuals of the GP algorithm (the programs) are not represented by structure trees but by linear chromosomes. These linear solutions represent lists of computer instructions which are executed linearly.

Linear GP chromosomes are more similar to those of conventional GAs; however, their size is usually not fixed so that a GP population is likely to contain chromosomes of different sizes which is usually not the case with conventional GA approaches. On the one hand this of course brings along the loss of the advantages mentioned in Section 2.2.1.4, but on the other hand this schema easily enables the representation of *stack-based* programs, *register-based* programs, and *machine code*.

- In general, a stack is a data structure based on the "last in first out" principle. If a program instruction is to be evaluated, it takes (pops) its arguments from the stack, performs the calculation, and writes back the result by adding (pushing) it back to the top of the stack. A chromosome in *stack-based GP* represents exactly such a stack-based program by storing the program instructions in a list and using a stack for executing the program. A typical example can be seen in Perkins' article "Stack-Based Genetic Programming" [Per94]; a recent implementation has for example been presented in [HRv07].

- *Register-based* and *machine code* GP are essentially similar [LP02]: In both cases data are stored in (a rather small number of) registers, and instructions read data from and write results back to these registers. Initially, a program's inputs are written to registers, and after executing the program the results are given in one or more registers. The main difference between these two GP approaches is the following:

  - Programs in register-based GP (as also those of any other kind of GP system) have to be interpreted, i.e., they are executed indirectly or compiled before execution.
  - On the contrary, programs in machine code GP consist of real hardware machine instructions; thus, these programs can be executed directly on a computer. The execution of machine code GP programs is therefore a lot faster than the evaluation of programs in traditional implementations.
    Nordin's Compiling Genetic Programming System (CGPS) [Nor97] for example presents an implementation of machine code GP.

2.2.3.2 Graphical Genetic Programming

Parallel Distributed Graphical Programming (PDGP, [Pol97], [Pol99b]) is a form of GP in which programs are represented as graphs representing functions and terminals as nodes; links between those nodes define the flow of control and results. PDGP defines a fixed layout for the nodes whereas the connections between them and the referenced functions are evolved by the GP process. PDGP enables a high degree of parallelism as well as an efficient and effective reuse of partial results; furthermore, it has been shown that it performs better than conventional tree-based GP on a number of benchmark problems.

Figure 2.6 shows the graphical representation of an exemplary program in PDGP (adapted from [Pol99b]).

2.3 Basic Steps of the GP-Based Problem Solving Process

2.3.1 Preparatory Steps

Before the GP process can be started there are several preparatory steps that have to be executed. As explained in Section 2.2.1.1, the *function* and *terminal* sets (F and T, respectively) have to be determined. Furthermore, as in any GA application, a *fitness measurement function* also has to be established so that a solution candidate can be evaluated and its fitness can be measured (either explicitly or implicitly).

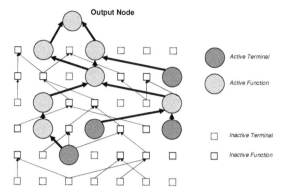

FIGURE 2.6: Intron-augmented representation of an exemplary program in PDGP [Pol99b].

In addition to these preparations that directly affect the construction and management of individuals of the GP population, there are also some things to be done regarding the execution of the GP algorithm:

- Parameters that control the GP run have to be set,

- a termination criterion has to be defined, and

- a result designation method has to be defined (as explained later in Section 2.3.4).

These preparations in fact have to be done for any genetic algorithm; a similar summary is for example given in [KIAK99]. Figure 2.7 summarizes the major preparatory steps for the basic GP process.

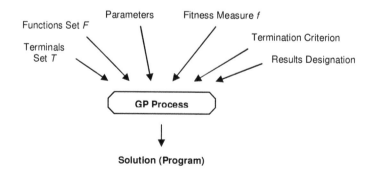

FIGURE 2.7: Major preparatory steps of the basic GP process.

2.3.2   Initialization

At the beginning of each GA and GP execution, the population is initialized arbitrarily before the intrinsic evolutionary process can be started. This initialization can be done either completely at random or using certain (problem-specific) heuristics.

For hierarchical program structures as used in GP the random initialization utilizes a maximum initial tree depth D_{max}. As introduced in [Koz92b] and for example reflected on in [ES03], there are two possibilities for creating random initial programs:

- Full method: Nodes at depth $d < D_{max}$ point to randomly chosen functions from function set F, and nodes at depth $d = D_{max}$ are randomly chosen terminals (from terminal set T);

- Grow method: Nodes at depth $d < D_{max}$ become either a function or a terminal (randomly chosen from $F \cup T$), and nodes at depth $d = D_{max}$ are again randomly chosen terminals (from T).

The so-called ramped half-half GP initialization method, proposed by Koza [Koz92b], has meanwhile become one of the most frequently used GP initialization approaches [ES03]. Both methods, grow and full, are hereby applied, each delivering parts of the initial population.

Still, there is research work going on regarding this issue of finding optimal initialization techniques as it is a fact that the use of different initialization strategies can lead to very different overall results (as for example demonstrated in [HHM04]). For example, there are approaches that produce initial populations that are generated adequately distributed in terms of tree size and distribution within the search space [GAMRRP07].

2.3.3   Breeding Populations of Programs

After preparing the GP process and initializing the population, the genetic process can be started. As it is the case in any GA, new individuals (programs) are created using recombination and mutation, tested, and become a part of the new population. Fitter individuals have a bigger chance to succeed in creating children of their own; thus, optimization happens during the run of the evolutionary algorithm. Unfit programs (and with them also their genetic material) wither out of the population.

As populations cannot grow infinitely in most applications, new programs somehow have to replace old ones that die off. There are in fact several ways how this replacement can be done:

- Generational replacement: The entire population is replaced by its descendants. This corresponds to generations changes in nature when for example annual plants or animals die in winter whereas their eggs (hopefully) survive; thus, the next generation of the species is founded.

- Steady state replacement: New individuals are produced continuously, and the removal of old individuals also happens continuously. Analogies in nature are obvious as this is more or less how for example human evolution happens.

- Selection of replaced programs: The individuals removed can be either chosen from the unfit ones (worst replacement), from the older ones (replacement with aging), or at random (random replacement), for example.

This whole procedure is graphically displayed in Figure 2.8 (adapted from [LP02]).

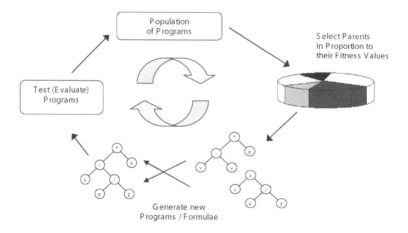

FIGURE 2.8: The genetic programming cycle [LP02].

In fact, the whole genetic programming process involves more than what is displayed in Figure 2.8: The preparatory steps summarized in Section 2.3.1 also have to be considered, and of course a validation of the results produced has to be done that might lead to a re-formulation of the pre-conditions. A more comprehensive overview of the GP process is given in Figure 2.9.

The execution of the GP cycle is – as GP is an extension to the GA – similar to the cyclic execution of the GA: Solutions are selected from the population, by crossing them they become parents, mutation is applied with a rather small probability, and thus a new offspring is produced. In the generational replacement scheme this is repeated until the next generation's population is complete; in the steady state scheme there is no generational cycle but this procedure is also repeated over and over again. The whole procedure is repeated until some pre-defined termination criterion is met (see Section 2.3.4 for details).

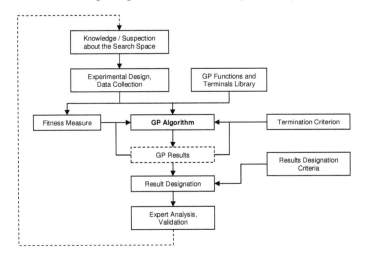

FIGURE 2.9: The GP-based problem solving process.

In fact, there is a veritable difference in the descriptions of this cyclic work-flow for GAs and for GP regarding the offspring creation scheme applied[5]:

- In GAs, crossover and mutation are used sequentially, i.e., both are applied (with mutation having a rather small probability).

- In GP, crossover or mutation (or a simple copy action) are executed independently; each time a new offspring is to be created, one of these variants is chosen probabilistically.

In fact, some researchers even recommend the GP-like offspring creation schema for all evolutionary computation systems (as for example given by Eick, see [Eic07]).

2.3.4 Process Termination and Results Designation

In general, the termination criteria of genetic algorithms are also applicable for genetic programming. A termination criterion might monitor the number of generations and terminate the algorithm as soon as a given limit is reached. Problem-specific criteria are also used frequently, i.e., the algorithm is termi-nated as soon as a problem-specific success predicate is fulfilled. In practice, one may manually monitor and manually terminate the run when the values of fitness for numerous successive best-of-generation individuals appear to have reached a plateau [KKS+03b].

---

[5]The GA workflow was described in detail in Chapter 1; the GP workflow as it is summarized here is also described in further detail in [Koz92b], [KKS+03b], and [ES03], for example.

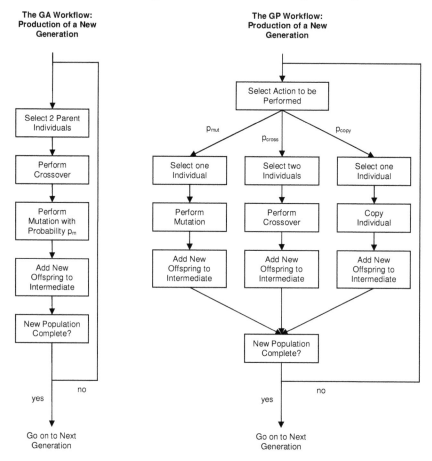

FIGURE 2.10: GA and GP flowcharts: The conventional genetic algorithm and genetic programming.

After terminating the algorithm it comes to the designation of the result returned by the algorithm. Normally, the single best-so-far individual is then harvested and designated as the result of the run [KKS+03b]. As we will see in Chapter 11 there are applications (as for example data-based structure identification) in which this is not the optimal strategy. In this case the use of a validation data set V is suggested, i.e., a data collection that was not used during the GP training phase; we eventually test the programs on V and pick the one that performs best on V.

2.4   Typical Applications of Genetic Programming

As genetic programming is a domain-independent method, there is an enormous number of applications for which it has been used for automatically producing solutions of high quality. Here we give a very short summary of exemplary problem classes which have been used for demonstrating GP's power in automatically learning programs for solving problems for more than 15 years, namely the automated learning of multiplexer functions (Section 2.4.1), the artificial ant (2.4.2), and symbolic regression (2.4.3). Finally, in Section 2.4.4 we give a short list of various problems for which GP has proven to be able to produce high quality results.

2.4.1   Automated Learning of Multiplexer Functions

The automated learning of functions requires the development of compositions of functions that can return correct values of functions after seeing only a relatively small number of specific examples; these training samples are combinations of values of the function associated with particular combinations of arguments.

The problem of learning Boolean multiplexer functions has become famous as a benchmark application for genetic programming since Koza's work on it for example presented in [Koz89] and [Koz92b]. The input to a Boolean k-multiplexer function is a bit-string consisting of k address bits a_i and 2^k data bits d_i; normally, the bits are thereby aligned following the form $[a_{k-1} \ldots a_1 a_0 d_{2^k-1} \ldots d_1 d_0]$. The value returned by the multiplexer function is the value of the particular data bit that is addressed by the k address bits. For example, let k be 3 and the three address bits $a_2 a_1 a_0 = 101$, then the multiplexer singles out data bit d_5 to be its output.[6] The abstract black box model of the Boolean multiplexer with three address bits and $2^3 = 8$ data bits as well as the concrete addressing of data bit d_5 is displayed in Figure 2.11.

A solution to this problem obviously has to be a function that uses input information a and d and calculates a Boolean return value. Thus, the terminal has $(k + 2^k)$ elements which correspond to the inputs to the multiplexer; in the case of $k = 3$ the terminal set $T = \{A_0, A_1, A_2, D_0, D_1, \ldots, D_7\}$. The functions used contain Boolean functions and the conditional function, i.e., $F = \{AND, OR, NOT, IF\}$. The evaluation of a solution candidate is done by applying the formula to all possible input bit combinations and counting the number of correct output values. As there are $(k + 2^k)$ inputs to the Boolean multiplexer, the number of possible input combinations is (2^{k+2^k}); in the case of $k = 3$, the number of possible input combinations is 2048.

---

[6]Data bit d_5 is in fact the sixth data bit since if $a_2 a_1 a_0 = 000$ data bit d_0 is addressed, so the indices of these data bits are zero-based.

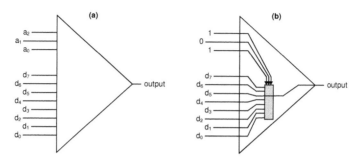

FIGURE 2.11: The Boolean multiplexer with three address bits; (a) general black box model, (b) addressing data bit d_5.

Koza was able to show that GP is able to solve the 3-address multiplexer problem 100% correctly [Koz92b]; this optimal result is shown in Figure 2.12. Of course, various test series have been documented in which GP was used for solving problem with multiplexers with more address bits in numerous publications.

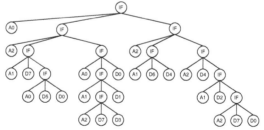

FIGURE 2.12: A correct solution to the 3-address Boolean multiplexer problem [Koz92b].

2.4.2　The Artificial Ant

The artificial ant problem ([CJ91a], [CJ91b], [JCC+92]) has also been a frequently used benchmark problem for GP since Koza's application [Koz92b]; meanwhile, it has become a well-studied problem in the GP community (see for example [LW95], [IIS98], [Kus98], [LP98], and [LP02]).

In short, the problem is to navigate an artificial ant on a grid consisting of

32×32 cells. The grid is toroidal so that if the ant moves off the edge of the grid, it reappears and continues on the opposite edge. On this grid, "food" units are distributed (normally along a trail); each time the ant enters a square containing food, the ant eats it. At the beginning of the ant's wanderings it starts at cell $(0, 0)$ facing in a particular direction (east, e.g.); at each time step, the ant is able to move forward in the direction it is facing, to turn right, or to turn left. The goal is to find a program that is able to navigate the ant so that as many food items as possible are eaten in a certain number of time units. The program can use the following:

- Three operations are available, namely Move, Left, and Right which let the ant move ahead, turn left, or turn right, respectively; these operations are used as terminals in the GP process.

- The sensing function IfFoodAhead investigates the cell the ant is currently facing and then executes the first child operation if food is ahead or the second child action otherwise.

- Additionally, two more functions are available: Prog2 and Prog3 take two and three arguments (operations), respectively, which are executed consecutively.

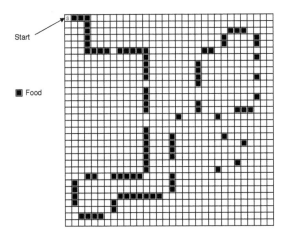

FIGURE 2.13: The Santa Fe trail.

The most frequently used trail is the so-called "Santa Fe trail" designed by Christopher Langton. This trail is displayed in Figure 2.13 (adapted from [LP02]); the ant is allowed to wander around the map for 600 time units. This problem is in fact considered a hard problem for GP; thorough explanations for this statement are for example given by Langdon and Poli in "Why ants are hard" ([LP98] and [LP02]). What makes it so hard is not that it is difficult to find correct solutions but rather to find these efficiently and significantly better than random search. As is listed in [LP98], the smallest solutions that solve the Santa Fe trail problem (i.e., those that provide programs that let the ant eat all food packets) are of length eleven[7]; one of them is exemplarily shown in Figure 2.14.

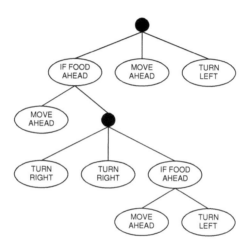

FIGURE 2.14: A Santa Fe trail solution. The black points represent nodes referencing to the `Prog3` function.

Even though it is a very "simple" problem, the artificial ant problem still provides a good basis for many theoretical investigations in GP such as building blocks and schema analysis [LP02], operators discussions ([LS97] or [IIS98], e.g.), further algorithmic development [CO07], and many other research activities.

2.4.3 Symbolic Regression

In short, symbolic regression is the induction of mathematical expressions on data. The key feature of this technique is, as Keijzer summarized in [Kei02], that the object of search is a symbolic description of a model, not just a set of coefficients in a pre-specified model. This is in sharp contrast with other

---

[7]In fact, there are 2,554,416 possible programs with length 11, but only 12 (i.e., 0.00047%) of them are successes. For programs of length 14 this ratio is approximately 0.0007%, for bigger program sizes (up to 200 – 500) it levels off between 0.0001% and 0.0002% [LP98].

methods of regression, including linear regression, polynomial approaches, or also artificial neural networks (ANNs), where a specific model is assumed and often only the complexity of this model can be varied.

The main goal of regression in general is to determine the relationship of a dependent (target) variable t to a set of specified independent (input) variables x. Thus, what we want to get is a function f that uses x and a set of coefficients w such that

$$t = f(x, w) + \epsilon \tag{2.1}$$

where ϵ represents the error (noise) term.

The form of f is usually pre-defined in standard regression techniques as for example linear regression (f_{LinReg}) and ANNs (f_{ANN}):

$$f_{LinReg}(x, w) = w_0 + w_1 x_1 + \ldots + w_n x_n \tag{2.2}$$

$$f_{ANN}(x, w) = w_0 \cdot g(w_1 x) \tag{2.3}$$

In linear regression, w is the set of coefficients $w_0, w_1, \ldots, w_n$. In ANNs we usually use an auxiliary transfer function g (which normally is a sigmoid function as for example the logistic function $\frac{1}{1+e^{-t}}$); the coefficients w are here called weights and include the weights from the hidden nodes to the output layer (w_0) and those from the input nodes to the hidden nodes (w_1) [Kei02].

In contrast to this, the function f which is searched for is not of any pre-specified form when applying genetic programming to symbolic regression. Instead, low-level functions are used and combined to more complex formulas during the GP process. Given a set of functions $f_1, \ldots, f_u$, the overall functional form induced by genetic programming can take a variety of forms. Usually, standard arithmetical functions such as addition, subtraction, multiplication, and division are in the set of functions f, but also trigonometric, logical, and more complex functions could be included.

An exemplary composed function therefore could be:

$$f(x, w) = f_1(f_4(x_1), f_5(x3, w1), f_4(f_2(x_1, w_2)), x_2)$$

or, by filling in some concrete functions for the abstract symbols f and w we could get:

$$f_1(x) = +(*(0.5, x), 1) \equiv 0.5 * x + 1$$

$$f_2(x) = +(2, *(x, x)) \equiv 2 + x * x$$

When it comes to evaluating solution candidates in a GP-based symbolic regression algorithm, the formulas have to be evaluated on a certain set of evaluation data X yielding the estimated values E. These estimated values are then compared to the original values T, i.e., those which are known from data retrieval (experiments) or calculated by applying the original formula to X.

For example, let f_{target} be the target function

$$f_{target}(x) = -(*(0.5, *(x, x)), 2) \equiv 0.5 * x^2 - 2 \tag{2.4}$$

and the functions f_1 and f_2 solution candidates. Furthermore, let the input data X be

$$X = [-5, -4, \ldots, +4, +5]. \tag{2.5}$$

Thus, by evaluating f_{target}, f_1, and f_2 on X we get T, E_1, and E_2:

$$T = [10.5, 6, 2.5, 0, -1.5, -2, -1.5, 0, 2.5, 6, 10.5] \tag{2.6}$$

$$E_1 = [-1.5, -1, -0.5, 0, 0.5, 1, 1.5, 2, 2.5, 3, 3.5] \tag{2.7}$$

$$E_2 = [27, 18, 11, 6, 3, 2, 3, 6, 11, 18, 27] \tag{2.8}$$

By crossing f_1 and f_2, these become parent functions (*parent formula 1* and *2*) and we could for example get the *child formula* f_3:

$$f_3(x) = +(*(0.5, *(x, x)), 1) \equiv 0.5 * x * x + 1 \tag{2.9}$$

and by evaluating it on X we get E_3:

$$E_3 = [13.5, 9, 5.5, 3, 1.5, 1, 1.5, 3, 5.5, 9, 13, 5] \tag{2.10}$$

Graphical displays of the formulas f_1, f_2, and f_3 (labeled as parent and child functions) and their evaluations are given in the Figures 2.16 and 2.15, respectively.

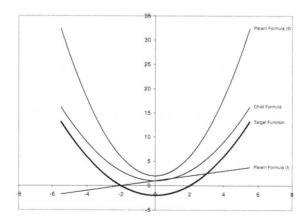

FIGURE 2.15: A symbolic regression example.

The task of GP in symbolic regression thus is to find a composition of the functions, input variables, and coefficients that minimizes the error of the function with respect to the desired target values. There are several ways how to measure this error, one of the simplest and probably most frequently

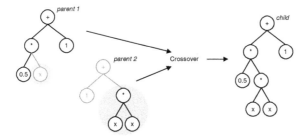

FIGURE 2.16: Exemplary formulas.

used ones being the mean squared error (mse) function; the mean squared error of the vectors A and B each containing n values is calculated as

$$mse(A, B) = \frac{1}{n} * \sum_{k=1}^{n} (A_k - B_k)^2; \; |A| = |B| = n \tag{2.11}$$

So, we can calculate the fitness of f_1, f_2, and f_3 as $mse(E_1, T)$, $mse(E_2, T)$, and $mse(E_3, T)$, respectively yielding

$$fitness(f_1) = 26.0 \tag{2.12}$$
$$fitness(f_2) = 100.5 \tag{2.13}$$
$$fitness(f_3) = 9.0 \tag{2.14}$$

Whereas the search for formulas that minimize a given error function (or maximize some other given fitness function) is the major goal of GP-based regression, the shape and the size of the solution could also be integrated into the fitness estimation function. The number and values of coefficients used is another issue that is tackled in the optimization process; the search process is also free whether to consider certain input variables or not, and thus it is able to perform variables selection (possibly leading to dimensionality reduction) [Kei02].

2.4.4 Other GP Applications

Finally we shall here give a short list of problems for which GP has proven to be able to produce high quality results - this list of course comes without the claim of completeness.

Koza can be for sure seen as one of the pioneers of applying GP to a variety of different problems: In [Koz92b], [Koz94], [KIAK99], and [KKS+03b] he reports (together with co-authors) on the GP-based solving of problems for example in classification, regression, pattern recognition, computational molecular biology, emergent behavior, cellular automata, sorting networks, design of topology and component sizing for complex hardware structures (such as analog electrical circuits, controllers, and antenna), and many others. Many of those results can be considered human-competitive results, some even being patentable new inventions created by GP.

In hardware design, for example, one of the problem situations explained in [KIAK99] is the automated design of amplifiers. In general, an amplifier is a circuit with one input and one output which multiplies the voltage of its input signal by a certain factor (the so-called voltage amplification factor) over a specified range of frequencies. The goal then is to realize such an amplifier only using resistors, capacitors, inductors, transistors, and power sources; the functions set used thus includes component creating functions for creating digital gates, inductors, transistors, power supplies, and resistors. Solution candidates are tree structures representing complete hardware entities which can be displayed in a way which we are used to.

Of course, there is a vast number of other fields of applications for genetic programming. Numerous applications of GP to problems of practical and scientific importance have for example also been documented in the conference proceedings of the GECCO, CEC, or EuroGP conferences ([CP$^+$03a], [CP$^+$03b], [D$^+$04a], [D$^+$04b], [B$^+$05], [K$^+$06], [T$^+$07], [KOL$^+$04], [KTC$^+$05], [CTE$^+$06], or [E$^+$07], e.g.). Please see the GP bibliography (Section 2.8) for a short list of sources of publications on those.

2.5   GP Schema Theories

As we have summarized *how* genetic programming works, we shall now turn our minds towards investigations *why* it works so well. Holland's work in the mid-1970s produced the well-known GA schema theorem; schemata have since then been frequently used to demonstrate how and why GAs work. In fact, as is summarized in [PMR04], in the 1990s interest in GA theory shifted towards exact microscopic Markov chain models possibly with aggregated states. However, after the work of Stephens and collaborators in the late 1990s on exact schema theories based on the notion of dynamic building blocks and the connection highlighted by Vose between his model and a different type of exact schema-based model, it is now clear that Markov-chain and schema-based models are, when exact, just different representations of the same thing.

Genetic programming theory has had a "difficult childhood," as Poli et al. stated in [PMR04]: After some early works on approximate GP schema theorems, it took quite some time until schema theories could be developed that give exact formulations for expected frequencies of schemata at the next generation.

In this section we give a rough overview of these GP schema theorems: After summarizing early work on GP schema theories in Section 2.5.1, which see schemata as components of programs, we give an introduction to rooted tree GP schema theories (Section 2.5.2) and an exact GP schema theory (Section 2.5.3). Finally, in Section 2.5.4 we summarize the GP schema theory concept.

The classification of schemata given in this section follows the grand concepts of [LP02], Chapters 3–6.

2.5.1   Program Component GP Schemata

First attempts to explain why GP works were given by Koza; in short, he gave an informal argument showing that Holland's schema theorem would work also for GP as described in [Koz92b], pp. 116–119. In Koza's definition, a schema is defined as a set of program subtrees (S-expressions); a schema can so be used for defining a subspace of the program trees search space by collecting all programs that include all subtrees given by the schema. For example, the schema H=[(+ x 3), y] includes the programs *(y,+(x,3)) and *(+(y,3),+(2,+(x,3))) as they both include (at least) one occurrence of the S-expressions (+ x 3) and y. This example is displayed graphically in Figure 2.17.

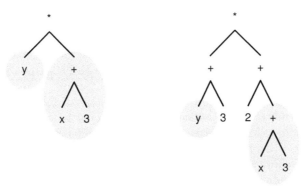

FIGURE 2.17: Programs matching Koza's schema H=[(+ x 3), y].

The first probabilistic model of GP that can be considered a mathematical formulation of a schema theorem for GP [LP02] was given by Altenberg in [Alt94a]. Also assuming very large populations, the neglection of mutation, and the application of proportional selection, he was able to calculate the frequency of a program at the next generation. Altenberg used a schema concept in which schemata are subexpressions and not, as in Koza's work, collections of subexpressions.

O'Reilly formalized Koza's work on schemata ([O'R95], [OO94]) and derived a schema theorem for GP with proportional selection and crossover (but without mutation). The main difference to Koza's approach was that she defined schemata as collections of subtrees and tree fragments; tree fragments in this context are trees with at least one leaf being a "don't care" symbol ('#'). O'Reilly was also able to calculate the frequency of a program at the next generation; unfortunately, the frequency depends on the shape, the size, and the composition of the trees containing the schemata investigated. Thus, frequencies are given rather as lower bounds than as concrete values. As O'Reilly argued in the discussion of her result, no hypotheses can be made on the basis of this theorem regarding the real propagation and the use of building blocks in GP.

Another approach was investigated by Whigham: He produced a definition of schemata for context free grammars and the related schema theorem which was published for example in [Whi95], [Whi96b], and [Whi96a]. Based on his definition of schemata he was able to give equations for the probabilities of disruption of schemata by crossover and mutation. Like in O'Reilly's work, also in Whigham's theorem the propagation of the components of schemata from one generation to the next is described.

In all these early attempts GP schemata were used for modeling how components (or groups of components) propagate within the population and how the number of these instances can vary over time.

2.5.2   Rooted Tree GP Schema Theories

In rooted tree GP schema theory, a schema can be seen as a set of points of the search space that share some syntactic feature. This can be defined in the following way [PMR04]: Let F be the set of functions used, and T the set of terminals. Syntactically, a GP schema is then defined as a tree composed of functions from the set $F \cup \{=\}$ and terminals from $T \cup \{=\}$; the primitive $=$ here means "don't care" and stands for a single terminal or function. Semantically, H is the set of programs that have the same shape and the same labels for the non-"$=$" nodes as the tree representation of H.

A simple example is given in Figure 2.18: Let F be defined as $F = \{+, -, *\}$ and T as $T = \{x, y, z\}$, and the schema H given as $*(=, = (x, =))$. For example, the programs $*(y, *(x, x))$, $*(z, +(x, z))$, and $*(x, -(x, z))$ are program members of H, i.e., they are included in H's semantics.

Rosca proposed this kind of schemata in [Ros97] (using the symbol '#' instead of '='). He formulated his schema theorem so that it became possible to calculate a lower bound for a schema's frequency at the next generation. As a matter of fact, here also schemata divide the space of programs into subspaces containing programs of different sizes and shapes.

Contrary to this, the following fixed-size-and-shape theory for GP was developed by Poli and Langdon ([PL97c], [PL97a]):

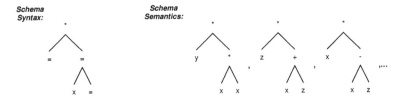

FIGURE 2.18: The rooted tree GP schema $*(=,= (x,=))$ and three exemplary programs of the schema's semantics.

Under the assumption that fitness proportional selection is applied, the probability of a program h sampling the schema H to be selected is

$$Pr\{h \in H\} = \frac{m(H,t)f(H,t)}{M\overline{f}(t)} \tag{2.15}$$

where $m(H,t)$ denotes the number of programs matching the schema H at generation t, $f(H,t)$ the mean fitness of programs matching H, M the population size, and $\overline{f}(t)$ the mean fitness of the programs in the population.

The main idea is that the probability of the disruption of a schema can be estimated. Let $D_c(H)$ be the event "H is disrupted when a program h matching H is crossed over with a program $\hat{h}$"; as is described in full detail in cite [LP02], the probability of such a disruption caused by one-point crossover can be formulated as

$$Pr\{D_c(H)\} \le p_{diff}(t)\left(1 - \frac{m(G(H),t)f(G(H),t)}{M\overline{f}(t)}\right)$$
$$+ \frac{\mathcal{L}(H)}{N(H)-1}\frac{m(G(H),t)f(G(H),t) - m(H,t)f(H,t)}{M\overline{f}(t)} \tag{2.16}$$

where $G(H)$ is the shape of all programs matching the schema H (which is called the *hyperspace* of H), and $\mathcal{L}(H)$ the defining length of H; p_{diff} is the probability of the disruption of schema H by crossing h (matching H) with program $\hat{h}$ that has different shape than h, i.e., which is not in $G(H)$: $p_{diff}(t) = Pr(D_c(H)|\hat{h} \notin G(H))$.

When it comes to point mutation, a schema H will survive mutation only if all of its $\mathcal{O}(H)$ defining nodes are not modified. Thus, the probability of H being disrupted by mutation $Pr\{D_m(H)\}$ is dependent on the probability of a node to be altered (p_m):

$$Pr\{D_m(H)\} = 1 - (1-p_m)^{\mathcal{O}(H)} \tag{2.17}$$

The overall formula uses these partial results and finally gives the expected number of programs matching schema H at generation $t+1$:

$$E[m(H,t+1)] \ge MPr\{h \in H\}(1-Pr\{D_m(H)\})(1-p_{xo}Pr\{D_c(H)\}) \tag{2.18}$$

By substituting (2.15), (2.16), and (2.17) in (2.18) we get the final overall formula for the lower bound of individuals sampling H at generation $t+1$ in generational GP with fitness proportional selection, one-point crossover, and point mutation as it is given in [LP02].

This GP schema theorem, produced by generalizing Holland's GA schema theorem, thus gives a pessimistic lower bound for the expected number of copies of a schema in the next generation. In the next chapter we will summarize an exact GP schema theory, produced by generalizing an exact GA schema theorem and using the concept of hyperschemata.

2.5.3   Exact GP Schema Theory

In the previous section we have summarized pessimistic GP schema theory based on generalization of Holland's GA schema theorem. As Langdon and Poli summarize in [LP02], the usefulness of these schema theorems has been widely criticized (see [CP94], [Alt94b], [FG97], [FG98], or [Vos99], e.g.). In order to overcome its main drawbacks, namely that they are pessimistic and only give lower bounds for the expected numbers of instances for a given schema at the next generation, more exact schema theorems for GAs and GP had to be developed. These are going to be summarized in this section: After explaining the main idea of Stephen and Waelbroeck's GA schema theory, the hyperschema concept is summarized, and finally, on the basis of these hyperschemata, exact GP schema theorems.

An exact GA schema theorem has been developed by the end of the last millennium ([SW97], [SW99]): The total transmission probability α of a schema H is defined so that $\alpha(H,t)$ is the probability that at generation t the individuals of the GA's population will match H. Assuming a crossover probability p_{xo}, $\alpha(H,t)$ is calculated as:

$$\alpha(H,t) = (1 - p_{xo})p(H,t) + \frac{p_{xo}}{N-1} \sum_{i=1}^{N-1} p(L(H,i),t)p(R(H,i),t) \qquad (2.19)$$

with $L(H,i)$ and $R(H,i)$ being the left and right parts of schema H, respectively, and $p(H,t)$ the probability of selecting an individual matching H to become a parent. The "left" part of a schema H is thereby produced by replacing all elements of H at the positions from the given index i to N with "don't care" symbols (with N being the length of the bit strings); the "right" part of a schema H is produced by replacing all elements of H from position 1 to i with "don't care." The summation sums over all positions from 1 to $N-1$, i.e., over all possible crossover points. A generalization of this theorem to variable-length GAs has also been constructed [SPWR02].

After the publication of this exact GA schema theory, immediately the question came to mind whether it would be possible to extend pessimistic GP schema theories towards an exact GP schema theorem [LP02]. In fact, it was: Poli developed an exact GP schema theorem (see [Pol99a], [Pol00c],

[Pol00b], [Pol00a], e.g.), a theorem which was then generalized by Poli and McPhee to become known as Poli and McPhee's Exact GP Schema Theorem ([PM01b], [PM01a], [PRM01], [PM01b], [Pol01], [PM03a], [PM03b], [PMR04], and [LP02]).

Assuming equal size and shape for GP programs, (2.19) can be also used for describing the transmission probability of a fixed-size-and-shape GP schema. In the presence of one-point crossover, the transmission probability for a GP schema H at generation t, $\alpha(H,t)$, can be thus given as

$$\alpha(H,t) = (1 - p_{xo})p(H,t) + \frac{p_{xo}}{N(H)} \sum_{i=1}^{N-1} p(l(H,i),t)p(u(H,i),t) \qquad (2.20)$$

with $l(H,i)$ and $u(H,i)$ being the lower and upper parts (building blocks) of schema H, respectively, and $N(H)$ the number of nodes in the schema (which is assumed to have the same size and shape as all other programs in the population). $l(H,i)$ is defined as the schema produced by replacing all nodes above cutting point i with "don't care" symbols, and $u(H,i)$ as the schema produced by replacing all nodes below cutting point i with "don't care" symbols. In analogy to (2.19), the summation in (2.20) sums over all possible crossover points.

Exemplary l and u schemata for the schema H $= +(*(=,x),=)$ are shown in Figure 2.19.

In order to generalize this exact GP schema theorem so that it can be applied to populations of programs of different sizes and shapes, a more general schema approach is used, namely the GP hyperschema concept.

A GP hyperschema represents a set of schemata in the same way as a schema represents a set of program trees (which is why it is called "hyperschema"). This can be defined in the following way [PMR04]: Let F be the set of functions used, and T the set of terminals. Syntactically, a GP schema is then defined as a tree composed of functions from the set $F \cup \{=\}$ and terminals from $T \cup \{=, \#\}$. The primitives = and # here mean "don't care"; = stands for exactly one node, whereas # stands for any valid subtree.

Examples are shown in Figure 2.20: Let F be defined as $F = \{+, -, *\}$, T as $T = \{x, y, z\}$, and the hyperschema H given as $*(\#, = (x, =))$. The three exemplary programs $*(y, *(x, *))$, $*(*(x, y), +(x, z))$, and $*(*(*(x, y), y), +(x, z))$ are a part of H's semantics.

In analogy to $l(H,i)$ and $u(H,i)$ defined above and sketched in Figure 2.19, the hyperschemata building blocks $L(H,i)$ and $U(H,i)$ are defined in the following way: $L(H,i)$ is the hyperschema obtained by replacing all nodes on the path between crossover point i and the root of hyperschema H with = nodes, and all subtrees connected with those nodes with # nodes. $U(H,i)$ is the hyperschema obtained by replacing the subtree below crossover point i with a # node [PMR04].

As examples might here also help to make this concept clearer, Figure 2.21 shows an exemplary schema $H = +(*(=, x), =)$ and potential hyperschema

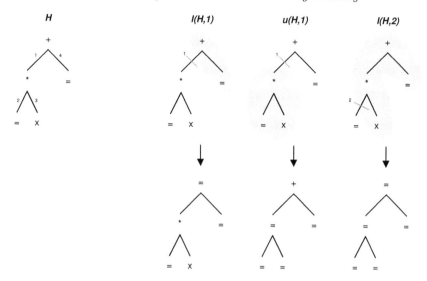

FIGURE 2.19: The GP schema H = +(*(=,x),=) and exemplary u and l schemata. Cross bars indicate crossover points; shaded regions show the parts of H that are replaced by "don't care" symbols.

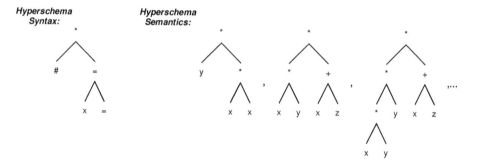

FIGURE 2.20: The GP hyperschema *(#,= (x, =)) and three exemplary programs that are a part of the schema's semantics.

building blocks. As for example shown in the second column, $L(H,1)$ is constructed by turning all nodes between crossover point 1 and the root (in this case only the root node) into = nodes, and all subtrees of the so modified nodes become # nodes. $U(H,1)$ is in column 3 constructed by replacing the subtree under crossover point 1 into a # node. And finally, as can be seen in column 4, $L(H,2)$ is again constructed by turning all nodes from crossover point 2 to the root into = nodes, and all subtrees of the so modified nodes become # nodes.

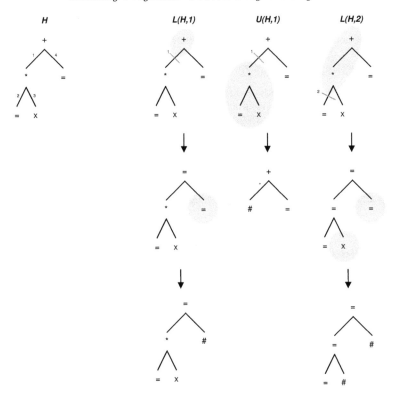

FIGURE 2.21: The GP schema $H = +(*(=,x),=)$ and exemplary U and L hyperschema building blocks. Cross bars indicate crossover points; shaded regions show the parts of H that are modified.

Using hyperschemata, it is possible to formulate a general, exact GP schema theorem for populations of programs of any size or shape. The total transmission probability of a fixed-size-and-shape GP schema H is, for GP with one-point crossover and no mutation, given as

$$\alpha(H,t) = (1 - p_{xo})p(H,t)+ \qquad (2.21)$$

$$p_{xo}\sum_{h_1}\sum_{h_2}\frac{p(h_1,t)p(h_2,t)}{\mathrm{NC}(h_1,h_2)}\sum_{i\in C(h_1,h_2)}\delta(h_1 \in L(H,i))\delta(h_2 \in U(H,i))$$

where $\mathrm{NC}(h_1, h_2)$ is the number of nodes in the tree fragment representing the common region of the programs h_1 and h_2, $C(h_1, h_2)$ is the set of indices of the crossover points in the common region of h_1 and h_2, and $\delta(x)$ is a function that returns 1 if x is true and 0 otherwise. The first two summations sum over all individuals in the population, i.e., we sum over all possible pairs of programs; the second summation sums over all indices of crossover points of the common region of the respective programs pair.

This GP schema theorem is called the "Microscopic Exact GP Schema Theorem" in the sense that it is necessary to consider each member of the population.

Via several transformations and lemmata (which are not given here) it is finally possible to formulate the "Macroscopic Exact GP Schema Theorem":

$$\alpha(H,t) = (1 - p_{xo})p(H,t)+ \qquad\qquad (2.22)$$

$$p_{xo}\sum_{j}\sum_{k}\frac{1}{\mathrm{NC}(G_j,G_k)}\sum_{i\in C(G_j,G_k)} p(L(H,i)\cap G_j,t)p(U(H,i)\cap G_k,t))$$

where $G(H)$ denotes the schema that is obtained by replacing all nodes in a schema H by "don't care" symbols[8]; the sets $L(H,i)\cap G_j$ and $U(H,i)\cap G_k$ are either schemata (of fixed size and shape), or the empty set $\emptyset$.

Thus, using this theorem (2.5.3), it is at last possible to give the exact transmission probability of a schema for genetic programming under one-point crossover and no mutation; an exact schema theorem for GP is established. We have here omitted lots of transformation steps and proofs; for these, the interested reader is for example referred to [PM03a], [PM03b], [LP02], or [PMR04].

An overview of the development of approximate and exact schema theorems for GAs and GP is graphically shown in Figure 2.22 (as given in [PMR04]).

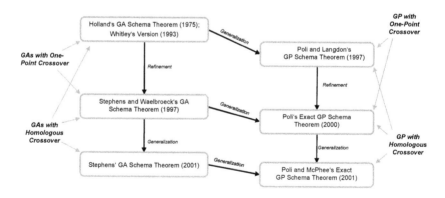

FIGURE 2.22: Relation between approximate and exact schema theorems for different representations and different forms of crossover (in the absence of mutation).

---

[8]$G(H)$ is called the *hyperspace* of H.

2.5.4   Summary

Until the development of the GP schema theorems described in this section, GP theory was typically considered scarce, approximate, and not terribly useful [PM01c]. The facts, that GP is relatively young and that building theories for variable size structures are very complex, are considered the reasons for this.

Significant breakthroughs, which have been summarized in this section, have fundamentally changed this understanding; after the development of GP schema theorems, we now have an exact GP theory based on schema and hyperschema concepts.

2.6   Current GP Challenges and Research Areas

Of course, theoretical work on GP was by far not finished after the development of GP schema theorems. Even though they shall not be discussed in detail here, we still want to line out a selection of current research areas in GP theory.

For example, operators design for GP has been discussed in numerous publications; extensive analysis of initialization, crossover, and mutation operators can be found in [Lan99], [ES03], or [LN00], for example.

The genetic programming search space has been subject to theoretical analysis (see [LP98], [LP02], e.g.). Experimental exploration of the GP search space by random sampling can be used for comparing GP to random search or other search techniques. Additionally, hypotheses have been stated regarding minimum and maximum tree depth.

As has already been mentioned before, a Markov model for GAs has been formulated by Vose, see [NV92], [VL91], and [Vos99] for explanations. In short, a GA is modeled as a Markov chain; selection, mutation, and crossover are incorporated into an explicitly given transition matrix, thus the method is complete, and no special assumptions are made which restrict populations or population trajectories.

This GA Markov model could also be extended to GP using the schema GP theory described in the previous section, which gives exact formulas for computing the probability that reproduction and recombination create any specific program. A GP Markov chain model is then easily obtained by plugging this ingredient into a minor extension of Vose's model of GAs [PMR04]; in fact, an alternative approach for describing the dynamics of evolutionary algorithms is provided by this theory.

One fact has been known for genetic programming since some of its first applications and has been frequently reported: Programs in genetic programming populations tend to grow in size ([Ang94], [Lan95], [NB95], [SFD96],

[AA05], [Ang98], [TH02]). "Redundancy," "introns," and, probably most frequently used as well as with the most negative connotation, "bloat" have (amongst others) been used since then as names for this tendency. In principle, it means that introns, i.e., code which has no effect on the performance of the program containing it, grow during the GP process; it is in fact a phenomenon also known from natural evolution [WL96].

Of course, this seems to be an unwanted phenomenon and does not conform to "Occam's Razor", a law attributed to the 14th-century Francisian friar William of Ockham. This law is also known as the "law of parsimony," the Latin principle "entia non sunt multiplicanda praeter necessitatem" meaning that "entities should not be multiplied beyond necessity" is also often quoted. In principle, this law demands the selection of exactly that theory that postulates the fewest entities and introduces the fewest assumptions (of course, in case if there are multiple competing theories which are considered equal in other respects). Argumentations pointing out how and why GP does or does not fulfill Occam's law can be found in [Dro98] and [LP97], for example.[9]

Examples for bloat are given in Figure 2.23: In the left example, the left subtree will always return $(x - (0 * y + x)) = x - x = 0$ and since the multiplication of 0 with any other value always results in 0, the result of the whole program will always be 0 regardless of the values of x, y, and z. In fact, the whole right subtree becomes code that does not influence the whole program's evaluation. In the second example shown on the right part of Figure 2.23, A will always be smaller than $A + 4$; thus the condition of the root condition will always be fulfilled and "else"-branch will never be activated.

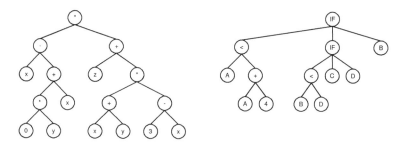

FIGURE 2.23: Examples for bloat.

In contrary to the examples in Figure 2.23, in which bloat is rather obvious,

---

[9] Especially "The Myth of Occam's Razor" [W18], a paper written by Thorburn in 1918, is worth reading in this context as it discusses the origins of the principle. For more discussions on Occam's razor and its reception in philosophy and science the interested reader is referred to [Jac94], [Nol97], [Pop92], or [RF99].

there are also of course examples in which it can be seen that GP will not always automatically produce rather simple results.

In their article entitled "Fitness Causes Bloat" [LP97], Langdon and Poli showed that fitness-based selection seems to be responsible for the solutions' growth in size; fitness-based parent selection therefore leads to code bloat. In this context bloat has also been ironically described as "survival of the fattest."

According to [Zha97], [Zha00], and [LP02], approaches used for preventing or at least decreasing bloat include, but are not restricted to the following anti-bloat techniques:

- Size and/or depth limitations: The growth of programs is limited, programs are not allowed to become bigger in size and/or depth (where the size of a program is normally the size of its structure tree and its depth the depth of its structure tree). Size limits are nowadays commonly used, see for example [KIAK99].

- Incorporation of program size in the selection process: An also often used technique to combat bloat is to include some preference for smaller programs in the criterion used to select programs for reproduction; this additional factor to selection is also called parsimony pressure. Examples and analysis can be for example found in [Kin93], [Zha97], [Zha00], [SF98], and [SH98]

- Incorporation of program size in evaluation: The size of a program could of course also be incorporated in its evaluation. It might also be included as one of the goals which the GP population tries to reach ([LN00], [EN01]).

- Genetic operators: Besides selection and evaluation, several crossover and mutation operators have been proposed which are designed so that they combat bloating, see for example [Ang98], [PL97b], or [Lan00].

Often we see another tendency of GP that does not fulfill Occam's law, namely that it is prone to producing programs that are overspecified. This means that programs that are too complex for the problem at hand and that much simpler programs could fulfill the given task as well; especially in data-based modeling this phenomenon is also known as "overfitting." We shall come back to this topic in Chapter 11.

Another field of GP research is the development of practical guides for ideal parameter settings for GP. As we find in [SOG04], for example, GP researchers and practitioners are often frustrated by the lack of theory available to guide them in selecting key algorithm parameters; GP population sizes, for example, run from ten to a million members or more, but at present there is no practical guide to knowing when to choose which size. [SOG04] here gives a population-sizing relationship depending on tree size, solution complexity, problem difficulty, and building block expression probability.

Furthermore, numerous other theoretical topics are widely discussed in the GP community, lots of them directly connected to well known problems (or rather challenges) with GP. Selected ones are to be mentioned in the next chapters.

As a part of the conclusions of [LP02], Langdon and Poli demand that GP users might like to consider how their GP populations are evolving, whether they are converging, and, if so, whether they are converging in the right direction. At the present, many GP packages offer only few possibilities to monitor populations. As we are going to demonstrate in later chapters, this is exactly what we try to accomplish by investigating dynamics in the populations of our GA and GP implementations.

2.7   Conclusion

In this chapter, genetic programming has been summarized and described as a powerful extension to the genetic algorithm. In fact, GP is more than a GA extension: It can be rather seen as the art of evolving computer programs and as a generic concept for the automated programming of computers.

After describing GP basics and a variety of applications for GP, we have summarized theoretical concepts for GP-based on schemata and hyperschemata. Problems and challenges in the context of GP have also been discussed.

In the following chapters we shall now come back to algorithmic developments in GAs. We will especially concentrate on enhanced algorithmic concepts which have been developed in order to support crossover-based evolutionary algorithms in their intention to combine those parts of chromosomes that define high quality solutions; these advanced concepts can of course also be used with GP.

In Chapter 11 we then come back to GP and its application to data-based system identification; we also demonstrate the effects of these algorithmic enhancements in GP.

2.8   Bibliographic Remarks

There are numerous books, journals, and articles available that survey the field of genetic programming. In this section we summarize some of the most important ones. Representatively, the following books are widely considered very important sources of information about GP:

- J. R. Koza et al.: *Genetic Programming I - IV* ([Koz92b], [Koz94], [KIAK99], [KKS+03b]): A series of books on theory and praxis of genetic programming by John Koza and varying co-authors

- W. Banzhaf et al.: *Genetic Programming – An Introduction* [BNKF98]

- W. Langdon: *Genetic Programming and Data Structures* [Lan98]

- W. Langdon and R. Poli: *Foundations of Genetic Programming* [LP02]

The following journals are dedicated to either theory and applications of genetic programming or evolutionary computation in general:

- *Genetic Programming and Evolvable Machines* (Springer Netherlands)

- *IEEE Transactions on Evolutionary Computation* (IEEE)

- *Evolutionary Computation* (MIT Press)

Moreover, several conference and workshop proceedings include papers related to genetic programming. Some examples are the following ones:

- *Genetic and Evolutionary Computation Conference (GECCO)*, a recombination of the *International Conference on Genetic Algorithms* and the *Genetic Programming Conference*

- *Congress on Evolutionary Computation (CEC)*

- *Parallel Problem Solving from Nature (PPSN)*

- *European Conference on Genetic Programming (EuroGP)*

Of course there is lots of GP-related information available on the internet including theoretical background and practical applications, course slides, and source code. Probably the most comprehensive overview of publications in GP is *The Genetic Programming Bibliography* which is maintained by Langdon, Gustavson, and Koza and available at http://www.cs.bham.ac.uk/~wbl/biblio/.

Finally, publications of the *Heuristic and Evolutionary Algorithms Laboratory (HEAL)* (including several articles on GAs and GP) are available at http://www.heuristiclab.com/publications/.

Chapter 3

Problems and Success Factors

DOI: 10.1201/9781420011326-4

3.1 What Makes GAs and GP Unique among Intelligent Optimization Methods?

In contrast to trajectory-based heuristic optimization techniques such as simulated annealing or tabu search, and also in contrast to population-based heuristics which perform parallel local search as for example the conventional variants of evolution strategies (ES without recombination), genetic algorithms and genetic programming operate under fundamentally different assumptions.

A neighborhood-based method usually scans the search space around a current solution in a predefined neighborhood in order to take moves to more promising directions, and are therefore often confronted with the problem of getting stuck in a local, but not global optimum of a multimodal solution space.

What makes GAs and GP unique compared to neighborhood-based search techniques is the crossover procedure which is able to assemble properties of solution candidates which may be located in very different regions of the search space. In this sense, the ultimate goal of any GA or GP is to assemble and combine the essential genetic information (i.e., the alleles of a globally optimal or at least high quality solution) step by step. This information is initially scattered over many individuals and must be merged to single chromosomes by the final stage of the evolutionary search process. This perspective, which is under certain assumptions stated in the variants of the schema theory and the according building block hypothesis, should ideally hold for any GA or GP variant. This is exactly the essential property that has the potential to make GAs and GP much more robust against premature stagnation in local optimal solutions than search algorithms working without crossover.

3.2   Stagnation and Premature Convergence

The fundamental problem which many meta-heuristic optimization methods aim to counteract with various algorithmic tricks is the stagnation in a locally, but not globally optimal solution. As stated previously, due to their methodology GAs and GP suffer much less from this problem.

Unfortunately, also users of evolutionary algorithms using crossover frequently encounter a problem which, at least in its effect, is quite similar to the problem of stagnating in a local, but not global optimum. This drawback, in the terminology of GAs called premature convergence, occurs if the population of a GA reaches such a suboptimal state that the genetic solution manipulation operators (crossover and mutation) are no longer able to produce offspring that outperform their parents (as discussed for example in [Fog94], [Aff03]). In general, this happens mainly when the genetic information stored in the individuals of a population does not contain that genetic information which would be necessary to further improve solution quality.

Several methods have been proposed to combat premature convergence in genetic algorithms (see [LGX97], [Gao03], or [Gol89], e.g.). These include, for example, the restriction of the selection procedure, the operators, and the according probabilities as well as the modification of the fitness assignment. However, all these methods are heuristic per definition, and their effects vary with different problems and even problem instances. A critical problem in studying premature convergence therefore is the identification of its occurrence and the characterization of its extent. Srinivas and Patnaik [SP94], for example, use the difference between the average and maximum fitness as a standard to measure genetic diversity, and adaptively vary crossover and mutation probabilities according to this measurement.

Classical Measures for Diversity Maintenance

The term "population diversity" has been used in many papers to study premature convergence (e.g., [SFP93], [YA94]) where the decrease of population diversity (i.e., a homogeneous population) is considered as the primary reason for premature convergence. The basic approaches for retarding premature convergence discussed in GA literature aim to maintain genetic diversity. The most common techniques for this purpose are based upon pre-selection [Cav75], crowding [DeJ75], or fitness-sharing [Gol89]. The main idea of these techniques is to maintain genetic diversity by the preferred replacement of similar individuals [Cav75], [DeJ75] or by the fitness-sharing of individuals which are located in densely populated regions [Gol89]. While methods based upon those discussed in [DeJ75] or [Gol89] require some kind of neighborhood measure depending on the problem representation, the approach given in [Gol89] is additionally quite restricted to proportional selection.

Limitations of Diversity Maintenance

In basic GA literature the topic of premature convergence is considered to be closely related to the loss of genetic variation in the entire population ([SFP93], [YA94]). In the opinion of the authors this perspective, which mainly stems from natural evolution, should be considered in more detail for the artificial evolutionary process as being performed by an GA or GP. In natural evolution the maintenance of genetic diversity is of major importance as a rich gene pool enables a certain species to adapt to changing environmental conditions. In the case of artificial evolution, the environmental conditions, for which the chromosomes are to be optimized, are represented in the fitness function which usually remains unchanged during the run of an algorithm. Therefore, we do not identify the reasons for premature convergence in the loss of genetic variation in general but more specifically in the loss of what we call essential genetic information, i.e., in the loss of alleles which are part of a global optimal solution. Even more specifically, whereas the alleles of high quality solutions are desired to remain in the gene pool of the evolutionary process, alleles of poor solutions are desired to disappear from the active gene pool in order to strengthen the goal-directedness of evolutionary search.

Therefore, in the following we denote the genetic information of the global optimal solution (which is unknown to the algorithm) as essential genetic information. If parts of this essential genetic information are missing or get lost, premature convergence is already predetermined in a certain way as only mutation (or migration in the case of parallel GAs) is able to regain this genetic information.

A very essential question about the general performance of a GA is whether or not good parents are able to produce children of comparable or even better fitness – after all, the building block hypothesis implicitly relies on this. Unfortunately, this property cannot be guaranteed easily for GA applications in general: The disillusioning fact here is that the user has to take care of an appropriate encoding in order to make this fundamental property hold.

Reconsidering the basic functionality of a GA, the algorithm selects two above average parents for recombination and sometimes (with usually rather low probability) mutates the crossover result. The resulting chromosome is then considered as a member of the next generation and its alleles are therefore part of the gene pool for the ongoing evolutionary process.

Reflecting the basic concepts of GAs, the following questions and associated problems arise:

- Is crossover always able to fulfill the implicit assumption that two above-average parents can produce even better children?

- Which of the available crossover operators is best suited for a certain problem in a certain representation?

- Which of the resulting children are "good" recombinations of their parents chromosomes?

- What makes a child a "good" recombination?

- Which parts of the chromosomes of above-average parents are really worth being preserved?

Conventional GAs are usually not always able to answer these questions in a satisfactory way, which should ideally hold for any GA application and not only for a canonical GA in the sense of the schema theorem and the building block hypothesis. These observations constitute the starting point for generic algorithmic enhancements as stated in the following chapters. The preservation of essential genetic information, widely independent of the actually applied representation and operators, plays a main role. These advanced evolutionary algorithm techniques called offspring selection, relevant alleles preserving genetic algorithm (RAPGA), and SASEGASA will be exemplarily compared to a conventional GA in Chapter 7 and extensively analyzed in the experimental part of the book on the basis of various problems.

Chapter 4

Preservation of Relevant Building Blocks

DOI: 10.1201/9781420011326-5

4.1 What Can Extended Selection Concepts Do to Avoid Premature Convergence?

The ultimate goal of the extended algorithmic concepts described in this chapter is to support crossover-based evolutionary algorithms, i.e., evolutionary algorithms that are ideally designed to function as building-block assembling machines, in their intention to combine those parts of the chromosomes that define high quality solutions. In this context we concentrate on selection and replacement which are the parts of the algorithm that are independent of the problem representation and the according operators. Thus, the application domain of the new algorithms is very wide; in fact, offspring selection and the RAPGA (a special variant of adaptive population sizing GA) can be applied to any application that can be treated by genetic algorithms (of course also including genetic programming).

The unifying purpose of the enhanced selection and replacement strategies is to introduce selection after reproduction in a way that checks whether or not crossover and mutation were able to produce a new solution candidate that outperforms its own parents. Offspring selection realizes this by claiming that a certain ratio of the next generation (pre-defined by the user) has to consist of child solutions that were able to outperform their own parents (with respect to their fitness values). The RAPGA, the second newly introduced selection and replacement strategy, ideally works in such a way that new child solutions are added to the new population as long as it is possible to generate unique and successful offspring stemming from the gene pool of the last generation. Both strategies imply a self-adaptive regulation of the actual selection pressure that depends on how easy or difficult it is at present to achieve evolutionary progress. An upper limit for the selection pressure provides a good termination criterion for single population GAs as well as a trigger for migration in parallel GAs.

4.2   Offspring Selection (OS)

As already discussed at length, the first selection step chooses the parents for crossover either randomly or in any other well-known way as for example roulette-wheel, linear-rank, or some kind of tournament selection strategy. After having performed crossover and mutation with the selected parents, we introduce a further selection mechanism that considers the success of the apparently applied reproduction. In order to assure that the progression of genetic search occurs mainly with successful offspring, this is done in such a way that the used crossover and mutation operators are able to create a sufficient number of children that surpass their parents' fitness. Therefore, a new parameter called success ratio ($SuccRatio \in [0,1]$) is introduced. The success ratio is defined as the quotient of the next population members that have to be generated by successful mating in relation to the total population size. Our adaptation of Rechenberg's success rule ([Rec73], [Sch94]) for genetic algorithms says that a child is successful if its fitness is better than the fitness of its parents, whereby the meaning of "better" has to be explained in more detail: Is a child better than its parents, if it surpasses the fitness of the weaker parent, the better parent, or some kind of weighted average of both?

In order to answer this question, we have borrowed an aspect from simulated annealing: The threshold fitness value that has to be outperformed lies between the worse and the better parent and the user is able to adjust a lower starting value and a higher end value denoted as comparison factor bounds; a comparison factor ($CompFactor$) of 0.0 means that we consider the fitness of the worse parent, whereas a comparison factor of 1.0 means that we consider the better of the two parents. During the run of the algorithm, the comparison factor is scaled between the lower and the upper bound resulting in a broader search at the beginning and ending up with a more and more directed search at the end; this procedure in fact picks up a basic idea of simulated annealing.

In the original formulation of the SASEGASA (which will be described in Chapter 5) we have defined that in the beginning of the evolutionary process an offspring only has to surpass the fitness value of the worse parent in order to be considered as "successful"; as evolution proceeds, the fitness of an offspring has to be better than a fitness value continuously increasing between the fitness values of the weaker and the better parent. As in the case of simulated annealing, this strategy gives a broader search at the beginning, whereas at the end of the search process this operator acts in a more and more directed way. Having filled up the claimed ratio ($SuccRatio$) of the next generation with successful individuals using the success criterion defined above, the rest of the next generation ($(1 - SuccRatio) \cdot |POP|$) is simply filled up with individuals randomly chosen from the pool of individuals that were also created by crossover, but did not reach the success criterion. The actual selection pressure $ActSelPress$ at the end of generation i is defined by the quotient of individuals that had to be considered until the success ratio was reached and the number of individuals in the population in the following way:

$$ActSelPress = \frac{|POP_{i+1}| + |POOL_i|}{|POP_i|} \tag{4.1}$$

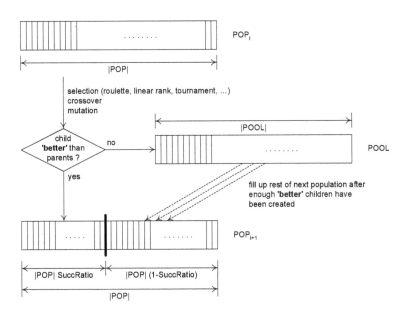

FIGURE 4.1: Flowchart of the embedding of offspring selection into a genetic algorithm. This figure is displayed with kind permission of Springer Science and Business Media.

Figure 4.1 shows the operating sequence of the concepts described above.

An upper limit of selection pressure (*MaxSelPress*) defines the maximum number of offspring considered for the next generation (as a multiple of the actual population size) that may be produced in order to fulfill the success ratio. With a sufficiently high setting of *MaxSelPress*, this new model also functions as a detector for premature convergence:

If it is no longer possible to find a sufficient number (SuccRatio · |POP|) of offspring outperforming their own parents even if (MaxSelPress · |POP|) candidates have been generated, premature convergence has occurred.

As a basic principle of this selection model, higher success ratios cause higher selection pressures. Nevertheless, higher settings of success ratio, and therefore also higher selection pressures, do not necessarily cause premature convergence. The reason for this is mainly that the new selection step does not accept clones that emanate from two identical parents per definition. In conventional GAs such clones represent a major reason for premature convergence of the whole population around a suboptimal value, whereas the new offspring selection works against this phenomenon (see Chapters 7, 10, and 11).

With all strategies described above, finally a genetic algorithm with the additional offspring selection step can be formulated as stated in Algorithm 4.1. The algorithm is formulated for a maximization problem; in case of minimization problems the inequalities have to be changed accordingly.

Algorithm 4.1 Definition of a genetic algorithm with offspring selection.

Initialize total number of iterations $nrOfIterations \in \mathbb{N}$
Initialize actual number of iterations $i = 0$
Initialize size of population $|POP|$
Initialize success ratio $SuccRatio \in [0, 1]$
Initialize maximum selection pressure $MaxSelPress \in \,]1, \infty[$
Initialize lower comparison factor bound $LowerBound \in [0, 1]$
Initialize upper comparison factor bound $UpperBound \in [LowerBound, 1]$
Initialize comparison factor $CompFactor = LowerBound$
Initialize actual selection pressure $ActSelPress = 1$
Produce an initial population POP_0 of size $|POP|$

while $(i < nrOfIterations) \wedge (ActSelPress < MaxSelPress)$ **do**
    Initialize next population POP_{i+1}
    Initialize pool for bad children $POOL$

    while $(|POP_{i+1}| < (|POP| \cdot SuccRatio)) \wedge (((|POP_{i+1}| + |POOL|) < (|POP| \cdot MaxSelPress))$ **do**
        Generate a child from the members of POP_i based on their fitness values using crossover and mutation

        Compare the fitness of the child c to the fitness of its parents par_1 and par_2 (without loss of generality assume that par_1 is fitter than par_2)
        if $f_c \leq (f_{par_2} + |f_{par_1} - f_{par_2}| \cdot CompFactor)$ **then**
            Insert child into $POOL$
        else
            Insert child into POP_{i+1}
        end if
    end while
    $ActSelPress = \frac{|POP_{i+1}| + |POOL|}{|POP|}$

    Fill up the rest of POP_{i+1} with members from $POOL$
    while $|POP_{i+1}| \leq |POP|$ **do**
        Insert a randomly chosen child from $POOL$ into POP_{i+1}
    end while
    Adapt $CompFactor$ according to the given strategy
    $i = i + 1$
end while

For a detailed analysis of the consequences of offspring selection the reader is referred to Chapter 7 where the characteristics of a GA incorporating offspring selection will be compared to the characteristics of a conventional GA on the basis of a benchmark TSP.

4.3 The Relevant Alleles Preserving Genetic Algorithm (RAPGA)

Assuming generational replacement as the underlying replacement strategy the most essential question at generation i is which parts of genetic information from generation i should be maintained in generation $i + 1$ and how this could be done most effectively applying the available information (chromosomes and according fitness values) and the available genetic operators selection, crossover, and mutation.

The here presented variant of enhanced algorithmic concepts based upon GA-solution manipulation operators aims to achieve this goal by trying to bring out as much progress from the actual generation as possible and losing as little genetic diversity as possible at the same time.

This idea is implemented using ad hoc population size adjustment in the sense that potential offspring generated by the basic genetic operators are accepted as members of the next generation if and only if they are able to outperform the fitness of their own parents and if they are new in the sense that their chromosome consists of a concrete allele alignment that is not represented yet in an individual of the next generation. As long as new and (with respect to the definition given previously) "successful" individuals can be created from the gene pool of the actual generation, the population size is allowed to grow up to a maximum size. A potential offspring which is not able to fulfill these requirements is simply not considered for the gene pool of the next generation.

Figure 4.2 represents the gene pool of the alleles at a certain generation i and Figure 4.3 illustrates how this genetic information can be used in order to generate a next population $i + 1$ of a certain size which may be smaller or larger than that of the actual population i. Whether the next population becomes smaller or larger depends on the success of the genetic operators crossover and mutation in the above stated claim to produce new and successful chromosomes.

For a generic, stable, and robust realization of these RAPGA ideas some practical aspects have to be considered:

- The algorithm should offer the possibility to use different settings also for conventional parent selection, so that the selection mechanisms for the two parents do not necessarily have to be the same. In many exam-

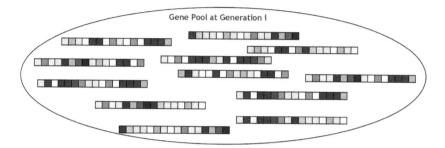

FIGURE 4.2: Graphical representation of the gene pool available at a certain generation. Each bar represents a chromosome with its alleles representing the assignment of the genes at the certain loci.

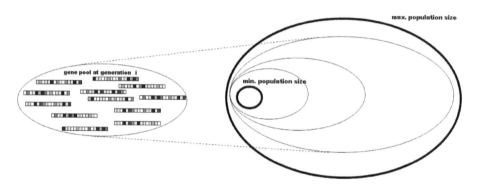

FIGURE 4.3: The left part of the figure represents the gene pool at generation i and the right part indicates the possible size of generation $i+1$ which must not go below a minimum size and also not exceed an upper limit. These parameters have to be defined by the user.

ples a combination of proportional (roulette wheel) selection and random selection has already shown a lot of potential (for example in combination with GP-based structure identification as discussed in [AWW08], e.g.). The two different selection operators are called male and female selection. It is also possible and reasonable in the context of the algorithmic concepts described here to disable parent selection totally, as scalable selection pressure comes along with the selection mechanisms after reproduction. This can be achieved by setting both parent selection operators to random.

- Due to the fact that reproduction results are only considered in case they are successful recombinations (and maybe mutations) of their parents' chromosomes, it becomes reasonable to use more than one crossover operator and more than one mutation operator at the same time. The reason for this possibility is given by the fact that only successful off-spring chromosomes are considered for the ongoing evolutionary process; this allows the application of crossover and mutation operators which do not produce good results mostly as long as they are still able to generate good offspring at least sometimes. On the one hand the insertion of such operators increases the average selection pressure and therefore also the average running time, but on the other hand these operators can help a lot to broaden evolutionary search and therefore retard premature convergence. If more than one crossover and mutation operator is allowed, the choice occurs by pure chance which has proven to produce better results than a preference of more successful operators [Aff05].

- As indicated in Figure 4.3, a lower as well as an upper limit of population size are still necessary in order to achieve efficient algorithmic performance. In case of a missing upper limit the population size would snowball especially in the first rounds which is inefficient; a lower limit of at least 2 individuals is also necessary as this indicates that it is no more possible to produce a sufficient amount of chromosomes that are able to outperform their own parents and therefore acts as a good detector for convergence.

- Depending on the problem at hand there may be several possibilities to fill up the next population with new individuals. If the problem representation allows an efficient check for genotypical identity, it is recommendable to do this and accept new chromosomes as members for the next generation if there is no structurally identical individual included in the population yet. If a check for genotypical identity is not possible or too time-consuming, there is still the possibility to assume two individuals are identical if they have the same fitness values as an approximative identity check. However, the user has to be aware of the fact that this assumption may be too restrictive in case of fitness landscapes with identical fitness values for a lot of different individuals; in such cases it is of course advisable to check for genotypical identity.

- In order to terminate the run of a certain generation in case it is not possible to fill up the maximally allowed population size with new successful individuals, an upper limit of effort in terms of generated individuals is necessary. This maximum effort per generation is the maximum number of newly generated chromosomes per generation (no matter if these have been accepted or not).

- The question, whether or not an offspring is better than its parents, is answered in the same way as in the context of offspring selection.

Figure 4.4 shows the typical development of the actual population size during an exemplary run of RAPGA applied to the ch130 benchmark instance of

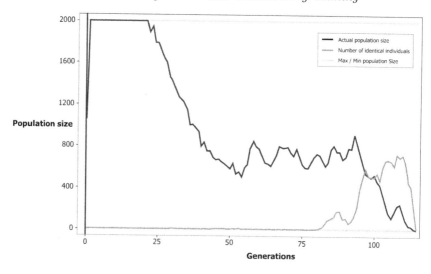

FIGURE 4.4: Typical development of actual population size between the two borders (lower and upper limit of population size) displaying also the identical chromosomes that occur especially in the last iterations.

the traveling salesman problem taken from the TSPLib [Rei91]. More sophisticated studies analyzing the characteristics of RAPGA will be presented in Chapter 7.

4.4 Consequences Arising out of Offspring Selection and RAPGA

Typically, GAs operate under the implicit assumption that parent individuals of above average fitness are able to produce better solutions as stated in Holland's schema theorem and the related building block hypothesis. This general assumption, which ideally holds under the restrictive assumptions of a canonical GA using binary encoding, is often hard to fulfill for many practical GA applications as stated in the questions of Chapter 3 which shall be rephrased and answered here in the context of offspring selection and RAPGA:

ad 1. *Is crossover always able to fulfill the implicit assumption that two above-average parents can produce even better children?*

Unfortunately, the implicit assumption of the schema theorem, namely that parents of above average fitness are able to produce even better children, is not accomplished for a lot of operators in many theoretical as well as practical applications. This disillusioning fact has several reasons: First, a lot of operators tend to produce offspring solution candidates that do not meet the implicit or explicit constraints of certain problem formulations. Commonly applied repair strategies included in the operators themselves or applied afterwards have the consequence that alleles of the resulting offspring are not present in the parents which directly counteracts the building block aspect. In many problem representations it can easily happen that a lot of highly unfit child solution candidates arise even from the same pair of above average parents (think of GP crossover for example, where a lot of useless offspring solutions may be developed, depending on the concrete choice of crossover points). Furthermore, some operators have disruptive characteristics in the sense that the evolvement of longer building block sequences is not supported. By using offspring selection (OS) or the RAPGA the necessity that almost every trial is successful concerning the results of reproduction is not given any more; only successful offspring become members of the active gene pool for the ongoing evolutionary process.

ad 2. *Which of the available crossover operators is best suited for a certain problem in a certain representation?*

For many problem representations of certain applications a lot of crossover concepts are available where it is often not clear a priori which of the possible operators is suited best. Furthermore, it is often also not clear how the characteristics of operators change with the remaining parameter settings of the algorithm or how the characteristics of the certain operators change during the run of the algorithm. So it may easily happen that certain, maybe more disruptive operators perform quite well at the beginning of evolution whereas other crossover strategies succeed rather in the final (convergence) phase of the algorithm. In contrast to conventional GAs, for which the choice of usually one certain crossover strategy has to be done in the beginning, the ability to use more crossover and also mutation strategies in parallel is an important characteristic of OS-based GAs and the RAPGA as only the successful reproduction results take part in the ongoing evolutionary process. It is also an implicit feature of the extended algorithmic concepts that when using more operators in parallel only the results of those will succeed which are currently able to produce successful offspring which changes over time. Even the usage of operator concepts that are considered evidentially weak for a certain application can be beneficial as long as these operators are able to produce successful offspring from time to time [Aff05].

ad 3. *Which of the resulting children are "good" recombinations of their parents' chromosomes?*

Both OS and RAPGA have been basically designed to answer this question in a problem independent way. In order to retain generality, these algorithms have to base the decision if and to which extent a given reproduction result is able to outperform its own parents by comparing the offspring's fitness with the fitness values of its own parent chromosomes. By doing so, we claim that a resulting child is a good recombination (which is a beneficial building block mixture) worth being part of the active gene pool if the child chromosome has been able to surpass the fitness of its own parents in some way.

ad 4. *What makes a child a "good" recombination?*

Whereas question 3 motivates the way, how the decision may be carried out whether or not a child is a good recombination of its parent chromosomes, question 4 intuitively asks why this makes sense. Generally speaking, OS and RAPGA direct the selection focus after reproduction rather than before reproduction. In our claim this makes sense, as it is the result of reproduction that will be part of the gene pool and that has to keep the ongoing process alive. Even parts of chromosomes with below average fitness may play an important role for the ongoing evolutionary process, if they can be combined beneficially with another parent chromosome which motivates gender specific parent selection [WA05b] as is for example applied in our GP experiments shown in the practical part (Chapter 11) of this book. With this gender specific selection aspect, which typically selects one parent randomly and the other one corresponding to some established selection strategy (proportional, linear-rank, or tournament strategies) or even both parents randomly, we decrease selection pressure originating from parent selection and balance this by increasing selection pressure after reproduction which is adjusted self-adaptively depending on how easy or difficult it is to achieve advancement.

ad 5. *Which parts of the chromosomes of parents of above-average fitness are really worth being preserved?*

Ideally speaking, exactly those parts of the chromosomes of above-average parents should be transferred to the next generation that make these individuals above average. What may sound like a tautology at the first view cannot be guaranteed for a lot of problem representations and corresponding operators. In these situations, OS and RAPGA are able to support the algorithm in this goal which is essential for the building block assembling machines GAs and GP.

Chapter 5

SASEGASA – More than the Sum of All Parts

DOI: 10.1201/9781420011326-6

The concept of offspring selection as described in Chapter 4 is very well suited to be transferred to parallel GA concepts. In the sense of parallel GA nomenclature, our proposed variant called SASEGASA (which stands for self-adaptive segregative genetic algorithm with simulated annealing aspects) is most closely related to the class of coarse-grained parallel GAs. The well-known island model supports global search by taking advantage of the steady pulsating interplay between breadth search and depth search supported by the forces of genetic drift and migration.

SASEGASA acts differently by allowing the certain subpopulations to drift a lot longer through the solution space; exactly until premature convergence is detected in each of the subpopulations, which will be denoted as local premature convergence. Then the algorithm aims very carefully to bring together the essential genetic information evolved in the certain demes individually.

Concretely, the following main distinguishing features can be pointed out comparing SASEGASA to a coarse-grained island model GA:

Dynamic Migration Intervals

Migration happens no longer in predefined fixed intervals but at those points in time when local premature convergence is detected in the certain subpopulations. The indicator of local premature convergence is the exceeding of a certain amount of selection pressure which can be measured in an offspring selection GA. New genetic information is then added from adjacent subpopulations that suffer from local premature convergence themselves, but have evolved different alleles due to the undirected forces of genetic drift. By this strategy the genetic search process can be initiated again until local premature convergence is detected next time.

From Islands to Growing Villages

The most important difference from SASEGASA to the well known island model is given by the fact that in case of SASEGASA the size of the subpopulations is slowly growing by decreasing the number of subpopulations. Therefore, the migration aspect of SASEGASA can rather be associated with a village-town-city model than with an island model. By this means the certain (at the beginning rather small) villages can drift towards different regions of the search space until they are all prematurely converged. Then the total number of subpopulations is decreased by one and the individuals are regrouped as sketched in Figure 5.1; then the new subpopulations evolve independently until local premature convergence is detected again for each subpopulation. So the initially rather small villages become larger and larger towns, finally forming a large panmictic population. The main idea of this strategy is that the essential alleles which may be shared over many different villages can slowly and carefully be collected in a single population resulting in a high quality solution. As a consequence, parallelization is no more that efficient as in the island model, due to the changing number of subpopulations. More sophisticated communication protocols between the involved CPUs become necessary for efficient parallel implementations.

5.1   The Interplay of Distributed Search and Systematic Recovery of Essential Genetic Information

When applying GAs to higher dimensional problems in combinatorial optimization, it happens that genetic drift also causes alleles to fix to suboptimal properties, which causes a loss of optimal properties in the entire population. This effect is especially observable, if attributes of a global optimal solution with minor influence on the fitness function, are "hidden" in individuals with a bad total fitness. In that case parental selection additionally promotes the drop out of those attributes.

This is exactly the point where considerations about multiple subpopulations, that systematically exchange information, come into play:

Splitting the entire population into a certain number of subpopulations (demes) causes the separately evolving subpopulations to explore different genetic information in the certain demes due to the stochastic nature of genetic drift. Especially in the case of multimodal problems the different subpopulations tend to prematurely converge to different suboptimal solutions. The idea is that the building blocks of a global optimal solution are scattered in the single subpopulations, and the aim is to develop concepts to systematically bring together these essential building blocks in one population in order to make it possible to find a global optimal solution by crossover. In contrast to the various coarse- and fine-grained parallel GAs that have been discussed in the literature (good reviews are given in [AT99] and [Alb05] for instance), we have decided to take a different approach by letting the demes grow together step by step in case of local premature convergence. Even if this property does not support parallelization as much as established parallel GAs, we have decided to introduce this concept of migration as it proved to support the localization of global optimal solutions to a greater extent [Aff05]. Of course, the concept of self-adaptive selection pressure steering is essential, especially immediately after the migration phases, because these are exactly the phases where the different building blocks of different subpopulations have to be unified. In classical parallel GAs it has been tried to achieve this behavior just by migration which is basically a good, but not very efficient idea, if no deeper thoughts about selection pressure are spent at the same time.

As first experiments have already shown, there also should be a great potential in equipping established parallel GAs with our newly developed self-adaptive selection pressure steering mechanisms which leads to more stability in terms of operators and migration rates, automated detection of migration interval, etc.

5.2 Migration Revisited

In nature the fragmentation of the population of a certain species into more subpopulations of different sizes is a commonly observable phenomenon. Many species have a great area of circulation of various environments which leads to the formation of subpopulations. An important consequence of the population structure is the genetic differentiation of subpopulations, i.e., the shift of allele frequencies in the certain subpopulations. The reasons for genetic differentiation are:

- Local adjustment of different genotypes in different populations

- Genetic drift in the subpopulations

- Random differences in the allele frequency of individuals which build up a new subpopulation

The structure of the population is hierarchically organized in different layers[1]:

- Individual

- Subpopulation

- Local population

- Entire population (species)

---

[1]The concept of hierarchical population structures has been introduced by Wright [Wri43].

An important goal of population genetics is the detection of population structures, the analyses of consequences and the location of the layer with most diverse allele frequencies. In this context a deeper consideration of genetic drift and its consequences is of major interest. The aspect of local adaptation of different genotypes in different populations should give useful hints for multi-objective function optimization or for optimization in changing environments.

One consequence of the population structure is the loss of heterozygosity (genetic variation). The Swedish statistician and geneticist Wahlund [HC89] described that genetic variation rises again, if the structure is broken up and mating becomes possible in the entire population. The SEGA and the SASEGASA algorithm, which will be described later, systematically take advantage of this effect step by step.

When talking about migration it is essential to consider some distinction concerning the genetic connection between the subpopulations which mainly depends on the gene flow (the exchange of alleles between subpopulations). Migration, the exchange of individuals, causes gene flow if and only if the exchanged individuals produce offspring. The most important effect of migration and gene flow is the introduction of new alleles into the subpopulations. In that sense migration has effects similar to mutation, but can occur at much higher rates. If the gene flow between subpopulations is high, they become genetically homogeneous; in the case of little gene flow, the subpopulations may diverge due to selection, mutation, and drift. Population genetics provides a set of models for the theoretical analysis of gene flows. The most popular migration models of population genetics are the mainland-island model, which considers migration in just one direction, and the island model, that allows migration in both directions. As discussed in parallel GA theory, the migration rate is an essential parameter for the description of migration. In population genetics the migration rate describes the ratio of chromosomes migrating among subpopulations.

5.3 SASEGASA: A Novel and Self-Adaptive Parallel Genetic Algorithm

We have already proposed several new EA-variants. The first prototype of this new class of evolutionary search which considers the concept of controllable selection pressure ([Aff01c], [Aff02]) for information exchange between independently evolving subpopulations has been introduced with the Segregative Genetic Algorithm (SEGA) [Aff01a], [Aff01b]. Even if the SEGA is already able to produce very high quality results in terms of global solution quality, selection pressure has to be set by the user which is a very time consuming and difficult challenge. Further research, which aimed to introduce self-adaptive selection principles for the steering of selection pressure ([AW03], [AW04a]), resulted in the so-called SASEGASA-algorithm ([AW03], [Aff05]), which already represents a very stable and efficient method for producing high quality results without introducing problem-specific knowledge or local search.

So far parallelism has "only" been introduced for improving global solution quality and all experiments have been performed on single-processor machines. Nevertheless, there is nothing to be said against transforming the concepts evolved in the parallel GA community to also improve the quantitative performance of our new methods. Empirical studies have shown that the theoretical concepts of the SASEGASA-algorithm [AW03] allow global solution quality to be steered up to the highest quality regions by just increasing the number of demes involved. The algorithm turned out to find the global optimum for all considered TSP benchmarks as well as for all considered benchmark test functions up to very high problem dimensions [Aff05].

Therefore an enormous increase of efficiency can be expected when applying concepts of supercomputing, allowing us to also attack much higher dimensional theoretical and practical problems efficiently in a parallel environment. Because of the problem independency of all newly proposed theoretical concepts there is no restriction to a certain class of problems that allows the attack of all problems for which GA theory (and also GP theory) offers adequate operators.

5.3.1   The Core Algorithm

In principle, the SASEGASA introduces two enhancements to the basic concept of genetic algorithms. Firstly, the algorithm makes use of variable selection pressure, as introduced as offspring selection (OS) in Chapter 4, in order to self-adaptively control the goal-orientedness of genetic search. The second concept introduces a separation of the population to increase the broadness of the search process so that the subpopulations are joined after local premature convergence has occurred. This is done in order to end up with a population including all genetic information sufficient for locating a global optimum.

The aim of dividing the whole population into a certain number of subpopulations (segregation) that grow together in case of stagnating fitness within those subpopulations (reunification) is to combat premature convergence which is the source of GA-difficulties. The basic properties (in terms of solution quality) of this segregation and reunification approach have already proven their potential in overcoming premature convergence [Aff01a], [Aff01b] in the so-called SEGA algorithm.

By using this approach of breadth search, essential building blocks can evolve independently in different regions of the search space. In the case of standard GAs those relevant building blocks are likely to disappear early on due to genetic drift and, therefore, their genetic information can not be provided at a later phase of evolution, when the search for a global optimum is of paramount importance.

However, within the SEGA algorithm there is no criterion to detect premature convergence, and there is also no self-adaptive selection pressure steering mechanism. Even if the results of SEGA are quite good with regard to global convergence [Aff01b], it requires an experienced user to adjust the selection pressure steering parameters, and as there is no criterion to detect premature convergence the dates of reunification have to be implemented statically.

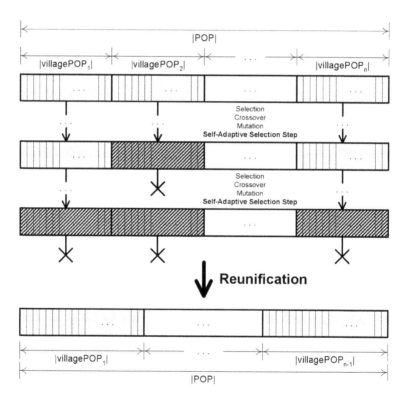

FIGURE 5.1: Flowchart of the reunification of subpopulations of a SASEGASA (light shaded subpopulations are still evolving, whereas dark shaded ones have already converged prematurely). This figure is displayed with kind permission of Springer Science and Business Media.

Equipped with offspring selection we have both: A self-adaptive selection pressure (depending on the given success ratio), as well as an automated detection of local premature convergence, if the current selection pressure becomes higher than the given maximal selection pressure parameter ($MaxSelPress$). Therefore, a date of reunification has to be set, if local premature convergence has occurred within all subpopulations, in order to increase genetic diversity again. Figure 5.1 shows a schematic diagram of the migration policy in the case of a reunification phase of the SASEGASA algorithm. The dark shaded subpopulations stand for already prematurely converged subpopulations. If all subpopulations are prematurely converged (dark shaded) a new reunification phase is initiated.

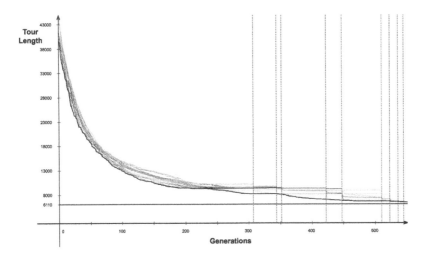

FIGURE 5.2: Quality progress of a typical run of the SASEGASA algorithm. This figure is displayed with kind permission of Springer Science and Business Media.

Figures 5.2 and 5.3 show typical shape of the fitness curves and selection pressure progresses of a SASEGASA test run. The number of subpopulations is in this example set to 10. The vertical lines indicate dates of reunification. In the quality diagram (Figure 5.2) the lines give the fitness value of the best member of each deme; the best known solution is represented by the horizontal line. In the selection pressure diagram (shown in Figure 5.3) the lines stand for the actual selection pressure in the certain demes, as the actual quotient of evaluated solution candidates per round (in a deme) to the subpopulation size. The lower horizontal line represents a selection pressure of 1 and the upper horizontal line represents the maximum selection pressure. If the actual selection pressure of a certain deme exceeds the maximum selection pressure, local premature convergence is detected in this subpopulation and evolution is stopped in this deme (which can be seen in the constant value of the corresponding fitness curve) until the next reunification phase is started (if all demes are prematurely converged).

With all the above described strategies, the complete SASEGASA algorithm

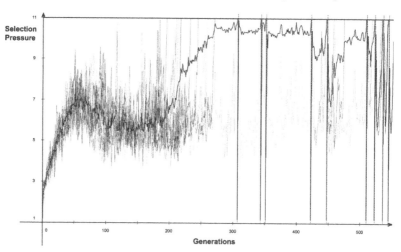

FIGURE 5.3: Selection pressure curves for a typical run of the SASEGASA algorithm. This figure is displayed with kind permission of Springer Science and Business Media.

can be stated as described in Figure 5.4.

Again, similar as in the context of offspring selection, it should be pointed out that a corresponding genetic algorithm is unrestrictedly included in SASEGASA, when the number of subpopulations (villages) is set to 1 and the success ratio is set to 0 at the beginning of the evolutionary process. Moreover, the introduced techniques also do not use any problem-specific information.

5.4   Interactions among Genetic Drift, Migration, and Self-Adaptive Selection Pressure

Using all the introduced generic algorithmic concepts combined in SASEGASA, it becomes possible to utilize the interactions between genetic drift and the SASEGASA specific dynamic migration policy in a very advantageous way in terms of achievable global solution quality:

Initially a certain number of subpopulations evolve absolutely independently from each other until no further evolutionary improvement is possible when using a genetic algorithm with offspring selection in the subpopulations, i.e., until local premature convergence is detected in all subpopulations.

As a matter of principle, it is also the case that primarily those alleles are fixed in the certain subpopulations which currently influence the fitness

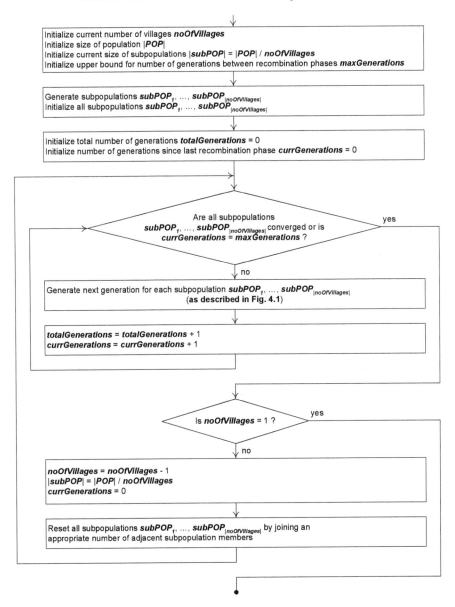

FIGURE 5.4: Flowchart showing the main steps of the SASEGASA. This figure is displayed with kind permission of Springer Science and Business Media.

function to a greater extent. The rest of the essential alleles with a currently low influence on the fitness value, which may be required for a global optimal solution later, are unlikely to be combined in any single subpopulation –

because basically these alleles are distributed over all subpopulations.

As after this first optimization stage all subpopulations are prematurely converged to solutions with comparable fitness values, genetic information that has not been considered before is suddenly taken into account by selection after a reunification phase. We so establish a step-by-step reunification of independently evolving subpopulations, which is triggered by the detection of convergence; in this way it becomes possible to systematically consider the essential alleles at exactly those points in time when they are in the position to become accepted in the newly emerging greater subpopulations. By means of this procedure smaller building blocks[2] of a global optimal solution are step-by-step enriched with new alleles, in order to evolve to larger and larger building blocks ending up in one single subpopulation, containing all genetic information being required for a global optimal solution.

With an increasing number of equally sized subpopulations, the probability that essential alleles are barred from dying off increases due to the greater total population size; this results in a higher survival probability of the alleles which are not yet considered.

As empirically demonstrated in Chapters 7, 10, and 11, the procedure described above makes it possible to find high quality solutions for more and more difficult problems by simply increasing the number of subpopulations. Of course this causes increasing computational costs; still, this growth in computational costs is not exponential, but linear.

---

[2]The notation of building blocks is considered in a more general interpretation than in Holland's definition.

Chapter 6

Analysis of Population Dynamics

DOI: 10.1201/9781420011326-7

There are several aspects of dynamics in populations of genetic algorithms that can be observed and analyzed. In this section we shall describe these aspects which we have concentrated on and which will also be analyzed for evaluating different algorithmic GA settings on various problem instances:

- In Section 6.1 we describe how we analyze which individuals of the population succeed in passing their genetic information on to the next generation.

- In Section 6.2 we give a summary of approaches for analyzing the diversity among populations of GAs using some kind of similarity measure for solution candidates. We use these concepts to measure how diverse the individuals of populations are as well as how similar populations of multi-population GAs become during runtime.

Furthermore, in Chapter 10 we will analyze the dynamics of population diversity over time for the combinatorial optimization problems TSP (see Section 10.1) and CVRP (see Section 10.2) on the basis of GA variants considered in this book.

6.1 Parent Analysis

In the context of conventional GAs, parent selection is normally responsible for selecting fitter individuals more often than those that are less fit. Thus, fitter individuals are supposed to pass on their genetic material to more members of the next generation.

When using offspring selection, several additional aspects have to be considered. As only those children survive this selection step that perform better than their parents to a certain degree, we cannot guarantee that fitter parents succeed more often than less fit ones.

This is why we document the parent indices of all successful offspring for each generation step. So we can analyze whether all parts of the population are considered for effective propagation of their genetic information, if whether only better ones or rather bad ones are successful.

Formally, in parent analysis we analyze the genetic propagation of parents P to their children C calculating the propagation count pc for each parent as the number of successful children it was able to produce by being crossed with other parents or mutation:

$$isParent(p, c) = \begin{cases} 1 & : & p \in c.Parents \\ 0 & : & otherwise \end{cases} \tag{6.1}$$

$$\forall(p \in \mathbf{P}) : \mathbf{pc}(\mathbf{p}) = \sum_{\mathbf{c} \in \mathbf{C}} \mathbf{isParent}(\mathbf{p}, \mathbf{c}) \tag{6.2}$$

In addition, we can optionally weight the propagation count for each potential parent by weighting it with the similarity of the parent and its children (supposing the availability of a similarity function sim which can be used for calculating the similarity of solution candidates):

$$\forall(p \in \mathbf{P}) : \mathbf{pc'}(\mathbf{p}) = \sum_{\mathbf{c} \in \mathbf{C}} \mathbf{isParent}(\mathbf{p}, \mathbf{c}) * \mathbf{sim}(\mathbf{p}, \mathbf{c}) \tag{6.3}$$

This kind of population dynamics analysis shall be used later in Section 11.3, where we will see how enhanced selection models affect genetic propagation in data-based modeling using genetic programming.

6.2 Genetic Diversity

In this section we describe the measures which we use to monitor the diversity and population dynamics with respect to the genetic make-up of solution candidates using some kind of similarity measurement function that estimates the mutual similarity of solution candidates.

As we know that similarity measures do not have to be symmetric (see Section 9.4 for examples and explanations), we can alternatively use the mean value of the two possible similarity calls and so define a symmetric similarity measurement:

$$symmetricAnalysis \Rightarrow sim(m_1, m_2) = \frac{sim(m_1, m_2) + sim(m_2, m_1)}{2} \tag{6.4}$$

6.2.1 In Single-Population GAs

In the context of single-population GAs we are mainly interested in the similarity among the individuals of the population: For each solution s of the population P we calculate the mean and the maximum similarity with all other individuals in the population:

$$meanSim(s, P) = \frac{1}{|P| - 1} \sum_{s2 \in P, s2 \neq s} sim(s, s2) \qquad (6.5)$$

$$maxSim(s, P) = max_{(s2 \in P, s2 \neq s)}(sim(s, s2)) \qquad (6.6)$$

The mean values of all individuals' similarity values are used for calculating the average mean and average maximum similarity measures for populations:

$$meanSim(P) = \frac{1}{|P|} \sum_{s \in P} meanSim(s, P) \qquad (6.7)$$

$$maxSim(P) = \frac{1}{|P|} \sum_{s \in P} maxSim(s, P) \qquad (6.8)$$

6.2.2 In Multi-Population GAs

In the context of parallel evolution of populations in genetic algorithms, which is summarized in Section 1.7, we can apply the population diversity analysis for each population separately; in the following we will describe a multi-population specific diversity analysis.

Basically, a solution s is compared to all solutions in another population P' which does not include s, and $multiPopSim(s, P')$ is equal to the maximum of the so calculated similarity values:

$$s \notin P' \Rightarrow multiPopSim(s, P') = max_{(s2 \in P')}(sim(s, s2)) \qquad (6.9)$$

So we can calculate the multi-population similarity of a solution with respect to a set of populations PP as the average $multiPopSim$ of the solution to all populations except the "own" one:

$$(s \in P \wedge P \in PP) \Rightarrow PP' = \{P' : P' \in PP \wedge P' \neq P\}, \qquad (6.10)$$

$$multiPopSim(s, PP) = \frac{1}{|PP'|} \sum_{P' \in PP'} multiPopSim(s, P'), \qquad (6.11)$$

Finally, a population's $multiPopSim$ value is equal to all its solutions' multi-population similarity values with respect to the whole set of populations:

$$multiPopSim(P, PP) = \frac{1}{|P|} \sum_{s \in P} multiPopSim(s, PP) \qquad (6.12)$$

6.2.3   Application Examples

As we are aware that these formulas for calculating the genetic diversity in populations might seem a bit confusing, we shall here give application examples and some graphical illustrations of the results so that the main ideas become a bit clearer.

For this purpose we have decided to let a genetic algorithm search for an optimal solution for a specific instance of the traveling salesman problem, namely the 130 cities problem *ch130* taken from the TSPLIB [Rei91].

The first algorithm tested was a conventional GA with 100 solution candidates, order crossover, and 5% mutation rate; it was executed over 10,000 generations, and the similarity for each pair of solutions in the population was measured every 10 iterations.

We hereby calculate the similarity of TSP solutions in the following way: As each solution is given as a path, we can find out which city is visited after another one for each step in the tour. These edges of the cities graphs are considered for calculating the similarity of tours: The proportion of edges that are common in both tour graphs represents the similarity of TSP solution candidates. In a more formal way we can use the following edge definition[1]:

$$isedge([i,j], t) \Leftrightarrow$$
$$\exists k : (t[k] = i \wedge t[k+1] = j) \vee (t[k_n] = i \wedge t[0] = j) \qquad (6.13)$$

where n is the number of cities in the given TSP problem instance (in the case of the *ch130* problem, $n = 130$), and define the similarity of two tours t_1 and t_2 as $sim(t_1, t_2)$ as

$$sim(t_1, t_2) = \frac{|\{[i,j] : (isedge([i,j], t_1) \wedge isedge([i,j], t_2))\}|}{n}. \qquad (6.14)$$

Figure 6.1 shows a graphical representation of the genetic diversity in the GA's population after 20 and 200 generations, shown on the left and right side of the figure. For each pair of solutions (at indices i and j) the similarity of these two solutions is represented by a square at position (i, j); light squares represent small similarity values, while dark squares indicate high similarities. Obviously the genetic diversity is high in the beginning of the GA run, but decreases very soon as the similarity of solutions becomes very high already after 200 rounds.

The histograms shown in Figure 6.2 sustain this impression: As most similarity values are rather low (below 0.25) in the beginning, i.e., after 20 iterations, most pairs of solutions show high similarity values between 0.5 and 0.8 after 200 generations.

---

[1]In the case of symmetric TSP instances $isedge([i,j], t)$ of course implies $isedge([j,i], t)$.

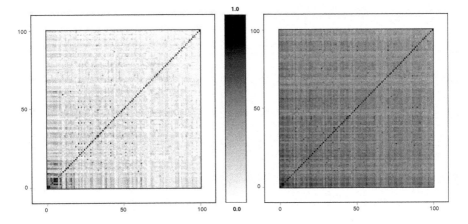

FIGURE 6.1: Similarity of solutions in the population of a standard GA after 20 and 200 iterations, shown in the left and the right charts, respectively.

Finally, Figure 6.3 shows the progresses of all solutions' average similarity values for the first 2,000 and 10,000 generations, respectively. As we can see clearly, most similarity values reach a very high level very soon, after a bit more than 500 iterations the overall average (shown as a black line) reaches 0.95; at the end of the GA run almost all pairs of solutions show a similarity of more than 0.9, the average being approximately 0.96.

For demonstrating the possibilities how to graphically represent multi-population specific similarities, we have set up a parallel GA with 4 populations each evolving exactly like the GA described for demonstrating single population similarity. Thus, the algorithm contains 4 populations each storing 100 solutions for the *ch130* problem, evolving over 10,000 generations by order crossover and 5% mutation.

Figure 6.4 graphically represents the multi-population specific similarity for each solution in the 4 given populations after 5,000 generations: In row i of bar j we give the maximum similarities of the ith solution in population j compared to all solutions of all other populations using Formula 6.9; the maximum similarities with other populations are given column wise. In each bar k we intentionally omit column k as the maximum similarity of a solution with its own population does not make sense in the context of the analysis of multi-population genetic diversity. For example, in column 1 of row 20 in bar 1 we represent the maximum similarity of solution 20 in population 1 with all solutions in population 2, and in column 2 of row 10 in hyper-column 2 we show the maximum similarity of solution 10 in population 2 with all solutions in population 3. Again, higher similarity values are represented by darker regions.

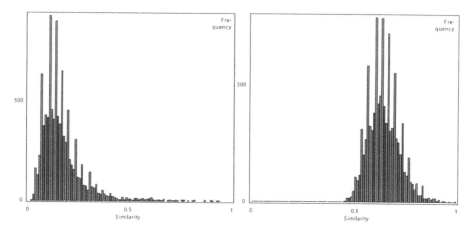

FIGURE 6.2: Histograms of the similarities of solutions in the population of a standard GA after 20 and 200 iterations, shown in the left and the right charts, respectively.

As we see in Figure 6.4, all solutions have rather high similarity with at least one solution of all other populations. This tendency is also shown in Figure 6.5 in which we give the average multi-population similarity values for each solution calculated using 6.11; the black line stands for the average of these values, which is equal to the overall average value calculated using 6.12. As we see, the populations become more and more similar to each other as the parallel GA is executed.

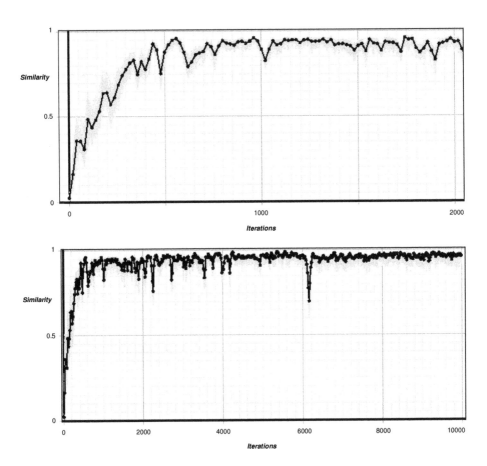

FIGURE 6.3: Average similarities of solutions in the population of a standard GA over for the first 2,000 and 10,000 iterations, shown in the upper and lower charts, respectively.

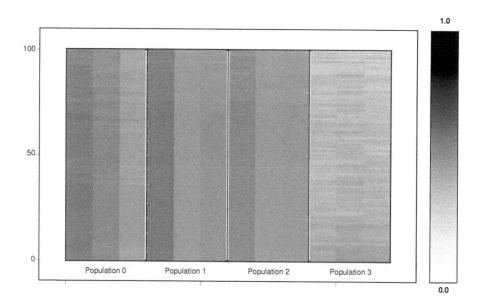

FIGURE 6.4: Multi-population specific similarities of the solutions of a parallel GA's populations after 5,000 generations.

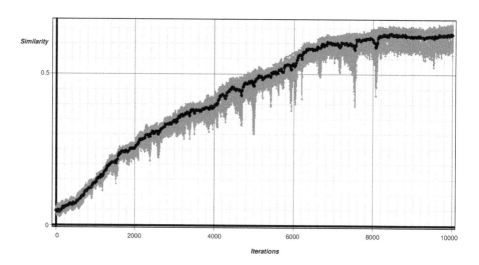

FIGURE 6.5: Progress of the average multi-population specific similarity values of a parallel GA's solutions, shown for 10,000 generations.

Chapter 7

Characteristics of Offspring Selection and the RAPGA

DOI: 10.1201/9781420011326-8

7.1 Introduction

In this chapter we will try to look inside the internal functioning of several GA variants already discussed in previous chapters. For this purpose we use the information about globally optimal solutions which is only available for well studied benchmark problems of moderate dimension. Of course, the applied optimization strategies (i.e., in our case variants of GAs) are not allowed to use any information about the global optimum; we just use this information for analysis purposes in order to obtain a better understanding of the internal functioning and the dynamics of the most relevant algorithmic concepts discussed so far.

A basic requirement for this (to a certain extent idealized) kind of analysis is the existence of a unique globally optimal solution which has to be known. Concretely, we aim to observe the distribution of the alleles of the global optimal solution over the generations in order to observe the ability of the certain algorithmic variants to preserve and possibly regain essential genetic material during the run of the algorithm.

The main aim of this book in general and especially of this chapter is not to give a comprehensive analysis of many different problem instances, but rather to highlight the main characteristics of the certain algorithm variants. For this kind of analysis as given in this chapter we have chosen the traveling salesman problem (TSP), mainly because it is a well known and well analyzed combinatorial optimization problem and a lot of benchmark problem instances are available. We here concentrate on the *ch130* TSP instance taken from the TSPLib [Rei91], for which the unique globally optimal tour is known; the characteristics of the global optimum of this 130 city TSP instance are exactly the 130 edges of the optimal tour which denote the essential genetic information as stated in Chapter 3. In contrast to the analyses described in Chapter 10, we here rather show results of single characteristical test runs in order to identify the essentially important algorithmic features; for statistically more significant tests the reader is referred to Chapter 10.

In a broader interpretation of the building block theory, these alleles should on the one hand be available in the initial population of a GA run, and on the other hand maintained during the run of the algorithm. If essential genetic information is lost during the run, then mutation is supposed to help regaining it in order to be able to eventually find the globally optimal solution (or at least a solution which comes very close to the global optimum). In order to observe the actual situation in the population we display each of the 130 essential edges as a bar indicating the saturation of each allele in the population so there are in total 130 bars. The disappearance of a bar therefore indicates the loss of the corresponding allele in the entire population, whereas a full bar indicates that the certain allele occurs in each individual (which is the desired situation at the end of an algorithm run). As a consequence, the relative height of a bar stands for the actual penetration level of the corresponding allele in the individuals of the population and the observation of the dynamic behavior allows observing the distribution of essential genetic information during the run.

Of course one has to keep in mind that this special kind of analysis can only be performed when the unique globally optimal solution is available; usually, this information is not available in real world applications where we can only observe genetic diversity which will be done in Chapters 10 and 11. Nevertheless, these somehow idealized conditions allow very deep insight into the dynamic behavior of certain algorithm variants which can also be extended to other more practically relevant problem situations.

In the following, the distribution of essential genetic information and its impact on achievable solution quality will be discussed for the standard GA, a GA variant including offspring selection as well as for the relevant alleles preserving GA (RAPGA) as introduced in Chapter 4.

7.2   Building Block Analysis for Standard GAs

For observing the distribution of essential alleles in a standard GA we have used the following test strategy: First, our aim was to observe the solution quality achievable with parameter settings that are quite typical for such kinds of GA applications (as given in Table 7.1) using the well known operators for the path representation, namely OX, ERX, and MPX; each algorithmic variant has been analyzed applying no mutation as well as mutation rates of 5% and 10%.

The following Figures 7.1, 7.2, and 7.3 show the fitness curves (showing best and average solution qualities of the GA's population as well as the best known quality) for a standard GA using OX (see Figure 7.1), ERX (Figure 7.2), and MPX (Figure 7.3), respectively; the parameter settings used for these experiments are given in Table 7.1.

Table 7.1: Parameters for test runs using a conventional GA.

Parameters for the conventional GA tests	
(Results are graphically presented in Figures 7.1, 7.2, and 7.3)	
Generations	20,000
Population Size	100
Elitism Solutions	1
Mutation Rate	0.00 or 0.05 or 0.1
Selection Operator	Roulette
Crossover Operator	OX (Fig. 7.1), ERX (Fig. 7.2) or MPX (Fig. 7.3)
Mutation Operator	Simple Inversion

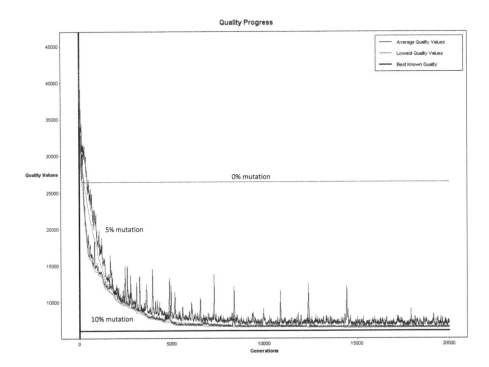

FIGURE 7.1: Quality progress for a standard GA with OX crossover for mutation rates of 0%, 5%, and 10%.

For the OX crossover, which achieved the best results with the standard parameter settings, the results are shown in Figure 7.1; it is observable that the use of mutation rates of 5% and 10% leads to achieving quite good results (about 5% to 10% worse than the global optimum), whereas disabling mutation leads to a rapid loss of genetic diversity so that the solution quality stagnates at a very poor level.

The use of more edge preserving crossover operators ERX and MPX (for which results are shown in Figures 7.2 and 7.3) shows different behavior in the sense that, applying the same parameter settings as used for the OX, the results are rather poor independent of the mutation rate. The reason for this is that these operators require more selection pressure (as for example tournament selection with tournament size 3); when applying higher selection pressures it is possible to achieve comparably good results also with ERX and MPX. Still, also when applying parameter settings which give good results with appropriate mutation rates, the standard GA fails dramatically when disabling mutation.[1]

When applying selection pressures which are sufficiently high to promote the genetic search process beneficially, mutation is absolutely necessary to achieve high quality results using a standard GA. Only if no alleles are fixed and therefore no real optimization process takes place, disabling mutation would not cause stagnation of evolutionary search.

Summarizing these aspects we can state for the SGA applied to the TSP that several well suited crossover operators[2] require totally different combinations of parameter settings in order to make the SGA produce good results. Considering the results achieved with the parameter setting as stated in Table 7.1, the use of the OX yields good results (around 10% worse than the global optimum) whereas the use of ERX and MPX leads to unacceptable results (more than 100% worse than the global optimum). On the contrary, tuning the residual parameters (population size, selection operator) for ERX or MPX would cause poor solution quality for OX.

Thus, an appropriate adjustment of selection pressures is of critical importance; as we will show in the following, self-adaptive steering of the selection pressure is able to make the algorithm more robust as selection pressure is adjusted automatically according to the actual requirements.

Figure 7.4 shows the distribution of the 130 essential alleles of the unique globally optimal solution over time for the overall best parameter constellation found in this section, i.e., the use of OX crossover with 5% mutation rate. In order to make the snapshots for the essential allele distribution within the SGA's population comparable to those captured applying a GA with offspring selection or the RAPGA, the timestamps are not given in iterations but rather

---

[1]In order to keep this chapter compact and on an explanatory level, detailed parameter settings and the corresponding analysis are not given here; interested readers are kindly invited to reproduce these results using HeuristicLab 1.1.

[2]OX, ERX, and MPX are all edge preserving operators and therefore basically suited for the TSP.

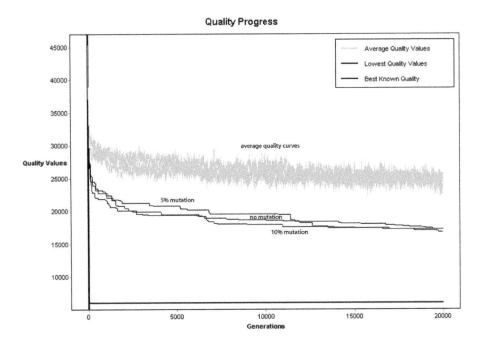

FIGURE 7.2: Quality progress for a standard GA with ERX crossover for mutation rates of 0%, 5%, and 10%.

in the number of evaluations (which is in the case of the SGA equal to the population size times the number of generations executed).

Until after about 10.000 evaluations, i.e., at generation 100, we can observe quite typical behavior, namely the rise of certain bars (representing the existence of edges of the global optimum). However, what happens between the 10.000th and 20.000th evaluation is that some of the essential alleles (about 15 in our test run) become fixed whereas the rest (here about $130 - 15 = 115$ in our test run) disappears in the entire population. As we can see in Figure 7.4, without mutation the genetic search process would already be over at that moment due to the fixation of all alleles; from now on mutation is the driving force behind the search process of the SGA.

The effects of mutation in this context are basically as follows: Sometimes high quality alleles are (by chance) injected into the population, and if those are beneficial (not even necessarily in the mutated individual), then a suited crossover operator is able to spread newly introduced essential allele information over the population and achieve a status of fixation quite rapidly. Thus, most of the essential alleles can be reintroduced and fixed approximately between the 20.000th and 2.000.000th evaluation.

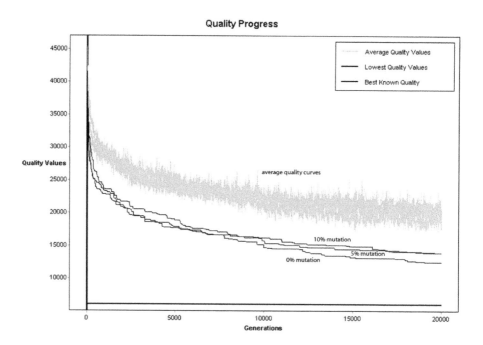

FIGURE 7.3: Quality progress for a standard GA with MPX crossover for mutation rates of 0%, 5%, and 10%.

However, even if this procedure is able to fulfill the function of optimization reasonably good when applying adjusted parameters, it has not much in common with the desired functioning of a genetic algorithm as stated in the schema theorem and the according building block hypothesis. According to this theory, we expect a GA to systematically collect the essential pieces of genetic information which are initially spread over the chromosomes of the initial population as reported for the canonical GA.[3] As we will point out in the next sections, GAs with offspring selection as well as RAPGA are able to considerably support a GA to function in exactly that way even under not so idealized conditions as required in the context of the canonical GA.

---

[3]This statement is in fact restricted to binary encoding, single point crossover, bit-flip mutation, proportional selection, and generational replacement; see Chapter 1 for further explanations.

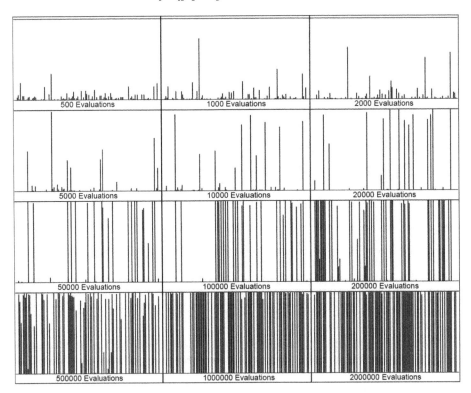

FIGURE 7.4: Distribution of the alleles of the global optimal solution over the run of a standard GA using OX crossover and a mutation rate of 5% (remaining parameters are set according to Table 7.1).

7.3   Building Block Analysis for GAs Using Offspring Selection

The aim of this section is to highlight some characteristics of the effects of offspring selection. In order to do so, we have chosen very strict parameter settings (no parental selection, strict offspring selection with 100% success ratio) which are given in Table 7.2. As the termination criterion of a GA with offspring selection is self-triggered, the effort of these test runs is not constant; however, the parameters are adjusted in a way that the total effort is comparable to the effort of the test runs for the SGA building block analyses discussed in Section 7.2.

Table 7.2: Parameters for test runs using a GA with offspring selection.

Parameter settings for the offspring selection GA runs	
(Results are graphically presented in Figures 7.5, 7.6, 7.7, and 7.8)	
Population Size	500
Elitism Solutions	1
Mutation Rate	0.00 or 0.05
Selection Operator	Random
Crossover Operator	OX , MPX, ERX, or combinations
Mutation Operator	Simple Inversion
Success Ratio	1.0
Comparison Factor Bounds	1.0
Maximum Selection Pressure	300

Similarly as for the SGA, we here take a look at the performance of some basically suited (edge preserving) operators. The results shown in Figures 7.5, 7.6, and 7.7 highlight the benefits of self-adaptive selection pressure steering introduced by offspring selection: Independent of the other parameter settings, the use of all considered crossover operators yields results near the global optimum.

As documented in the previous section we have observed that the standard GA heavily relies on mutation, when the selection pressure is adjusted at a level that allows the SGA to search in a goal oriented way. Therefore, we are now especially interested in how offspring selection can handle the situation when mutation is disabled. Figure 7.8 shows the quality curves for the use of the ERX crossover operator (which achieved the best results with 5% mutation) without mutation and the same settings for the remaining parameters. The remarkable result is that the result is practically as good as with mutation, which at the same time means that offspring selection does not rely on the genetic diversity regaining aspect of mutation. Furthermore, this also means that offspring selection is able to keep the essential genetic information which in the concrete example is given by the alleles of the globally optimal solution. When using offspring selection (in contrast to the SGA) the algorithm is not only able to keep the essential genetic information, but slowly merges the essential building blocks step by step which complies with the core statements of the building block theory and is not restricted to binary encoding or the use of certain operators.

This behavior of offspring selection, which is very important for practical applications, comes along with the property from which the method derived its name: Due to offspring selection only those children take part in the ongoing evolutionary process which were successful offspring of their own parents. Thus, one implicit assumption of the schema theory and the building block hypothesis holds, which would not be valid for a lot of practical applications: We enhance the evolutionary process in such a way that two parents (with above average qualities) are able to produce offspring of comparable or even

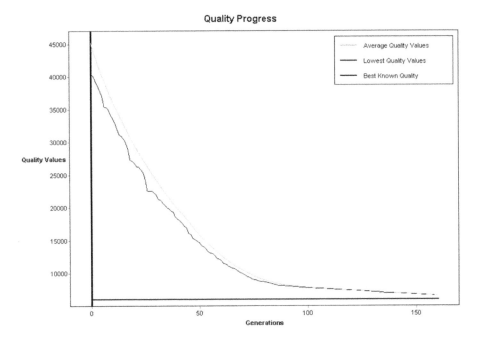

FIGURE 7.5: Quality progress for a GA with offspring selection, OX, and a mutation rate of 5%.

better fitness, and that exactly these children take part in the ongoing evolutionary process.

In previous publications we have even gone one step further: We have been able to show in [AW04b] and [Aff05] that even with crossover operators basically considered unsuitable for the TSP (as they inherit the position information rather than the edge information like CX or PMX) it becomes possible to achieve high quality results in combination with offspring selection. The reason is the sufficiency that a crossover operator is able to produce good recombinations from time to time (as only these are considered for the future gene pool); the price which has to be paid is that higher average selection pressure has to be applied, if the crossover operator is more unlikely to produce successful offspring.

For the path representation of the TSP, the characteristics of the crossover operators mentioned before are well analyzed. However, when it comes to practical applications it is often not known a priori, which of the possible crossover concepts will perform well. In this context a further interesting aspect of offspring selection becomes obvious: As only the successful crossover results are considered for the ongoing evolutionary process, we can simply

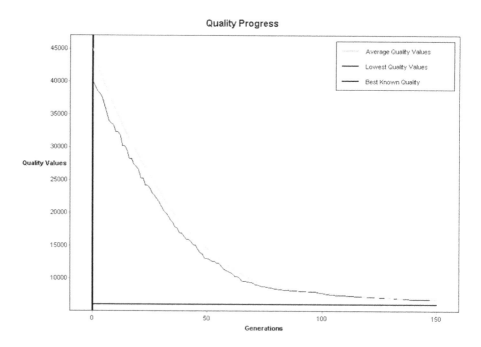

FIGURE 7.6: Quality progress for a GA with offspring selection, MPX, and a mutation rate of 5%.

apply several different crossover operators and select one of those at random for each crossover.

As a proof of concept for applying more than one crossover at the same time, we have repeated the previous test runs with OX, MPX, and ERX with the only difference that for these tests all crossover operators have been used. Figure 7.9 shows the quality curves (best and average results) for this test run and shows that the results are in the region of the global optimal solution and therefore at least as good as in the test runs before. A further question that comes along using multiple operators at once is their performance over time: Is the performance of each of the certain operators relatively constant over the run of the algorithm?

In order to answer this question, Figure 7.10 shows the ratio of successful offspring for each crossover operator used (in the sense of strict offspring selection which requires that successful children have to be better than both parents). Figure 7.10 shows that ERX performs very well at the beginning (approximately until generation 45) as well as in the last phase of the run (circa from gen. 75). In between (approximately from generation 45 to generation 75), when the contribution of ERX is rather low, MPX shows significantly

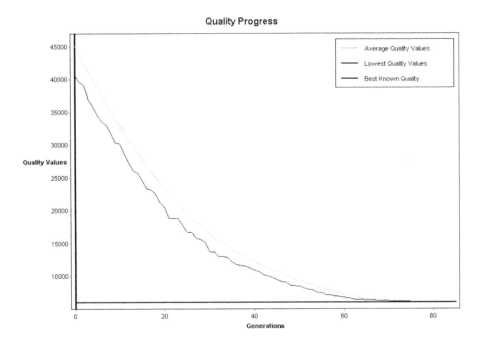

FIGURE 7.7: Quality progress for a GA with offspring selection, ERX, and a mutation rate of 5%.

better performance. The performance of OX in terms of its ability to generate successful offspring is rather mediocre during the whole run showing very little success in the last phase. The analysis of reasons of the behavior of the certain operators over time would be an interesting field of research; anyway, it is already very interesting to observe that the performance characteristics of the operators are changing over time to such an extent.

For a more detailed observation of the essential alleles during the runs of the GA using offspring selection we show the allele distribution for the ERX crossover, which achieved slightly better results than the other crossover operators, in Figure 7.11. However, the characteristics of the distribution of essential alleles are quite similar also for the other crossover operators when using offspring selection. As a major difference in comparison to the essential allele distributions during a standard GA, we can observe that the diffusion of the essential alleles is established in a rather slow and smooth manner. The essential alleles are neither lost nor fixed in the earlier stages of the algorithm, so the bars indicating the occurrence of the certain essential allele (edges of the optimal TSP path) in the entire population are growing steadily until almost all of them are fixed by the end of the run. This behavior not

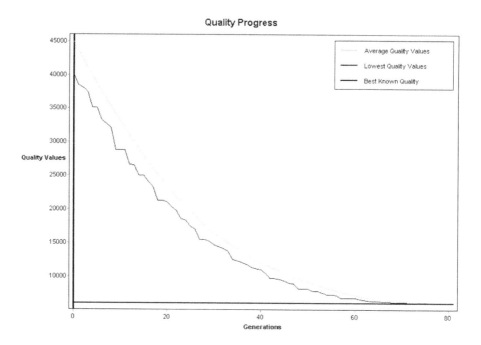

FIGURE 7.8: Quality progress for a GA with offspring selection, ERX, and no mutation.

only indicates a behavior in accordance with the building block hypothesis, but also implies that the algorithm performance no more relies on mutation to an extent as observed for the corresponding SGA analyses. In order to confirm this assumption we have repeated the same test without mutation and indeed, as it can be seen by a comparison of Figures 7.11 and 7.12, the saturation behavior of the essential building blocks is basically the same, no matter if mutation is used or not. This is a remarkable observation as it shows that offspring selection enables a GA to collect the essential building blocks represented in the initial population and compile high quality solutions very robustly in terms of parameters and operators like mutation, selection pressure, crossover operators, etc. This property is especially important when exploring new fields of application where suitable parameters and operators are usually not known a priori.

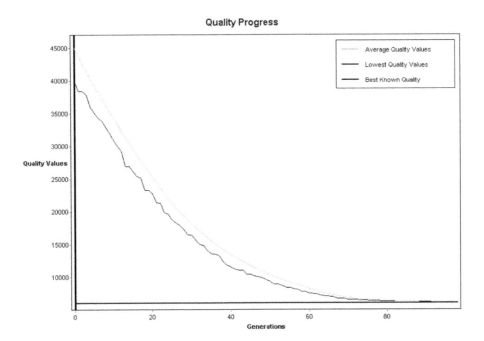

FIGURE 7.9: Quality progress for a GA with offspring selection using a combination of OX, ERX, and MPX, and a mutation rate of 5%.

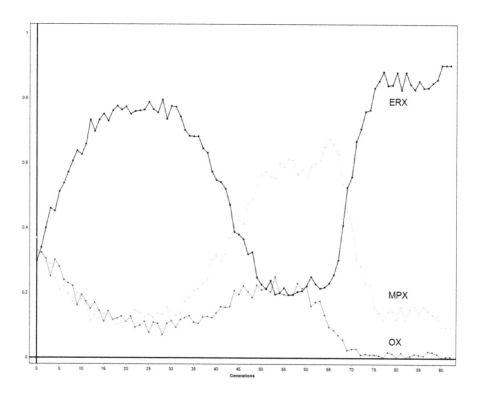

FIGURE 7.10: Success progress of the different crossover operators OX, ERX, and MPX, and a mutation rate of 5%. The plotted graphs represent the ratio of successfully produced children to the population size over the generations.

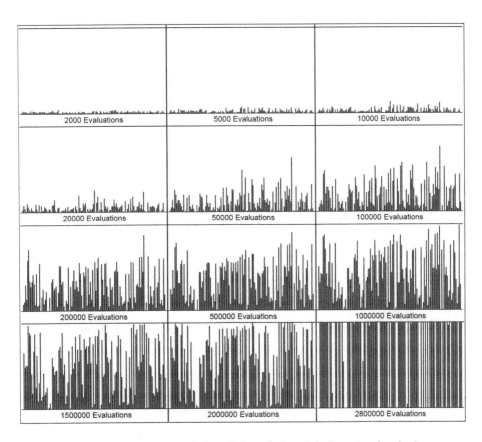

FIGURE 7.11: Distribution of the alleles of the global optimal solution over the run of an offspring selection GA using ERX crossover and a mutation rate of 5% (remaining parameters are set according to Table 7.2).

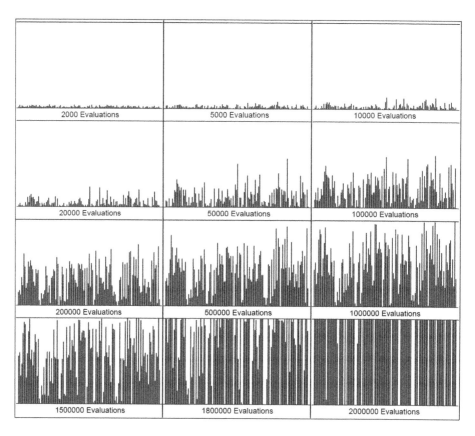

FIGURE 7.12: Distribution of the alleles of the global optimal solution over the run of an offspring selection GA using ERX crossover and no mutation (remaining parameters are set according to Table 7.2).

7.4 Building Block Analysis for the Relevant Alleles Preserving GA (RAPGA)

Similar to the previous section we aim to highlight some of the most characteristic features of the relevant alleles preserving GA as introduced in Section 4.3. The parameter settings of the RAPGA as given in Table 7.3 are also described in Section 4.3.

Table 7.3: Parameters for test runs using the relevant alleles preserving genetic algorithm.

Parameters for the RAPGA tests	
(Results presented in Fig.7.13, Fig.7.14, Fig.7.15, Fig.7.16, and Fig.7.17)	
Max. Generations	1,000
Initial Population Size	500
Mutation Rate	0.00 or 0.05
Elitism Rate	1
Male Selection	Roulette
Female Selection	Random
Crossover Operators	OX
	ERX
	MPX
	combined (OX, ERX, and MPX)
Mutation Operator	Simple Inversion
Minimum Population Size	5
Maximum Population Size	700
Twin Exclusion	true
Check Structural Identity	true
Effort	20,000
Comparison Factor Bounds	1 to 1
Attenuation	0

The main characteristics of the RAPGA are quite similar to a GA using offspring selection. The most important aspects of offspring selection are implicitly included in RAPGA; additionally, the RAPGA also introduces adaptive population size adjustment in order to support offspring selection to exploit the available genetic information in the actual population to the maximum in terms of achieving new (in order to maintain diversity) and even better (the offspring selection aspect) solution candidates for the next generation. RAPGA is a rather young algorithmic idea which has been presented in [AWW07]. Nevertheless, there is evidence that the RAPGA is comparably generic and flexible as offspring selection has already proven to be for a wide range of GA and GP applications.

The following experiments are set up quite similar to the offspring selection experiments of the previous section. Firstly, the considered operators OX, MPX, and ERX as well as their combination (OX, MPX, ERX) are applied to the *ch130* benchmark TSP problem taken from the TSPLib. Then the most successful operator or operator combination, respectively, is also exemplarily considered without mutation in order to show that the RAPGA like offspring selection does not rely on mutation to such an extent as conventional GAs.

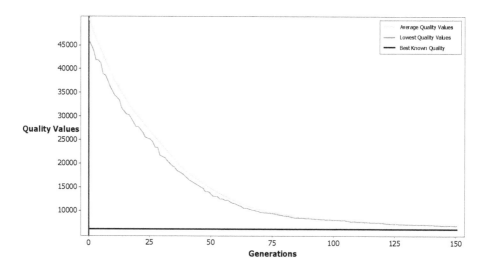

FIGURE 7.13: Quality progress for a relevant alleles preserving GA with OX and a mutation rate of 5%.

Already the experiments using OX (see Figure 7.13) and MPX (Figure 7.14) show good results (approximately 5% – 10% worse than the globally optimal solution) which are even slightly better than the corresponding offspring selection results. Even if only single test runs are shown in this chapter it has to be pointed out that the authors have taken care that characteristical runs are shown. Besides, as can be seen in the more systematical experiments of Chapter 10, especially due to the increased robustness caused by offspring selection and RAPGA the variance of the results' qualities is quite small.

Similar to what we stated for the OS analyses, also for the RAPGA the best results could be achieved using ERX (as shown in Figure 7.15) as well as using the combination of OX, ERX, and MPX (see Figures 7.16 and 7.17).

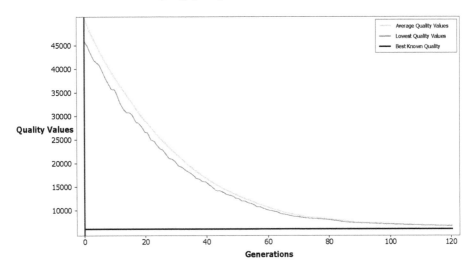

FIGURE 7.14: Quality progress for a relevant alleles preserving GA with MPX and a mutation rate of 5%.

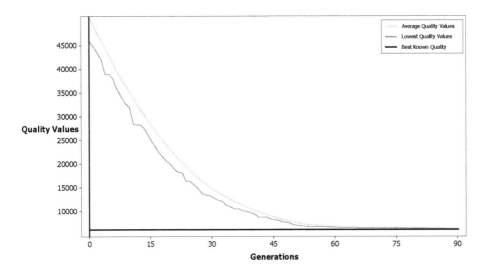

FIGURE 7.15: Quality progress for a relevant alleles preserving GA with ERX and a mutation rate of 5%.

The achieved results using these operators are about 1% or even less worse than the global optimal solution. In the case of the RAPGA the operator combination turned out to be slightly better than ERX (in 18 of 20 test runs). Therefore, this is the operator combination we have also considered for a detailed building block analysis without mutation as well as applying 5% mutation.

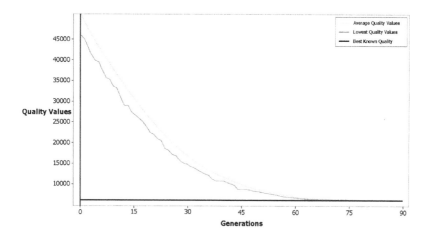

FIGURE 7.16: Quality progress for a relevant alleles preserving GA using a combination of OX, ERX, and MPX, and a mutation rate of 5%.

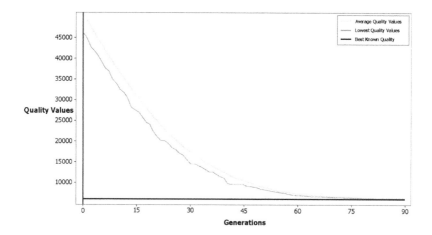

FIGURE 7.17: Quality progress for a relevant alleles preserving GA using a combination of OX, ERX, and MPX, and mutation switched off.

Barely surprising, the results of RAPGA with the operator combination consisting of OX, ERX, and MPX turned out to be quite similar to those achieved using offspring selection and the ERX operator. Due to the name giving aspect of essential allele preservation, disabling mutation (see Figures 7.16 and 7.17) has almost no consequences concerning achievable global solution quality. Even without mutation the results are just 1-2% worse than the global optimum. The distributions of essential alleles over the generations of the RAPGA run (as shown in Figure 7.18 and Figure 7.19) also show quite similar behavior as already observed in the corresponding analyses of the effects of offspring selection. Almost all essential alleles are represented in the first populations and their diffusion is slowly growing over the GA run, and even without mutation the vast majority of essential alleles is fixed by the end of the RAPGA runs.

Summarizing these results, we can state that quite similar convergence behavior is observed for a GA with offspring selection and the RAPGA, which is characterized by efficient maintenance of essential genetic information. As shown in Section 7.2, this behavior (which we would intuitively expect from any GA) cannot be guaranteed in general for GA applications where it was mainly mutation which helped to find acceptable solution qualities.

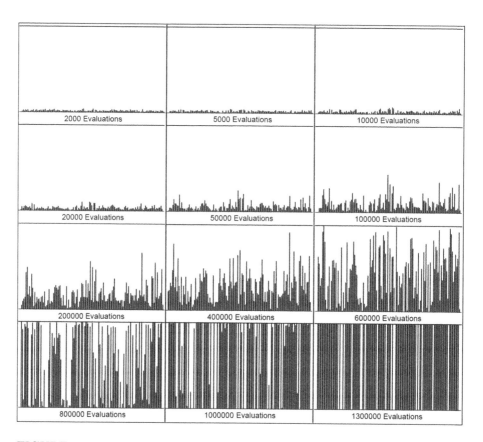

FIGURE 7.18: Distribution of the alleles of the global optimal solution over the run of a relevant alleles preserving GA using a combination of OX, ERX, and MPX, and a mutation rate of 5% (remaining parameters are set according to Table 7.3).

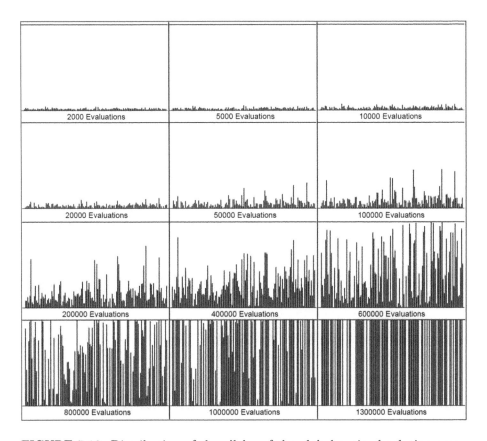

FIGURE 7.19: Distribution of the alleles of the global optimal solution over the run of a relevant alleles preserving GA using a combination of OX, ERX, and MPX without mutation (remaining are set parameters according to Table 7.3).

Chapter 8

Combinatorial Optimization: Route Planning

DOI: 10.1201/9781420011326-9

There are a great many of combinatorial optimization problems that genetic algorithms have been applied on so far. In the following we will concentrate on two selected route planning problems with a lot of attributes which are representative for many combinatorial optimization problems, namely the traveling salesman problem (TSP) and the vehicle routing problem (VRP).

The traveling salesman problem is certainly one of the classical as well as most frequently analyzed representatives of combinatorial optimization problems with a lot of solution methodologies and solution manipulation operators. Comparing the TSP to other combinatorial optimization problems, the main difference is that very powerful problem-specific methods as for example the Lin-Kernighan algorithm [LK73] and effective branch and bound methods are available that are able to achieve a global optimal solution in very high problem dimensions. These high-dimensional problem instances with a known global optimal solution are very well suited as benchmark problems for metaheuristics as for example GAs.

The VRP as well as its derivatives, the capacitated VRP (CVRP) and the capacitated VRP with time windows (CVRPTW) which will be introduced in this chapter, are much closer to practical problem situations in transport logistics, and solving them requires the handling of implicit and explicit constraints. There are also no comparable powerful problem-specific methods available, and metaheuristics like tabu search and genetic algorithms are considered the most powerful problem solving methods for VRP which is a different but not less interesting situation than handling the TSP problem.

8.1 The Traveling Salesman Problem

The TSP is quite easy to state: Given a finite number of cities along with the cost of travel between each pair of them, the goal is to find the cheapest way of visiting all the cities exactly once and returning to your starting point. Usually the travel costs are symmetric. A tour can simply be described by the order in which the cities are visited; the data consist of integer weights assigned to the edges of a finite complete graph and the objective is to find a Hamiltonian cycle, i.e., a cycle passing through all the vertices, of minimum total weight. In this context, Hamiltonian cycles are commonly called tours.

Already in the early 19th century the TSP appeared in literature [Voi31]. In the 1920s, the mathematician and economist Karl Menger [Men27] published it in Vienna; it reappeared in the 1930s in the mathematical circles of Princeton. In the 1940s, it was studied by statisticians (Mahalanobis, see [Mah40], e.g., and Jessen, see for instance [Jes42]) in connection with an agricultural application. The TSP is commonly considered the prototype of a hard problem in combinatorial optimization.

8.1.1 Problem Statement and Solution Methodology

8.1.1.1 Definition of the TSP

In a formal description the TSP is defined as the search for the shortest Hamiltonian cycle of a graph whose nodes represent cities. The objective function f represents the length of a route and therefore maps the set S of admissible routes into the real numbers $\mathbb{R}$ [PS82]:

$$f : S \to \mathbb{R}$$

The aim is to find the optimal tour $s^* \in S$ such that $f(s^*) \leq f(s_k), \forall s_k \in S$. In order to state the objective function f we have to introduce a distance matrix $[d_{ij}], d_{ij} \in \mathbb{R}^+$ whose entries represent the distance from a city i to a city j. In that kind of representation the cities are considered as the nodes of the underlying graph. If there is no edge between two nodes, the distance is set to infinity.

Using the notation given in [PS82], that $\pi_k(i)$ represents the city visited next after city i in a certain tour s_k, the objective function is defined as

$$f(s_k) = \sum_{i=1}^{n} d_{i\pi_k(i)} \tag{8.1}$$

By this definition the general asymmetric TSP is specified. By means of certain constraints on the distance matrix it is possible to define several variants of the TSP. A detailed overview about the variants of the TSP is given in [LLRKS85]. The most important specializations consider symmetry, the triangle-inequality, and Euclidean distances:

Symmetry

A TSP is defined to be symmetric if and only if its distance matrix is symmetric, i.e., if

$$d_{ij} = d_{ji}, \forall i, j \in 1, \ldots, n \tag{8.2}$$

If this set of equalities is not satisfied for at least one pair (i, j), for example if "one-way streets" occur, we denote the problem as an asymmetric TSP.

The Triangle Inequality

Symmetric as well as asymmetric TSPs can, but don't necessarily have to satisfy the triangle inequality:

$$d_{ij} \leq (d_{ik} + d_{kj}), \forall(i, j, k \in 1, \ldots, n) \tag{8.3}$$

i.e., that the direct route between two cities must be shorter than or as long as any route including another node in between.
A reasonable violation of the triangle inequality is possible especially when the entries in the distance matrix are interpreted as costs rather than as distances.

Euclidean Distances

As a rather important subset of symmetric TSP satisfying the triangle inequality we consider the so-called Euclidean TSP. For the Euclidean TSP it is mandatory to specify the coordinates of each node in the n-dimensional Euclidean space. For the 2-dimensional case the entries d_{ij} of the distance matrix are consequently given by the Euclidean distance

$$d_{ij} = \sqrt{(x_i - x_j)^2 + (y_i - y_j)^2} \tag{8.4}$$

whereby x_i and y_i denote the coordinates of a certain city i.
In contrast to most problems that occur in practice, many TSP benchmark tests use Euclidean TSP instances. Anyway, GA-based metaheuristics do not take advantage of the Euclidean structure and can therefore also be used for Non-Euclidean TSPs.

8.1.1.2   Versions of the TSP

Motivated by certain situations appearing in operational practice, some more variants of the TSP have emerged. Appreciable standardizations that will not be taken into further account within the scope of this book are the following ones:

Traveling Salesman Subtour Problems (TSSP)

In contrast to the TSP not all cities have to be visited in the context of the TSSP; only those cities are visited that are worth being visited which implies the necessity of some kind of profit function in order to decide if the profit is higher than the travel expenses. Vice versa, depending on the actual implementation, this can also be realized by the introduction of penalties for not visiting a certain node (city).

Postman Problems

For postman problems (e.g., [Dom90]) not certain sets of nodes (cities) have to be visited but rather given sets of edges (which can be interpreted as streets of houses) have to be passed at least once with the goal to minimize the total route length. Therefore, the aim is a suitable selection of the edges to be passed in a certain order for obtaining minimal cost.

Time Dependent TSP

In time dependent TSPs the cost of visiting a city j starting from a city i does not only depend on d_{ij} but also on the position in the total-route or, even more general, on the point of time a certain city is visited [BMR93].

Traveling Salesman Problem with Time Windows (TSPTW)

Like the TSP, the TSPTW is stated as finding an optimal tour for a set of cities where each city has to be visited exactly once. Additionally to the TSP, the tour must start and end at a unique depot within a certain time window and each city must be visited within its own time window. The cost is usually defined by the total travel distance and/or by the total schedule time (which is defined as the sum of travel time, waiting time, and service time) [Sav85].

8.1.1.3 Review of Optimal Algorithms

Total Enumeration

In principle, total enumeration is applicable to all integer optimization problems with a finite solution space: All points of the solution space S are evaluated by means of an objective-function storing the best solution so far. As the TSP has a worst case complexity of $\mathcal{O}(n!)$, total enumeration is only applicable to very small problem instances. For example, even for a rather small and simple 30-city symmetric TSP one would have to consider $\frac{(n-1)!}{2} = \frac{(29)!}{2}$ possible solutions which would require a computational time of about $1.4 * 10^{12}$ years assuming the use of a very powerful computer which can evaluate 100,000 million routes per second.

Integer Programming

In order to apply integer programming to the TSP it is mandatory to introduce a further $n \times n$ matrix $X = [x_{ij}]$ with $x_{ij} \in \{0, 1\}$ where x_{ij} indicates whether or not there is a connection from city i to city j. Thus, the optimization problem can be stated in the following way:
Find

$$\min(\sum_{i=1}^{n} \sum_{j=1}^{n} d_{ij} x_{ij}) \qquad (8.5)$$

such that

$$\sum_{j=1}^{n} x_{ij} = 1; \forall i \in \{1, \ldots, n\} \tag{8.6}$$

$$\sum_{i=1}^{n} x_{ij} = 1; \forall j \in \{1, \ldots, n\} \tag{8.7}$$

$$x_{ij} \geq 0; \forall i, j \in \{1, \ldots, n\} \tag{8.8}$$

These constraints ensure that each city has exactly one successor and is the predecessor of exactly one other city. The representation given above is also called an assignment problem and has firstly been applied to the TSP by Dantzig [DR59]. However, the assignment problem alone does not assure a unique Hamiltonian cycle, i.e., it is also possible that two or more subcycles exist which does not specify a valid TSP.

Hence, for the TSP it is necessary to state further conditions in order to define an assignment problem without subcycles, and therefore the integer property of the assignment problem does not hold any more. Similar to linear programming, in integer programming the admissible solution space can be restricted by the given constraints – but the corners of the emerging polyhedron won't represent valid solutions in general. In fact, only a rather small number of points inside the polyhedron will represent valid solutions of the integer program.

As described in [Gom63] Gomory tried to overcome this drawback by introducing the cutting-plane method that introduces virtual constraints, the so-called cutting-planes, in order to ensure that all corners of the convex polyhedron are integer solutions. The crux in the construction of suitable cutting planes is that it requires a lot of very problem-specific knowledge.

Grötschel's dissertation [Grö77] was one of the first contributions that considered a special TSP instance in detail and a lot of articles about the solution of specific TSP-benchmark problems have since then been published (as for example in [CP80]) with problem-specific cutting-planes.

Unfortunately, the methods for constructing suitable cutting-planes are far away from working in an automated way and require a well versed user. Therefore, the main area of application of integer programming for the TSP is the exact solution of some large benchmark problems in order to get reference problems for testing certain heuristics.

8.1.2 Review of Approximation Algorithms and Heuristics

During the last four decades a variety of heuristics for the TSP has been published; Lawler et al. have given a comparison of the most established ones in [LLRKS85]. Operations research basically distinguishes between methods that are able to construct new solutions routes, called route building heuristics or construction heuristics, and methods that assume a certain (valid) route in order to improve it, which are understood as route improving heuristics.

Nearest Neighbor Heuristics

The nearest neighbor algorithm [LLRKS85] is a typical representative of a route building heuristics. It simply considers a city as its starting point and takes the nearest city in order to build up the Hamiltonian cycle. At the beginning this strategy works out quite well whereas adverse stretches have to be inserted when only a few cities are left.

Figure 8.1 shows a typical result of nearest neighbor heuristics applied to a TSP instance that demonstrates its drawbacks.

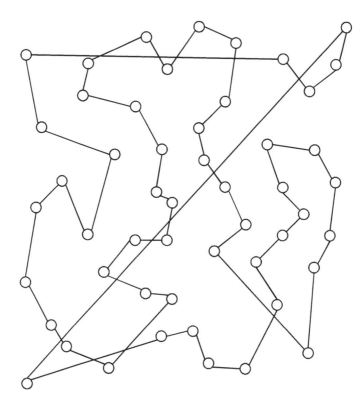

FIGURE 8.1: Exemplary nearest neighbor solution for a 51-city TSP instance ([CE69]).

Partitioning Heuristics

Applying partitioning heuristics to the TSP means splitting the total number of cities into smaller sets according to their geographical position. The emerging subsets are treated and solved as independent TSPs and the solution of the original TSP is given by a combination of the partial solutions.

The success rate of partitioning heuristics for TSPs very much depends on the size and the topological structure of the TSP. Partitioning heuristics do not perform well in the general case. Particularly suitable for partitioning heuristics are only higher dimensional Euclidean TSPs with rather uniformly distributed cities. Algorithms based on partitioning have been presented in [Kar77], [Kar79], and [Sig86].

Local Search

Typical representatives of route improving heuristics are the so-called k-change methods that examine a k-tuple of edges of a given tour and test whether or not a replacement of the tour segments effects an improvement of the actual solution quality.

A lot of established improvement methods are based upon local search strategies also causing the nomenclature "neighborhood search." The basic idea is to search through the surroundings of a certain solution s_i in order to replace s_i by an eventually detected "better" neighbor s_j.

The formal description of the neighborhood structure is given by $N \subseteq \mathcal{S} \times \mathcal{S}$ with $\mathcal{S}$ denoting the solution space. The choice of N is up to the user with the only restriction that the corresponding graph has to be connected and undirected, i.e., the neighborhood structure should be designed in a way that any point in the solution space is reachable and that s_i being a direct neighbor of s_j implies s_j being a direct neighbor of s_i.

Mainly for reasons of implementation the following formal definition has been established:

$$N \subseteq \mathcal{S} \times \mathcal{S}$$

with

$$N(s_i) := \{s_j \in \mathcal{S} \mid (s_i, s_j) \in N\}.$$

Choosing a neighborhood of larger size can cause problems concerning computational time whereas a rather small neighborhood increases the probability of getting stuck in a local optimum [PS82]. The search process for a better solution in the neighborhood is performed successively until no better solution can be detected. Such a point is commonly referred to as a local minimum with respect to a certain neighborhood structure and the neighborhood structure is termed definite if and only if any local optimum coincides with the global optimum (optima) s^* due to the neighborhood. Unfortunately, the verification of a definite neighborhood itself mostly is a NP-complete problem [PS82].

The 2-Opt Method

The most popular local edge-recombination heuristic is the 2-change replacement of two edges. In this context the neighborhood is defined in the following way:

> A tour s_i is adjacent (neighboring) to a tour s_j *if and only if s_j* can be derived from s_i by replacing two of s_i's edges.

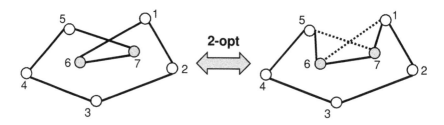

FIGURE 8.2: Example of a 2-change for a TSP instance with 7 cities.

Numbering the cities in the order they are visited with $c_1 \ldots c_n$ yields the following representation of two adjacent routes:

$$(c_1 \ldots c_i c_{i+1} \ldots c_j c_{j+1} \ldots c_n) \longleftrightarrow (c_1 \ldots c_i c_j \ldots c_{i+1} c_{j+1} \ldots c_n)$$

Figure 8.2 illustrates one possible 2-change operation for a small TSP instance. In this example (assuming that the left tour is transformed to the right tour) the two edges $5 - 7$ and $6 - 1$ are removed and the edges $5 - 6$ and $7 - 1$ are inserted in order to reestablish a valid tour.

Any route s_j can be derived from any other route s_i by at most $(n - 2)$ 2-change operations [AK89] and any solution s_i has exactly $\frac{n(n-1)}{2}$ neighboring (adjacent) solutions. For a symmetrical TSP (as indicated in the example of Figure 8.2) the number of neighboring solutions reduces to $\frac{n(n-2)}{2}$ [GS90].

Already half a century ago Croes [Cro58] published a solution technique for the TSP which is based upon the 2-change method: The algorithm has to check if an existing route s_i can be upgraded by the 2-change operator and perform it where applicable. This process is repeated until $f(s_j) \geq f(s_i)$ for all s_j that can be generated by using 2-change and the resulting route is called 2-optimal. Unfortunately, it is very unlikely that a 2-optimal tour is globally optimal.

The 3-Opt Method

The 3-opt method is very similar to the 2-opt method with the exception that not two but three edges are replaced. Considering a route with n nodes being involved $\frac{n(n-1)(n-2)}{2}$ different 3-change operations are possible [GS90].

$$(c_1 \ldots c_i c_{i+1} \ldots c_j c_{j+1} \ldots c_k c_{k+1} \ldots c_n) \longleftrightarrow (c_1 \ldots c_i c_{j+1} \ldots c_k c_{i+1} \ldots c_j c_{k+1} \ldots c_n)$$

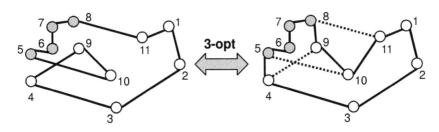

FIGURE 8.3: Example of a 3-change for a TSP instance with 11 cities.

Figure 8.3 illustrates one possible 3-change operation for a small TSP instance. In this example (assuming that the left tour is transformed to the right tour) the three edges $4-9$, $5-10$, and $8-11$ are removed and the edges $4-5$, $8-9$, and $10-11$ are inserted in order to reestablish a valid tour.

Also already half a century ago, Bock [Boc58] was the first one who applied the 3-opt method to the TSP. Similar as for the 2-opt method the final route was derived by successively applying 3-change operations terminates to a so-called 3-optimal solution. The probability to obtain a global optimal solution using the 3-opt method was empirically detected to be about $2^{-\frac{n}{10}}$ [Lin65].

The k-opt Method

In principle, the k-opt method is the consequential generalization of the methods described previously: k edges are replaced in a k-change neighborhood structure and a route is called k-optimal if it cannot be improved by any k-change. If $k = n$ then it is proven that the k-optimal solution is the global solution [PS82]. But as the complexity of locating a k-optimal solution is given by $\mathcal{O}(n^k)$ [GBD80], the computational effort is still enormous even for rather small values of k. A very efficient implementation for Euclidean traveling salesman problems is the Lin-Kernighan algorithm [LK73]. An efficient implementation is given in [Hel00].

8.1.3 Multiple Traveling Salesman Problems

The multiple traveling salesman problem (MTSP) describes a generalization of the TSP in the sense that there is not just one traveling salesman performing the whole tour but rather a set of salesmen, each serving a subset of the cities involved. Therefore, one of the cities has to be selected as the location for the depot representing the starting as well as the end point of all routes. So the MTSP is a combination of the assignment problem and the TSP. Usually a tour denotes the set of cities served by one traveling salesman and the number of tours is specified by m.

In literature there are mainly two definitions of the MTSP:

- In Bellmore's definition (given in [BH74]) the task is to find exactly m tours in such a way that each city in a tour and the depot are visited exactly once with the objective to minimize the total way.

- The second definition of the MTSP (as given in [Ber98], e.g.) does not postulate exactly but at most m routes and the goal is to minimize the total distance if each tour includes the depot and each city is visited exactly once in some tour.

At first sight the second definition seems more reasonable because there is no comprehensible reason why one should consider m tours if there is a solution involving only $(m-1)$ tours, for example. Still, one has to be aware of the fact that in the second definition with no additional constraints the solution will always be a single tour including all cities for any distance matrix fulfilling the triangle inequality.

8.1.4 Genetic Algorithm Approaches

Sequencing problems as for example the TSP are among the first applications of genetic algorithms, even if the classical binary representation as suggested in [Gol89] is not particularly suitable for the TSP because crossover hardly ever produces valid descendants.

In the following we will discuss some GA coding standards for the TSP as proposed in the relevant GA and TSP literature:

8.1.4.1 Problem Representations and Operators

Adjacency Representation

In the adjacency representation [GGRG85] a tour is represented as a list of n cities where city j is listed in position i if and only if the tour leads from city i to city j. Thus, the list

$$(7 \quad 6 \quad 8 \quad 5 \quad 3 \quad 4 \quad 2 \quad 1)$$

represents the tour

$$3 - 8 - 1 - 7 - 2 - 6 - 4 - 5.$$

In the adjacency representation any tour has its unique adjacency list representation. An adjacency list may represent an illegal tour. For example,

$$(3 \quad 5 \quad 7 \quad 6 \quad 2 \quad 4 \quad 1 \quad 8)$$

represents the following collection of cycles:

$$1 - 3 - 7, \quad 2 - 5, \quad 4 - 6, \quad 8$$

Obviously, the classical crossover operator(s) (single or n-point crossover) are very likely to return illegal tours for the adjacency representation. Therefore, the use of a repair operator becomes necessary.

Other operators for crossover have been defined and investigated for this kind of representation:

- Alternating Edge Crossover :
  The alternating edge crossover [GGRG85] chooses an edge from the first parent at random. Then, the partial tour created in this way is extended with the appropriate edge of the second parent. This partial tour is extended by the adequate edge of the first parent, etc. By doing so, the partial tour is extended by choosing edges from alternating parents. If an edge is chosen which would produce a cycle into the partial tour, then the edge is not added; instead, the operator randomly selects an edge from the edges which do not produce a cycle.
  For example, the result of an alternating edge crossover of the parents

$$(2 \; 3 \; 8 \; 7 \; 9 \; 1 \; 4 \; 5 \; 6) \qquad (7 \; 5 \; 1 \; 6 \; 9 \; 2 \; 8 \; 4 \; 3)$$

  could for example be

$$(2 \; 5 \; 8 \; 7 \; 9 \; 1 \; 6 \; 4 \; 3)$$

  The first edge chosen is $(1 - 2)$ included in the first parent's genetic material; the second edge chosen, edge $(2 - 5)$, is selected from the second parent, etc. The only randomly introduced edge is $7 - 6$ instead of $7 - 8$. Nevertheless, experimental results using this operator have been discouraging. The obvious explanation seems to be that good subtours are often disrupted by the crossover operator. Ideally, an operator ought to promote longer and longer high performance subtours; this has motivated the development of the following operator.

- Subtour Chunks Crossover:
  Using the subtour chunks crossover [GGRG85] an offspring is constructed from two parent tours in the following way: A random subtour of the first parent is chosen, and this partial tour is extended by choosing a subtour of random length from the second parent. Then, the partial tour is extended by taking subtours from alternating parents. If the use

of a subtour, which is selected from one of the parents, would lead to an illegal tour, then it is not added; instead, an edge is added which is randomly chosen from the edges that do not produce a cycle.

- Heuristic Crossover:
  The heuristic crossover [GGRG85] starts with randomly selecting a city for being the starting point of the offspring's tour. Then, the edges starting from this city are compared and the shorter of these two edges is chosen. Next, the city on the other side of the chosen edge is selected as a reference city. The edges which start from this reference city are compared and the shortest one is added to the partial tour, etc. If at some stage a new edge introduces a cycle into the partial tour, then the tour is extended with an edge chosen at random from the remaining edges which do not introduce cycles.

The main advantage of the adjacency representation is that it allows schemata analysis as described in [OSH87], [GGRG85], [Mic92]. Unfortunately, the use of all operators described above lead to poor results; in particular, the experimental results with the alternating edge operator have been very poor. This is because this operator often destroys good subtours of the parent tours. The subtour chunk operator which chooses subtours instead of edges from the parent tours performs better than the alternating edge operator. However, it still has quite a low performance because it does not take into account any information available about the edges. The heuristic crossover operator performs far better than the other two operators; still, the performance of the heuristic operator is not remarkable either [GGRG85].

Ordinal Representation

When using the ordinal presentation as described in [GGRG85] a tour is also represented as a list of n cities; the i-th element of the list is a number in the range from 1 to $n - i + 1$, and there an ordered list of cities serving as a reference point is also used.

The easiest way to explain the ordinal representation is probably by giving an example. Assume, for instance, that the ordered list L is given as

$$L = (1 \quad 2 \quad 3 \quad 4 \quad 5 \quad 6 \quad 7).$$

Now the tour

$$1 - 2 - 7 - 5 - 6 - 3 - 4$$

in ordinal representation is given as

$$T = (1 \quad 1 \quad 5 \quad 3 \quad 3 \quad 1 \quad 1).$$

This can be interpreted in the following way: The first member of T is 1, which means that in order to get the first city of the tour we take the first element of the list L and remove it from the list. So the partial tour is 1 at the beginning. The second element of T is also 1 so the second city of the route is 2 which is situated at the first position of the reduced list. After removing city 2 from the list, the next city to add is in position 5 according to T, which is city 7 in the again reduced list L, etc. If we proceed in this way until all elements of L are removed, we will finally find the tour $1 - 2 - 7 - 5 - 6 - 3 - 4$ with the corresponding ordinal representation $T = (1 \ \ 1 \ \ 5 \ \ 3 \ \ 3 \ \ 1 \ \ 1)$. The main advantage of this rather complicated ordinal representation lies in the fact that the classical crossover can be used. This follows from the fact that the i-th element of the tour representation is always a number in the range from 1 to $n - i + 1$. It is self-evident that partial tours to the left of the crossover point do not change whereas partial tours to the right of the crossover point are split in a quite random way and, therefore, the results obtained using ordinal representation have been generally poor (approximately in the dimension of the results with adjacency representation) [GGRG85], [LKM+99].

Path Representation

The path representation is probably the most natural representation of a tour. Again, a tour is represented as a list of n cities. If city i is the j-th element of the list, city i is the j-th city to be visited. Hence the tour

$$1 - 2 - 7 - 5 - 6 - 3 - 4$$

is simply represented by

$$(1 \ \ 2 \ \ 7 \ \ 5 \ \ 6 \ \ 3 \ \ 4).$$

Since the classical operators are not suitable for the TSP in combination with the path representation, other crossover and mutation operators have been defined and discussed. As this kind of representation will be used for our experiments in Chapter 10, we shall now discuss the corresponding operators in a more detailed manner:

- Partially Matched Crossover (PMX):
  The partially matched crossover operator has been proposed by Goldberg and Lingle in [GL85]. It passes on ordering and value information from the parent tours to the offspring tours: A part of one parent's string is mapped onto a part of the other parent's string and the remaining information is exchanged.

  Let us for example consider the following two parent tours

  $$(a \ \ b \ \ c \ \ d \ \ e \ \ f \ \ g \ \ h \ \ i \ \ j) \quad \text{and} \quad (c \ \ f \ \ g \ \ a \ \ j \ \ b \ \ d \ \ i \ \ e \ \ h).$$

  The PMX operator creates an offspring in the following way: First, it randomly selects two cutting points along the strings. As indicated in Figure 8.4, suppose that the first cut point is selected between the fifth

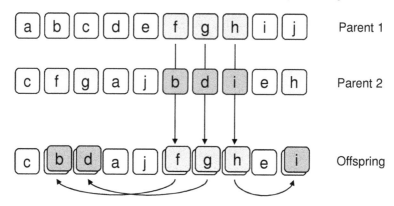

FIGURE 8.4: Example for a partially matched crossover (adapted from [Wen95]).

and the sixth element and the second one between the eighth and ninth string element. The substrings between the cutting points are called the mapping sections. In our example they define the mappings $f \leftrightarrow b$, $g \leftrightarrow d$, and $h \leftrightarrow i$. Now the mapping section of the first parent is copied into the offspring resulting

$$(\Box \quad \Box \quad \Box \quad \Box \quad \Box \quad f \quad g \quad h \quad \Box \quad \Box)$$

Then the offspring is filled up by copying the elements of the second parent; if a city is already present in the offspring then it is replaced according to the mappings. Hence, as illustrated in Figure 8.4, the resulting offspring is given by

$$(c \quad b \quad d \quad a \quad j \quad f \quad g \quad h \quad e \quad i)$$

The PMX operator therefore tries to keep the positions of the cities in the path representation; these are rather irrelevant in the context of the TSP problem where the most important goal is to keep the sequences. Thus, the performance of this operator for the TSP is rather poor, but we can easily imagine that this operator could perform well for other combinatorial optimization problems like the machine scheduling problem even if it has not been developed for such problem instances.

- Order Crossover (OX):
  The order crossover operator has been introduced by Davis in [Dav85]. For the first time it employs the essential property of the path representation, that the order of cities is important and not their position.

  It constructs an offspring by choosing a subtour of one parent preserving the relative order of the other parent. For example let us consider the

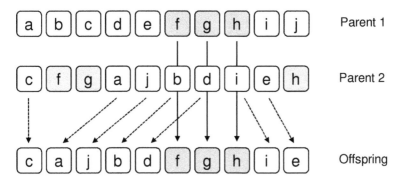

FIGURE 8.5: Example for an order crossover (adapted from [Wen95]).

following two parent tours

$$(a \quad b \quad c \quad d \quad e \quad f \quad g \quad h \quad i \quad j) \quad \text{and} \quad (c \quad f \quad g \quad a \quad h \quad b \quad d \quad i \quad e \quad j)$$

and suppose that we select the first cut point between the fifth and sixth position and the second cut point between the eighth and ninth position. For creating the offspring the tour segment between the cut points of the first parent is copied into it, which gives

$$(\square \quad \square \quad \square \quad \square \quad \square \quad f \quad g \quad h \quad \square \quad \square)$$

Then the selected cities of the first parent's tour segment are canceled from the list of the second parent and the blank positions of the child are filled with the elements of the shortened list in the given order (as illustrated in Figure 8.5), which gives

$$(c \quad a \quad b \quad d \quad i \quad f \quad g \quad h \quad e \quad j)$$

Since a much higher number of edges is maintained, the results are unequivocally much better compared to the results achieved using the PMX operator.

- Cyclic Crossover (CX):
  The cyclic crossover operator, proposed by Oliver et al. in [OSH87], attempts to create an offspring from the parents where every position is occupied by a corresponding element from one of the parents.

  For example, again consider the parents

  $$(a \quad b \quad c \quad d \quad e \quad f \quad g \quad h \quad i \quad j) \quad \text{and} \quad (c \quad f \quad g \quad a \quad h \quad b \quad d \quad i \quad e \quad j)$$

  and choose the first element of the first parent tour as the first element of the offspring. As node c can no longer be transferred to the child

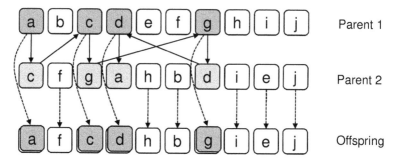

FIGURE 8.6: Example for a cyclic crossover (adapted from [Wen95]).

from the second parent, we visit node c in the first parent and transfer it to the offspring which makes it impossible for the first parent's node g to occupy the same position in the child. Therefore, g is taken from parent 2 and so on. This process is continued as long as possible, i.e., as long as the selected node is not yet a member of the offspring. In our example this is the case after four successful copies resulting in the following partial tour:

$$(a \ \square \ c \ d \ \square \ \square \ g \ \square \ \square \ \square).$$

The remaining positions can then simply be taken from one of the two parents; in this example, which is graphically illustrated in Figure 8.6, these are taken from from parent 2).

Oliver et al. [OSH87] concluded from theoretical and empirical results that the CX operator gave slightly better results than the PMX operator. Anyway, the results of both position preserving operators CX and PMX are definitely worse than those obtained with OX which fortifies our basic assumption that in the context of the TSP it is much more important to keep sequences rather than positions.

- Edge Recombination Crossover (ERX):
  Even if the main aim of the OX operator is to keep the sequence of at least one parent there are still quite a lot of new edges in the offspring.[1] Whitley et al. [WSF89] tried to overcome this drawback and came up with the edge recombination crossover which has been designed with the objective of keeping as many edges defined by the parents as possible. Indeed it can be shown that about $95\% - 99\%$ of each child's edges occur in at least one of the two respective parents [WSF89]. Therefore, the ERX operator for the first time represented an almost mutation-free

---

[1]In the present contents "new" means that those edges do not occur in any of the two parents.

crossover operator, but unfortunately this can only be achieved by a quite complicated and time consuming procedure:

The ERX operator is an operator which is suitable for the symmetrical TSP as it assumes that only the values of the edges are important and not their direction. Pursuant to this assumption, the edges of a tour can be seen as the carriers of heritable information. Thus, the ERX operator attempts to preserve the edges of the parents in order to pass on a maximum amount of information to the offspring whereby the breaking of edges is considered as an unwanted mutation. The problem that usually occurs with operators that follow an edge recombination strategy is that they often leave cities without a continuing edge [GGRG85] whereby these cities become isolated and new edges have to be introduced.

The ERX operator tries to avoid this problem by first choosing cities that have few unused edges; still, there has to be a connection with a city before it can be selected. The only edge that the ERX operator may fail to enforce is the edge from the final city to the initial city which inhibits the ERX operator of working totally mutation free. When constructing an offspring (descendant), we first have to construct a so-called "edge map" which gives the edges for each of the parents that start or finish in it. Then, the ERX works according to the following algorithm [WSF89]

1. Choose the initial city from one of the two parent tours. It might be chosen randomly or according to criteria outlined in step 4. This is the "current city."

2. Remove all occurrences of the "current city" from the left hand side of the edge map.

3. If the current city has entities in its edge list go to step 4; otherwise, go to step 5.

4. Determine which of the cities in the edge-list of the current city has the fewest entities in its own edge list. The city with the fewest entities becomes the "current city"; ties are broken at random. Proceed with step 2.

5. If there are no remaining unvisited cities, then terminate; otherwise randomly choose an unvisited city and continue with step 2.

We will explain the functioning of ERX on the basis of a small example which has also been used in [Wen95]. Consider for instance the tours

(1 2 3 4 5 6 7 8 9) and (4 1 2 8 7 6 9 3 5)

The edge map for our example parent tours is given in Table 8.1.

Table 8.1: Exemplary edge map of the parent tours for an ERX operator.

city	connected cities
1	9, 2, 4
2	1, 3, 8
3	2, 4, 9, 5
4	3, 5, 1
5	4, 6, 3
6	5, 9, 7
7	6, 8
8	7, 9, 2
9	8, 1, 6, 3

According to the procedure given before, we select city 1 as the initial city. The edge map of city one shows that cities 9, 2, and 4 are the candidates for becoming the next current city. As city 9 actually has 4 (8,1,6,3) further links, we have to decide between the cities 2 and 4 which both have 3 further links. Choosing city 4 as the next current city we obtain 3 and 5 as the next candidates, etc. Proceeding in that way, we might finally end up with the offspring tour

$$(1 \quad 4 \quad 5 \quad 6 \quad 7 \quad 8 \quad 2 \quad 3 \quad 9)$$

which for this special case of our example is totally mutation free, i.e., all edges of the offspring occur in at least one of the two parents.

As common sequences of the parent tours are not taken into account by the ERX operator an enhancement, commonly denoted as "enhanced edge recombination crossover (EERX)", has been developed [SMM+91]. The EERX additionally gives priority to those edges starting from the current city which are present in both parents.

For mutation in the context of applying genetic algorithms to the TSP, the $2 - change$ and $3 - change$ techniques have turned out to be very successful [WSF89]. A comprehensive review of mutation operators for the TSP is given in [LKM+99]. In the following some of the most important mutation operators are described which are also applied in the experimental part of this book:

- Exchange Mutation:
  The exchange mutation operator selects two cities of the tour randomly and simply exchanges them. In various publications the exchange mutation operator is also referred to as swap mutation, point mutation, reciprocal exchange, or order-based mutation [LKM+99].

- Insertion Mutation:
  The insertion mutation operator [Mic92] randomly chooses a city, removes it from the tour and inserts it at a randomly selected place. An

alternative naming for insertion mutation is position-based mutation [LKM+99].

- **Simple Inversion Mutation:**
  The simple inversion mutation operator [Hol75], which is used in the TSP experiments of the book, randomly selects two cut points and simply reverses the string between them.

- **Inversion Mutation:**
  The inversion mutation operator [Fog93] randomly selects a subtour, removes it, and inserts it in reverse order at a randomly chosen position. An alternative naming for inversion mutation is cut-inversion mutation [LKM+99].

8.2 The Capacitated Vehicle Routing Problem

In principle, the vehicle routing problem (VRP) is a m-TSP where a demand is associated with each city, and the salesmen are interpreted as vehicles each having the same capacity. A survey of the VRP is for example given in [Gol84]. During the later years a number of authors have "renamed" this problem the capacitated vehicle routing problem (CVRP). The sum of demands on a route cannot exceed the capacity of the vehicle assigned to this route; as in the m-TSP we want to minimize the sum of distances of the routes. Note that the CVRP is not purely geographic since the demand may be constraining.

The CVRP is the basic model for a number of vehicle routing problems:

If a time slot, in which customers have to be visited, is added to each customer, then we get the "vehicle routing problem with time windows" (VRPTW or CVRPTW). In addition to the capacity constraint, a vehicle now has to visit a customer within a certain time frame given by a ready time and due date. It is generally allowed that a vehicle may arrive before the ready time (in this case it simply waits at the customer's place), but it is forbidden to arrive after the due date. However, some models allow early or late servicing but with some form of additional cost or penalty. These models are denoted "soft" time window models (as for example in [Bal93]).

If customers are served from several depots, then the CVRP becomes the "multiple depots vehicle routing problem" (MDVRP); in this variant each vehicle starts and returns to the same depot. The problem can be solved by splitting it into several single depot VRP problems if such a split can be done effectively. Another variant of the CVRP is the "vehicle routing problem with length constraints" (VRPLC or CVRPLC). Here each route is not allowed to exceed a given distance; this variant is also known as the "distance constrained vehicle routing problem" (DVRP) in case there are no capacity restrictions and the length or cost is the only limiting constraint.

In the "split delivery" model the demand of a customer is not necessarily covered by just one vehicle but may be split between two or more. The solutions obtained in a split delivery model will always be at least as good as for the "normal" CVRP and we often might be able to utilize the vehicles better and thereby save vehicles.

Finally we shall also mention the "pickup and delivery" variant where the vehicles not only deliver items but also pick up items during the routes. This problem can be varied even more according to whether the deliveries must be completed before starting to pick up items or the two phases can be interleaved.

All of these problems have in common that they are "hard" to solve. For the VRPTW exact solutions can be found within reasonable time for some instances including up to about 100 customers. A review of exact methods for the VRPTW is given in Subsection 8.2.1.2.

Often the number of customers combined with the complexity of real-life data does not permit solving the problems exactly. In these situations it is commendable to apply approximation algorithms or heuristics. Both can produce feasible, but not necessarily optimal solutions; whereas a worst-case deviation is known for approximation algorithms, nothing is known a priori for heuristics. Some of these inexact methods will be reviewed in Subsection 8.2.1.3.

If the term "vehicle" is interpreted more loosely, numerous scheduling problems can also be modeled as CVRPs or VRPTWs. An example is the following one: For a single machine we want to schedule a number of jobs for which we know the flow time and the time to go from one running job to the next one. This scheduling problem can be regarded as a VRPTW with a single depot, a single vehicle, and the customers representing the jobs. The cost of changing from one job to another is equal to the distance between the two customers, and the time it takes to perform an action is the service time of the job. For a general description of the connection between routing and scheduling see for instance [vB95] or [CL98].

8.2.1 Problem Statement and Solution Methodology

8.2.1.1 Definition of the CVRP

In this section we present a mathematical formulation of the general vehicle routing problem with time windows (VRPTW or CVRPTW) as the (capacitated) vehicle routing problem ((C)VRP) is fully included within this definition under special parameter settings. The formulation is based upon the model defined by Solomon [SD88].

In this description the VRPTW is given by a fleet of homogeneous vehicles

$\mathcal{V}$, a set of customers $\mathcal{C}$, and a directed graph $\mathcal{G}$. The graph consists of $|\mathcal{C}| + 2$ vertices, whereby the customers are denoted as $1, 2, \ldots, n$ and the depot is represented by the vertices 0 (the "driving-out depot") and $n + 1$ (the "returning depot"). The set of vertices $0, 1, \ldots, n + 1$ is denoted as $\mathcal{N}$; the set of arcs $\mathcal{A}$ represents connections between customers and between the depot and customers, where no arc terminates in vertex 0 and no arc originates from vertex $n + 1$. With each arc (i, j), where $i \neq j$, we associate a cost c_{ij} and a time t_{ij}, which may include service time at customer i.

Each vehicle j has a capacity q_j and each customer i a demand d_i. Furthermore, each customer i has a time window $[a_i, b_i]$; a vehicle can arrive before a_i, but service does not start before a_i); however, the vehicle must arrive at the customer before b_i. In the general description, the depot also has a time window $[a_0, b_0] = [a_{n+1}, b_{n+1}]$, the scheduling horizon. Vehicles may not leave the depot before a_0 and must be back before or at time b_{n+1}.

It is postulated that q, a_i, b_i, d_i, and c_{ij} are nonnegative integers, while the t_{ij} values are assumed to be positive integers. Furthermore, it is assumed that the triangular inequality is satisfied for both c_{ij} values as well as the t_{ij} values.

This model contains two sets of decision variables, namely x and s. For each arc (i, j), where $i \neq j, i \neq n + 1, j \neq 0$, and each vehicle k we define x_{ijk} in the following way:

$$
x_{ijk} = \begin{cases} 0 & , \quad \text{if vehicle } k \text{ does not drive from vertex } i \text{ to vertex } j \\ 1 & , \quad \text{if vehicle } k \text{ drives from vertex } i \text{ to vertex } j \end{cases}
$$

The decision variable s_{ik} is defined for each vertex i and each vehicle k denoting the time vehicle k starts to service customer i. If the given vehicle k doesn't service customer i, then s_{ik} does not mean anything. We assume $a_0 = 0$ and therefore $s_{0k} = 0$ for all k.

The goal is to design a set of routes with minimal cost, one for each vehicle, such that

- each customer is serviced exactly once,

- every route originates at vertex 0 and ends at vertex $n + 1$, and

- the time windows and capacity constraints are complied with.

The mathematical formulation for the VRPTW is stated as follows [Tha95]:

$$\min \sum_{k \in \mathcal{V}} \sum_{i \in \mathcal{N}} \sum_{j \in \mathcal{N}} c_{ij} x_{ijk} \quad s.t. \tag{8.9}$$

$$\sum_{k \in \mathcal{V}} \sum_{j \in \mathcal{N}} x_{ijk} = 1 \quad \forall i \in \mathcal{C} \tag{8.10}$$

$$\sum_{i \in \mathcal{C}} d_i \sum_{j \in \mathcal{N}} x_{ijk} \leq q \quad \forall k \in \mathcal{V} \tag{8.11}$$

$$\sum_{j \in \mathcal{N}} x_{0jk} = 1 \quad \forall k \in \mathcal{V} \tag{8.12}$$

$$\sum_{i \in \mathcal{N}} x_{ihk} - \sum_{j \in \mathcal{N}} x_{hjk} = 0 \quad \forall h \in \mathcal{C}, \forall k \in \mathcal{V} \tag{8.13}$$

$$\sum_{i \in \mathcal{N}} x_{i,n+1,k} = 1 \quad \forall k \in \mathcal{V} \tag{8.14}$$

$$s_{ik} + t_{ij} - K(1 - x_{ijk}) \leq s_{jk} \quad \forall i, j \in \mathcal{N}, \forall k \in \mathcal{V} \tag{8.15}$$

$$a_i \leq s_{ik} \leq b_i \qquad \forall i \in \mathcal{N}, \forall k \in \mathcal{V} \tag{8.16}$$

$$x_{ijk} \in \{0, 1\} \qquad \forall i, j \in \mathcal{N}, \forall k \in \mathcal{V} \tag{8.17}$$

The constraint (8.10) states that each customer is visited exactly once, and (8.11) implies that no vehicle is loaded with more than its capacity allows. Equations (8.12), (8.13), and (8.14) ensure that each vehicle leaves depot 0, leaves again after arriving at a customer, and finally arrives at the depot $n+1$. Inequality (8.15) states that a vehicle k cannot arrive at j before $s_{ik} + t_{ij}$ if it is traveling from i to j, whereby K is a large scalar. Finally, constraints (8.16) ensure that the time windows are adhered to and (8.17) are the integrality constraints. In this definition an unused vehicle is modeled by driving the empty route $(0, n+1)$.

As already mentioned earlier, the VRPTW is a generalization of TSP and CVRP; in case the time constraints (8.15) and (8.16) are not binding, the problem becomes a CVRP. This can be achieved by setting $a_i = 0$ and $b_i = M$ (where M is a large scalar) for all customers i. In this context it should be noted that the time variables enable us to formulate the CVRP without subtour elimination constraints. If only one vehicle is available, then the problem is in fact a TSP.

8.2.1.2 Exact Algorithms

Almost all papers proposing an exact algorithm for solving the CVRP or the VRPTW use one or a combination of the following three principles:

- Dynamic programming

- Lagrange relaxation-based methods

- Column generation

The dynamic programming approach for the VRPTW was presented in [KRKT87]. This paper is inspired by an earlier publication [CMT81] where Christofides et al. used the dynamic programming paradigm to solve the CVRP.

Lagrange relaxation-based methods have been published in a number of papers using slightly different approaches. There are approaches applying variable splitting followed by Lagrange relaxation as well as variants applying the k-tree approach followed by Lagrange relaxation. In [FJM97] Fisher et al. presented a shortest path approach with side constraints followed by Lagrangean relaxation. The main problem, which consists of finding the optimal Lagrange multipliers that yield the best lower bounds, is solved by a method using both subgradient optimization and a bundle method. Kohl et al. [KM97] managed to solve problems of 100 customers from the Solomon test cases; among them some previously unsolved problems.

If a linear program contains too many variables to be solved explicitly, it is possible to initialize the linear program with a smaller subset of variables and compute a solution of this reduced linear program. Afterwards one has to check whether or not the addition of one or more variables, currently not in the linear program, might improve the solution; this check is commonly done by the computation of the reduced costs of the variables. An introduction to this method (commonly called "column generation method") can for example be found in [BJN+98].

Again, similar as for the TSP it takes well versed users in order to benefit from the mentioned exact algorithms - especially if they are applied to large problems.

Therefore, the main area of application of exact methods in the context of CVRP is to locate the exact solution of some large benchmark problems in order to get some reference-problems for testing certain heuristics which can easily be applied to practical problems of higher dimension.

8.2.1.3 Approximation Algorithms and Heuristics

The field of inexact algorithms for the CVRP has been very active - far more active than that of exact algorithms, and a long series of papers has been published over the recent years. Heuristic algorithms that build a set of routes from scratch are typically called route-building heuristics, while an algorithm that tries to produce an improved solution on the basis of an already available solution is denoted as route-improving.

The Savings Heuristic

The savings heuristic has been introduced in [CW64]. At the beginning of the algorithm, each of the n customers (cities) is considered to be delivered with an own vehicle. For every pair of two cities a so-called savings value is calculated; this value specifies the reduction of costs which is achieved when the two routes are combined. Then the routes are merged in descending order of their saving values if all constraints are satisfied. According to [Lap92] the time complexity of the savings heuristic is given as $\mathcal{O}(n^2 \log n)$.

A lot of papers based on savings heuristics have been published. Especially Gaskell's approach [Gas67] is appreciable in this context as it introduces a different weighting of the savings with respect to the length of the newly inserted route-part as well as the so-called parallel savings algorithm that not only examines a pair but rather a n-tuple of routes.

The Sweep Heuristic

The sweep heuristic has been introduced by Gillett and Miller [GM74]. It belongs to the so-called "successive methods" in the sense that the ultimate goal of this approach is not necessarily the location of the final solution but rather the generation of reasonable valid tours which can be optimized by some kind of route improving heuristic.

The fundamental idea of the sweep heuristic can be described as follows: Imagining a watch hand that is mounted at the depot, the sweep heuristic builds up the first tour starting from an arbitrary angle and takes the cities in the order the watch hand sweeps over them as long as all constraints are fulfilled. Then the next cluster is established in the same way.

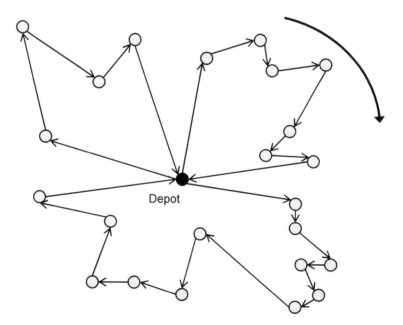

FIGURE 8.7: Exemplary result of the sweep heuristic for a small CVRP.

Figure 8.7 shows a possible solution achieved by a simple sweep heuristic. The more customers can be assigned to a route, the better the sweep heuristic typically performs. The time complexity of sweep heuristics is $\mathcal{O}(n \log n)$, which is equal to the complexity of a sorting algorithm.

The Push Forward Insertion Heuristic

In [Sol87] Solomon describes and evaluates three insertion heuristics for the VRPTW. Here a new route is started by a customer which minimizes a certain cost function cf_1. All unrouted customers are then evaluated for insertion between any two customers i and j of the partial route according to another cost function cf_2. If no feasible insertion is possible for all customers, cf_1 is evaluated again for all unrouted customers in order to determine the starting customer of a new route. Three different possible criteria for selecting the next customer are the following ones:

- Farthest customer from the depot first

- Customer with the earliest due date first

- Customer with the minimum equally weighted direct route-time and distance first

The third function basically describes the closest customer that will be directly reached in time. During the evaluation of their performance Solomon states that generally neither is better than the other. The farthest customer first criterion is suited for problems with shorter scheduling horizons, while selecting the customers regarding the earliest due date gives better results in situations where the scheduling horizons are longer, i.e., where vehicles have to visit more customers. The third alternative for cf_1 was not examined closer as it was not used in conjunction with the best performing alternative for cost function cf_2.

As Solomon notes, the three insertion heuristics that are described as alternatives for cf_2 are guided by both geographical and temporal criteria. The insertion heuristic, which is described first and termed $I1$, performed best in a number of test cases. Basically it extends the savings heuristic insofar as it takes into account the prolongation of the arrival time at the next customer. This function evaluates the difference between scheduling a customer directly and servicing it in an existing route between two customers. Mathematically it can be described as

$$I1(i, u, j) = \lambda t_{0u} - (\alpha_1(t_{iu} + t_{uj} - \mu t_{ij}) + \alpha_2(b_{j_u} - b_j)) \tag{8.18}$$

with the following restrictions: $\lambda, \mu, \alpha_1, \alpha_2 \geq 0$ and $\alpha_1 + \alpha_2 = 1$. In this case the VRP cost function c_{ij} equals to 1 for each pair of different customers i and j.

Tests with a choice of some configurations for λ, μ, α_1, and α_2 showed that good results were achieved with $\lambda = 2$, $\mu = 1$, $\alpha_1 = 0$, and $\alpha_2 = 1$, thus using only time-based savings instead of distance-based savings.

The name "push forward insertion heuristic" stems from a more efficient computation of the feasibility of an insertion. At each point, where a customer could be inserted, the time at which the vehicle would arrive later at the preceding customer is propagated sequentially through the route. As soon as this time becomes 0 the insertion is feasible as the remaining customers would not be serviced later than they already are. If the old partial route is feasible, then the new one thus will also be feasible. If the push forward value surpasses the due date at a customer, then an infeasible insertion is encountered and the rest of the route does not have to be checked. In the worst case this method still needs to perform the calculation for every customer in the tour. Feasibility regarding the capacity constraints, at least for the VRP variants without pickup & delivery, is easier to compute.

Solomon concludes that a hybridization of *I1* with a sweep heuristic could achieve excellent initial solutions with a reasonable amount of computation. Such an approach can be found in [TPS96] where cf_1 is a function taking into account three different properties: distance, due date, and the polar angle. The mathematical description reads

$$cf_1(u) = -\alpha t_{0u} + \beta b_u + \gamma \varphi_u \qquad (8.19)$$

with empirically derived weights $\alpha = 0.7$, $\beta = 0.2$, and $\gamma = 0.1$.

Other Methods

The problem of building one route at a time (which is done when using the heuristics described above) is usually that the routes generated in the latter part of the process are of worse quality because the last unrouted customers tend to be scattered over the geographic area. Potvin and Rousseou [PR93] tried to overcome this problem of the insertion heuristic by building several routes simultaneously where the initialization of the routes is done by using Solomon's insertion heuristic:

On each route the customer farthest away from the depot is selected as a "seed customer." Then, the best feasible insertion place for each unserviced customer is computed and the customer with the largest difference between the best and the second best insertion place is inserted. Even if this method works out better than the Solomon heuristic it is still quite far away from optimum. Russell elaborates further on the insertion approach in [Rus95].

Another approach built up upon the classical insertion heuristic is presented in [AD95]. Defined in a very similar way to the Solomon heuristics, every unrouted customer requests an offer and receives a price for insertion from every route in the schedule. Then unrouted customers send a proposal to the route with the best offer, and each route accepts the best proposal among the customers with the fewest number of alternatives. Therefore, more customers can be inserted in each iteration. If a certain threshold of routes is violated, then a certain number of customers is removed and the process is started again. The results of Antes and Derigs are comparable to those presented in [PR93]. As a matter of principle it has turned out that building several routes in parallel results in better solutions than building the routes one by one.

Similar to the route first schedule second principle mentioned previously, Solomon also suggests doing it the other way round in the "giant tour heuristic" [Sol86]. First all customers are scheduled in a giant route and then this route is divided into a number of routes. In the paper no computational results are given for the heuristic. Implementations of route-building heuristics on parallel hardware are reported for example in [FP93] and [Lar99].

8.2.2 Genetic Algorithm Approaches

Applying genetic algorithms to vehicle routing problems with or without time constraints is a rather young field of research and therefore, even if a lot of research work is done, no widely accepted standard representations or operators have yet been established. In the following we will in short discuss some of the more popular or promising proposals:

- A genetic algorithm for the VRPTW has been presented in [TOS94]. This algorithm uses the already mentioned cluster first route second method whereby clustering is done by a genetic algorithm while routing is done by an insertion heuristic. The GA works by dividing each chromosome into K divisions of N bits. The algorithm is based on dividing the plane by using the depot as the center and assuming the polar angle to each customer. Each of the divisions of a chromosome then represents the offset of the seed of a sector; the seeds here are polar angles that bound the sector and thereby determine the members of the sector.

- The genetic algorithm of Potvin and Bengio [PB96] operates on chromosomes of feasible solutions. The selection of parent solutions is stochastic and biased towards the best solutions. Two types of crossover, called RBX and SBX, are used. They rarely produce valid solutions and the results therefore have to undergo a repair phase as the algorithm only works with feasible solutions. The reduction of routes is often obtained by two mutation operators and the routes are optimized by local search every k iterations.
  The approach described in [PTMC02] is similar, but does not use trip delimiters in the actual representation. Instead, the routes that a solution is composed of are saved separately as ordered sets. The number of routes is not predefined and practically only limited by the number of customers. The crossover described in [PTMC02] is biased insofar as it uses distance information to decide where to insert a segment, though

an unbiased generic variant of this crossover has been defined later in [TMPC03]. A repair method is applied (after applying the crossover) which constructs a feasible solution by removing customers that are visited twice. Additionally, any route that has become too long (so that the demand or time window constraints would not be satisfied) is split before the violating customer. This algorithm has been applied on several benchmark instances of the CVRP and CVRPTW problem variants.

- An encoding similar to the TSP is used in [Zhu00] and [Pri04]. As described in these publications, a GA optimizes solutions for the problem using a path representation that does not include the depot. The representation is thus similar to the TSP and the subtours are identified deterministically whenever needed. [Pri04] describes a genetic algorithm that is applied to the DVRP using a splitting procedure that will find the optimal placement of the depots; this guarantees the feasibility of solutions in any case. Additionally, classic operators like the order crossover (as used for tackling the TSP) are applied without modifications. This approach does not rely solely on unbiased operators and is hybridized with a local search procedure, while [Zhu00] makes use of biased crossover operators in which distance information is used for determining the point where a crossing might occur. In this approach an initial population, consisting of individuals created by suited construction heuristics as well as randomly generated individuals, is also used. Additionally, the mutation probability is adjusted depending on the diversity in the population leaving a minimum probability of 6%.

- A cellular genetic algorithm has been proposed in [AD04]. It uses an encoding with unique trip delimiters such that the customers are assigned numbers from 0 to $(|\mathcal{C}| - 1)$ while the trip delimiters are numbers from $|\mathcal{C}|$ to $(|\mathcal{C}| + |\mathcal{V}| - 1)$. The representation of a solution thus consists of a string of consecutive numbers and is syntactically equal to a TSP path encoding. This allows the use of crossover operators known from the TSP such as the ERX. For mutating solutions the authors use insertion, swap, and inversion operators which are similar to relocate, exchange, and 2-opt as described below. The difference is that swap and inversion are used in an inter- as well as an intraroute way. There is also a local search phase which is conducted after every generation; in this phase all individuals are optimized by 2-opt and λ-interchanges. The best results have been achieved using both methods and setting $\lambda = 2$.

8.2.2.1 Crossover Operators

Sequence-Based Crossover (SBX)

The sequence-based crossover (SBX) operator has been proposed for the CVRP in [PB96], but it is applicable also to other VRP variants as well.

Generally, it assumes a path representation with trip delimiters, but can be applied on a representation without trip delimiters in which case the subtours have to be calculated first.

It works by breaking a subtour in each parent and linking the first part of one parent to the second part of another parent. The newly created subtour is then copied to a new offspring individual and completed with the genetic information of one of the parents. This behavior is exemplarily illustrated in Figure 8.8.

This operator is very likely to create ill-formed children with duplicate or unrouted customers; therefore the authors also propose a repair method which creates syntactically valid genetic representations. However, feasibility cannot be guaranteed in every case as it is not always possible to find a feasible insertion space for all unrouted customers. In such a case the offspring is discarded and the crossover is executed anew with a new pair of parents. It is stated in [PB96] that when applied on the Solomon benchmark set [Sol87] 50% of the offspring are infeasible.

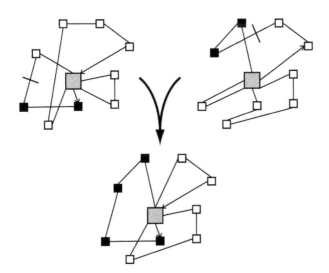

FIGURE 8.8: Exemplary sequence-based crossover.

Let us for example consider the following tours in path representation with trip delimiters where the depot is denoted as 0 and all other values represent customers.

$$(0\ 1\ 2\ 3\ 0\ 4\ 5\ 6\ 0)\quad\text{and}\quad(0\ 2\ 5\ 3\ 0\ 1\ 4\ 6\ 0)$$

In this case the SBX would randomly select two cut points in both solutions, for example at customer 2 in the first parent and customer 4 in the second one. Then the first half of the route in the first solution is concatenated with the second half of the route in the second solution yielding

$$(0 \ 1 \ 2 \ 6 \ 0)$$

This route then replaces the route with the selected customer in the first solution; thus the solution becomes

$$(0 \ 1 \ 2 \ 6 \ 0 \ 4 \ 5 \ 6 \ 0)$$

Obviously, now customer 6 is served twice, while customer 3 is not served at all and the repair procedure will have to correct this situation: First it will remove all duplicate customers in all routes except the new one which was just formed by the concatenation. This results in

$$(0 \ 1 \ 2 \ 6 \ 0 \ 4 \ 5 \ 0)$$

Then it will try to route all unserviced customers in that location in which the detour is minimal. For this example let us assume that this is after customer 5 and the final offspring solution thus is

$$(0 \ 1 \ 2 \ 6 \ 0 \ 4 \ 5 \ 3 \ 0)$$

Route-Based Crossover (RBX)

The route-based crossover (RBX) operator has also been proposed in [PB96]. It differs from the SBX insofar as subtours are not merged, but rather a complete subtour of one parent is copied to the offspring individual filling up the rest of the chromosome with subtours from the other parent. This procedure is exemplarily illustrated in Figure 8.9. Again, the operator does not produce feasible solutions in all cases and a repair procedure is needed.

Let us again consider the following tours in path representation with trip delimiters where 0 denotes the depot and all other values represent customers.

$$(0 \ 1 \ 2 \ 3 \ 0 \ 4 \ 5 \ 6 \ 0) \quad \text{and} \quad (0 \ 2 \ 5 \ 3 \ 0 \ 1 \ 4 \ 6 \ 0)$$

The RBX randomly selects a complete route in solution 1, in this case for example the first route:

$$(0 \ 1 \ 2 \ 3 \ 0)$$

This route then replaces a route in the second solution and thus we get

$$(0 \ 1 \ 2 \ 3 \ 0 \ 1 \ 4 \ 6 \ 0)$$

Obviously, now customer 1 is served twice, while customer 5 is not served at all. The same repair procedure as the one described for the SBX operator is applied here as well, resulting in the following possible final solution:

$$(0 \ 1 \ 2 \ 3 \ 0 \ 4 \ 6 \ 5 \ 0)$$

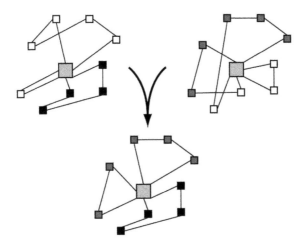

FIGURE 8.9: Exemplary route-based crossover.

Other Crossover Operators

The crossover that is introduced in [PTMC02] does not necessarily concatenate the opposite ends of two partial routes such as the SBX, but inserts a partial route of one solution into a good location of another solution. The fitness of such a location is determined by the distance between the customer which would precede the new partial route and the first customer of that partial route. Such an approach works well when solving the CVRP, but as is noted in [TMPC03] it does not help in the CVRPTW; this is in fact not surprising as distance alone might not be sufficient enough to determine the fitness of an insert when there are additional constraints which to a certain degree determine the shape of a route. Thus, a generic variant of the crossover is proposed which works similar to the RBX, except that it does not replace, but rather appends the new route to a given solution. Removing customers which are served twice then is the only necessary repair method. Such a removal also has the benefit that any solution remains feasible if it has been feasible before.

Other genetic algorithm approaches such as the one described in [Pri04] build on a path representation without trip delimiters as in the TSP. This allows the application of those crossover operators that have been mentioned in Section 8.1.4.1 without modifications. In [Zhu00] the PMX is compared to two new crossovers called "heuristic crossover" and "merge crossover" that take into account spatial and temporal features of the customers. However, these new operators achieve slightly better results only for those problem instances in which the customers are clustered, whereas for the other cases of randomly placed customers and a mix of random and clustered customers PMX showed better results.

8.2.2.2   Mutation Operators

Relocate

The relocate operator moves one of the customers within the solution string from one location to another randomly chosen one. An example of this behavior is illustrated in Figure 8.10.

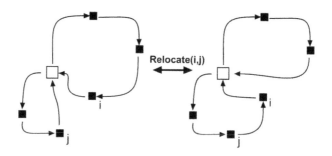

FIGURE 8.10: Exemplary relocate mutation.

Exchange

The exchange operator selects two customers within different tours and switches them so that they are served by the other vehicle, respectively. Both relocate and exchange operators are similar to a $(1, 0)$ and $(1, 1)$ λ-exchange defined by Osman [Osm93]. An example of the exchange behavior is shown in Figure 8.11.

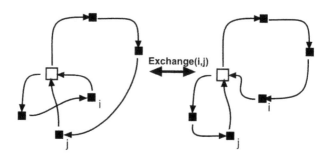

FIGURE 8.11: Exemplary exchange mutation.

2-Opt

The 2-opt operator selects two sites within the same route and inverts the route between them, so that the vehicle travels in the opposite direction. An example of this behavior is given in Figure 8.12.

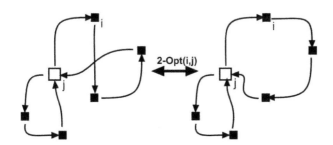

FIGURE 8.12: Example for a 2-opt mutation for the VRP.

2-Opt*

The 2-opt* operator behaves like a one point crossover operator in a tour: It first selects two customers in two different tours and creates two new tours; the first tour here consists of the first half of the first tour unified with the second half of the second tour, and the second tour consists of the first half of the second tour unified with the second half of the first tour. An example for the behavior of this operator is illustrated in Figure 8.13.

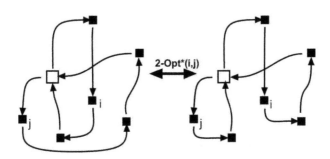

FIGURE 8.13: Example for a 2-opt* mutation for the VRP.

Or-Opt

The or-opt operator takes a number of consecutive customers, deletes them and inserts them at some point of the same tour. An example of this behavior is given in Figure 8.14.

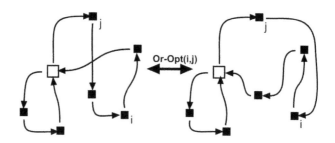

FIGURE 8.14: Example for an or-opt mutation for the VRP.

One Level Exchange (M1)

The one level exchange (M1) operator is mentioned in the context of the GA proposed in [PB96]. It tries to eliminate routes by inserting the customers into other routes while maintaining a feasible solution. This operator favors small routes with higher probability, because these are in general easier to remove. The probability is chosen such that a trip of size N is half as likely to be chosen as one of size $\frac{N}{2}$.

Two Level Exchange (M2)

The two level exchange (M2) operator looks one level deeper than the M1 operator as it removes routes from the whole route, but tries harder for each of the customers. After selecting a trip using the same bias towards smaller routes as the M1, each customer is tried to be inserted instead of another customer in a different route which in turn is tried to be inserted in any other place (except the originally selected trip). If such a feasible insertion is found, then the second selected customer is inserted and the first customer is inserted at the second customer's original place. This operator is more likely to find feasible insertion places, but requires quite a lot of computational effort; in the worst case the runtime complexity is $O(N^2)$.

Local Search (LSM)

Another operator which is also proposed in [PB96] applies several or-opt exchanges until a local optimum is reached. It first selects all possible combinations of three consecutive customers and tries to insert them in any other place while still maintaining the feasibility of the solution. If no solution can be found the process is repeated with two consecutive customers and finally with every single customer. Once a better solution has been found, the operator starts again with three consecutive customers and continues until no further improvement is possible. Local search methods can improve solutions on the one hand, but on the other hand they also might reduce the diversity in a population when applied to several individuals which could end in the same local minimum.

Chapter 9

Evolutionary System Identification

DOI: 10.1201/9781420011326-10

9.1 Data-Based Modeling and System Identification

9.1.1 Basics

In general, data mining is understood as the practice of automatically searching large stores of data for patterns. Nowadays, incredibly large (and quickly growing) amounts of data are collected in commercial, administrative, and scientific databases. Several sciences (e.g., molecular biology, genetics, astrophysics, and many others) produce extreme amounts of information which are often collected automatically. This is why it is impossible to analyze and exploit all these data manually; what we need are intelligent computer systems that can extract useful information (such as general rules or interesting patterns) from large amounts of observations. In short, "data mining is the non-trivial process of identifying valid, novel, potentially useful, and ultimately understandable patterns in data" [FPSS96].

One of the ways how genetic algorithms and, more precisely, genetic programming can be used in data mining is its application in data-based modeling. A given system is to be analyzed and its behavior described by a mathematical model; the process is therefore (especially in the context of modeling dynamic physical systems) called system identification [Lju99].

The principles have already been summarized in the GP introduction chapter, especially in Section 2.4.3 on symbolic regression, and they shall be repeated and extended in the following:

The main goal of regression is to determine the relationship of a dependent (target) variable t to a set of specified independent (input) variables x. Thus, what we want to get is a function f that uses x and a set of coefficients w such that

$$t = f(x, w) + \epsilon \tag{9.1}$$

where ϵ represents the error (noise) term.

Applying this procedure we assume that a model can be created with which it will also be possible to predict correct outputs for other data examples (test samples); from the training data we want to generalize to situations not known (or allowed to analyze) during the training phase.

When it comes to evaluating a model (i.e., a solution candidate in a GP-based modeling algorithm), the formula has to be evaluated on a certain set of evaluation (training) data X yielding the estimated values E. These estimated target values are compared to the original values T, i.e., those which are known from data retrieval (experiments) or calculated applying the original formula to X.

This comparison is done by calculating the error between original and calculated target values. There are several ways how to measure this error, one of the simplest and probably most frequently used one being the mean squared error (mse) function; the mean squared error of the vectors A and B each containing n values is calculated as

$$mse(A, B) = \frac{1}{n} * \sum_{k=1}^{n} (A_k - B_k)^2 \qquad (9.2)$$

Some of the major problems of data-based modeling are *noise* and *overfitting*:

- In common language, on the one hand we know noise as in general that what is heard, but on the other hand also as unwanted sound which is added to the audio signals that are of interest. Furthermore, the concept of noise is also known in image and video processing, where it is used more to describe unwanted signals that are rather disturbing. In the context of data-based modeling we often see that additional and somehow unwanted values are added to the original signals; this disturbing additional data is called noise.

- In machine learning, overfitting is understood as the exceeding fitting of models to given data. As already mentioned, data-based training of models is done using training data, i.e., sets of training examples of the functions which are searched for; the problem is that it can happen – especially in cases where too complex models are trained or the training process is executed too long – that the learner may adjust to very specific features or samples of the training data. Even a structurally inadequate model may fit to given training data perfectly if the model is complex enough.

  From the point of view of mathematical systems theory, we assume that a system Σ can be described by a function $\phi(\theta) : u \to y$, where u and y are the system's input and output, respectively, ϕ describes the structure of the function and θ denotes the vector of parameters. Data-based structure identification is supposed to find a function $\psi(\hat{\theta}) : u \to y$ that reproduces the system's output. The more parameters are stored in $\hat{\theta}$ the easier it becomes to reproduce the given training data, but it also becomes more probable that $\psi(\hat{\theta})$ represents not the basic behavior of Σ but rather the measured signal (which also includes noise). Of course, as we do not know the size of θ (or the structure of ϕ) in general, we cannot know when $\hat{\theta}$ becomes "too big."

Overfitting can also be seen as a violation of Occam's razor (see Section 2.6 for explanations on this); fitting too exactly to (noisy) training data might lead to a model whose ability to generalize is far worse than the general applicability of a simpler model.

Unfortunately there is no rule how to generally avoid overfitting as we often do not exactly know the complexity of the system whose behavior is to be modeled. However, there are several techniques that can help to avoid it: For instance, overfitting might cause a significant rise of the variances of the estimated parameter values $\hat{\theta}_i$, i.e., the parameter values estimated in independent identification runs diverge (which should of course not be the case if the structure of ψ and the size of $\hat{\theta}$ are correct); early stopping and the use of validation sets which are not included in the training data can also help to decrease the probability of overfitting.

Thus, accuracy (on training data) is not the only requirement for the result of the modeling process: Compact and (if possible) minimal models are preferred as they can be used in other applications easier. It is, of course, not easy to find models that ignore unimportant details and capture the behavior of the system that is analyzed; due to this challenging character of the task of system identification, modeling has been considered as "an art" [Mor91].

In the following section we are going to explain the problems of noise and overfitting using a simple example.

9.1.2 An Example

9.1.2.1 Learning Polynomial Models

Let us consider the following example: Let S be a system whose behavior is to be modeled using the input / output (target) training examples given in Table 9.1 (where X and Y values denote input and output data, respectively).

By looking at these values as they are shown in Figure 9.1 the suspicion is aroused that there might be a cubic connection between the X and Y values, distorted by additive noise. This is in fact correct: The data were generated using the model $y = x^3 - 100x + 100$ and adding noise (uniformly distributed in the interval $[-250; +250]$). This is why the original function $x^3 - 100x + 100$ is also depicted in Figure 9.1.

If we want to evaluate the original formula that was used for simulating the system ($x^3 - 100x + 100$), we can for example evaluate this model on all integral values for X in the range of the given training data (i.e., $-15, -14, \ldots$, 4, 5) and calculate the mean squared differences of these calculated values and

Table 9.1: Data-based modeling example: Training data.

X	Y	X	Y
-15	-1571.1605	-4	229.6581
-14	-1405.3919	-3	249.9523
-13	-644.6956	-2	518.4009
-12	-398.4149	-1	294.8873
-11	-69.9755	0	22.0334
-10	-87.4658	1	-193.7337
-9	126.4967	2	-146.7154
-8	227.3979	3	-294.5191
-7	309.4894	4	-179.5208
-6	522.4300	5	-353.2186
-5	474.8867		

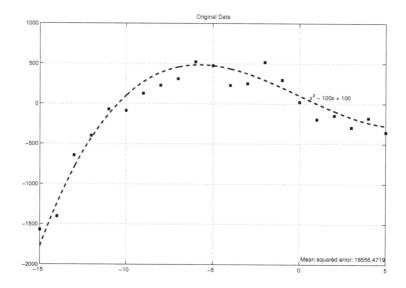

FIGURE 9.1: Data-based modeling example: Training data.

the given training target data for Y which yields 18,556.4719 – the "fitness" of the original formula therefore is approximately 18,556.

Now let us suppose that we do not know or suspect anything about the system or its order. We could therefore try for example polynomial approaches of order 2, 3, 10, and 20; thus, we assume model structures of the form

$$y = a_0 + a_1 x + a_2 x^2 + \ldots + a_n x^n \tag{9.3}$$

for a model of order n. The parameters $[a_0, a_1, a_2, \ldots, a_n]$ are now to be set so that the model fits the given training data as exactly as possible.

As we see in the Figures 9.2, 9.3, 9.4, and 9.5, the quadratic model performs fairly, the model of order 3 performs better on the given training data, and the models of order 10 and especially 20 perform even a lot better; the polynomial of order 20 is even able to explain the training data perfectly.

The quality of the so generated models of order 1, 3, 10, and 20 is approximately 244,218, 14,435, 6,605, and 0^1, respectively.

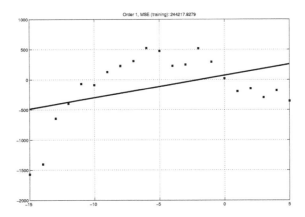

FIGURE 9.2: Data-based modeling example: Evaluation of an optimally fit linear model.

9.1.2.2 Testing Polynomial Models

Now let us assume that test data are available for evaluating the models; these test data are not included in the training data but rather used for estimating the quality of the models produced (and of the identification method itself). These test data are given in Table 9.2.

Now we see that the linear model performs even worse on the test data ($mse_{test} \approx 25 * 10^6$, see Figure 9.6); the cubic model, which performed a lot better in training, is much more accurate also on test data ($mse_{test} \approx 4.5 * 10^6$, see Figure 9.7)

---

[1] Minor inaccuracies are here due to numerical imprecisions.

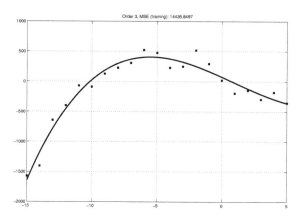

FIGURE 9.3: Data-based modeling example: Evaluation of an optimally fit cubic model.

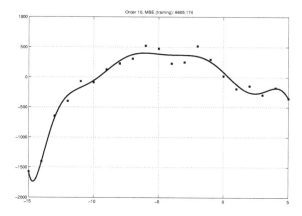

FIGURE 9.4: Data-based modeling example: Evaluation of an optimally fit polynomial model ($n = 10$).

So, does this trend go on and does better fit on training data guarantee better fit on test data? Analyzing the test performance of the models of order 10 and 20 the answer to this question obviously is: No. In Figure 9.8 we see that the polynomial model of order 10 predicts values out of the range of the given test data yielding a mean squared error value of $5 * 10^{16}$. The model of order 20 is not shown; its mean squared error on test data is $5.8 * 10^{34}$.

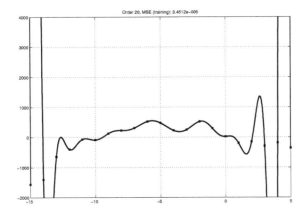

FIGURE 9.5: Data-based modeling example: Evaluation of an optimally fit polynomial model ($n = 20$).

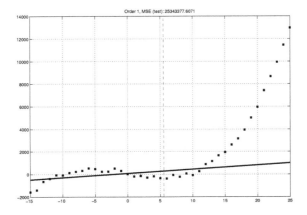

FIGURE 9.6: Data-based modeling example: Evaluation of an optimally fit linear model (evaluated on training and test data).

Summarizing this example we give an overview of training and test errors for the data and models mentioned above in Figure 9.9 (models of order 0 and 5 were created in the same way as the other models). This behavior is typical: As the number of parameters increases, the training errors decrease; in the

Table 9.2: Data-based modeling example: Test data.

X	Y	X	Y
6	-381.4362	16	2609.0386
7	-73.1285	17	3147.7311
8	-226.3715	18	3941.3802
9	60.7464	19	5006.4839
10	-84.9143	20	5957.1595
11	251.8633	21	7424.0707
12	877.4408	22	8664.473
13	1149.4064	23	9937.4536
14	1666.7466	24	11452.5263
15	1941.3963	25	12980.5208

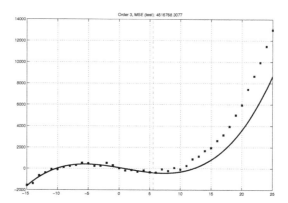

FIGURE 9.7: Data-based modeling example: Evaluation of an optimally fit cubic model (evaluated on training and test data).

beginning, test errors also tend to decrease[2], but after some time (as soon as overfitting happens), test errors start to increase with increasing training effort.

Please note that the training and test errors shown in Figure 9.9 are depicted on a logarithmic y-axis.

---

[2]In the summary chart displayed in Figure 9.9 we have intentionally omitted the training and test errors for $n = 2$. The reason is that it would have shown that in this particular case the test error for the quadratic model is a lot worse than for the linear as well as the cubic model; this would be correct, of course, but in this way it is easier to sketch the characteristic behavior of first decreasing and then increasing test errors as the number of parameters increases.

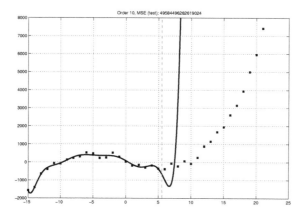

FIGURE 9.8: Data-based modeling example: Evaluation of an optimally fit polynomial model ($n = 10$) (evaluated on training and test data).

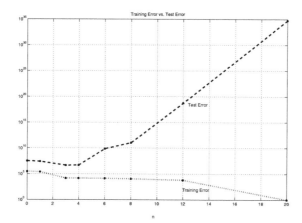

FIGURE 9.9: Data-based modeling example: Summary of training and test errors for varying numbers of parameters n.

9.1.2.3 Implementation

All data generation and modeling steps used here have been implemented in MATLAB®, Version 7.0 (R14); the source code representing the implementation of this example can be found on the website of this book.[3]

For the fitting of a polynomial of order n we first compose a matrix M as a concatenation of the input values (namely the x values of the training data, i.e., all values in $[-15; 5]$) potentiated by $0, 1, \ldots, n$:

$$Z = [X^0 X^1 \ldots X^n], X^k = [x_1^k x_2^k \ldots x_N^k]^T \tag{9.4}$$

where X^k is a column vector consisting of all N input values to the power of n.

Secondly, the training target values Y are, after transposing the matrices, divided by Z using the right matrix division function ($/$); this numerically solves the system of linear equations defined by the order of the model n, the input data Z, and the target values Y. Thus, we get the coefficients $a_0, a_1 \ldots a_n$ in the result of this division (as a vector p) and calculate the estimated target values $\hat{Y}$ (denoted in the source code as Yhat) by multiplying *poly* and Z; this represents the evaluation of the identified polynomial for each given sample.

The training and test qualities are calculated using the mean squared errors function, i.e., we calculate the sum of squared residuals and divide by the number of samples considered.

The data documented in Section 9.1.2 were generated using a noise range of 500.

9.1.3   The Basic Steps in System Identification

The following two phases in data-based modeling are often distinguished: Structural identification and parameter optimization.

- First, structural identification is hereby seen as the determination of the structure of the model for the system which is to be analyzed; physical knowledge, for example, can influence the decision regarding the mathematical structure of the formula. This of course includes the determination of the functions used, the order of the formula (in the case of polynomial approaches, e.g.), and, in the case of dynamical models, potential time lags for the input variables used.

  In the simple example given previously this step was the decision to use a polynomial modeling approach; for example, the decision to try a polynomial model $y = a_0 + a_1 x + a_2 x^2 + \ldots + a_n x^n$ of specific orders was the structural identification part. As we tried several polynomials of different orders we simply executed the procedure several times; this is exactly what is indicated by the feedback loop in Figure 9.10 (a).

- Parameter identification is then the second step: Based on training data, the parameters of the formula are determined (optimized) meaning that

---

[3]http://gagp2009.heuristiclab.com.

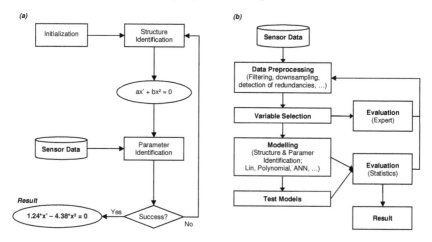

FIGURE 9.10: The basic steps of system identification ([BES01], [WER06]).

the coefficients and, if used, time lags are fixed.

Basically, this is what we did in the previous example by calculating the coefficients for the polynomials of different orders separately.

This separation is schematically shown in the left part (a) of Figure 9.10 (adapted from [BES01]).

Of course, the whole process of building models out of data includes more steps than those mentioned above. Especially data preprocessing is a very important issue, i.e., preparing data before it is used for the "real" modeling process. Data downsampling, filtering, and the removal of data without information can be applied in order to retrieve preprocessed data on which it is easier to efficiently generate appropriate models.

Variables selection is also often considered a key issue in data-based modeling: Those variables are selected from the pool of variables available which shall be used for the essential modeling process. For example, variables which do not include information (since they are constant in the whole data set, e.g.) or are redundant to other ones can be omitted for simplifying the modeling process. Variables selection can thereby be done using expert knowledge or statistical methods. Exhaustive statistical methods are available as well as sequential iterative forward or backward variable selection:

- Exhaustive search is executed by computing all possible combinations of variables and evaluating them; exactly that combination of channels will be selected which provides best approximation of measurement data. This method is able to provide an optimal solution (if the process is linear), but especially for higher dimensional problems (including big numbers of channels) it requires excessive computation time. In order to overcome this drawback, forward and backward selection can be used as alternatives even if they provide only suboptimal solutions.

- In sequential forward selection the algorithm sequentially derives the list of input channels. In the first step, only one input channel is considered where that channel is selected that minimizes the sum of squares errors. In the next step, another input channel is selected where once again that channel is chosen which minimizes the sum of squares errors; the algorithm iteratively adds more and more input channels until a predefined accuracy is reached and hence the algorithm terminates. Of course the results depend on the chosen basis functions.

- The main difference when applying backward selection is that the algorithm starts with all variables available in a set of selected variables and then iteratively removes variables that do not have a statistically measurable connection with the observed (measured) target values.

- Hybrid variants combining backward selection and a subsequent forward selection step have also been investigated for producing good results very efficiently.

These basic steps of the data driven modeling process are shown in the right part (b) of Figure 9.10.

As we see in both diagrams shown in Figure 9.10, the total system identification process based on measurement data is not finished as soon as models are created. A decision whether the model at hand is appropriate and fulfills the given quality requirements has to be made during a subsequent validation step. If this validation (often also called test phase[4]) fails, the process might be repeated starting again at the structural identification or data preprocessing step.

The major drawback of this classical approach is obvious: As the structure of the model has to be fixed before identifying parameters, thus it has to use *a priori* knowledge. However, there is a large number of applications in which the a priori model information is not available to the desired precision. For all these cases, several generic so-called "model free" approaches are widely used, ranging from simple static maps up to self-organizing neural networks; see for instance [dRLF+05] for ANN-based identification of a Diesel engine's NO_x emissions, [PP01] for a specific spectral analysis tool to describe the behavior of a plant or [THL94] for a neural network approach to optimal filtering of sensor data.

---

[4]Please note that in some cases the terms validation and test phase are used synonymously, but often (and also in the following test case documentations) the validation and test phase are separate model analysis phases. Detailed explanation is to come in the following sections.

In spite of the evident simplicity of generic approaches, the drawbacks are known as well: Over-parameterization, lack of extrapolation and often even of interpolation capabilities [dRLF⁺05], large data requirements, etc.

9.1.4 Data-Based Modeling Using Genetic Programming

Using Genetic Programming for data-based modeling has the advantage that we are able to design an identification process that automatically incorporates variables selection, structural identification, and parameters optimization in one process.

In GP, the function f which is searched for is not of any pre-specified form when applying genetic programming to data-based modeling; low-level functions are during the GP process combined to more complex formulas. Given a set of functions $f_1, \ldots, f_u$, the overall function induced by genetic programming can take a variety of forms. Usually, standard arithmetical functions such as addition, subtraction, multiplication, and division are in the set of functions f, but also trigonometric, logical, and more complex functions can be included.

Thus, the key feature of this technique is that the object of search is a symbolic description of a model, not just a set of coefficients in a pre-specified model. This is in sharp contrast with other methods of regression, including linear regression, polynomial approaches, or also artificial neural networks, where a specific structure is assumed and often only the complexity of this model can be varied.

Of course, data preprocessing and a separate validation / test phase are also parts of the GP-based modeling process; the main workflow is sketched in Figure 9.11.

In the following we are going to give an overview of our system identification implementation in HeuristicLab in Section 9.2 and discuss concepts developed for analyzing the similarity of mathematical models produced by GP in Section 9.4.

Two typical application scenarios for GP-based modeling are then analyzed using real-world test data as well as benchmark data:

- In Section 11.1 we bring basics and examples for time series analysis and the design of so-called virtual sensors;

- in Section 11.2 we demonstrate classification as a possible application for GP-based structure identification.

In both cases we discuss the effects of using enhanced concepts for GAs that have been discussed in the previous chapters as well as advanced GP concepts that are to be described in the following sections.

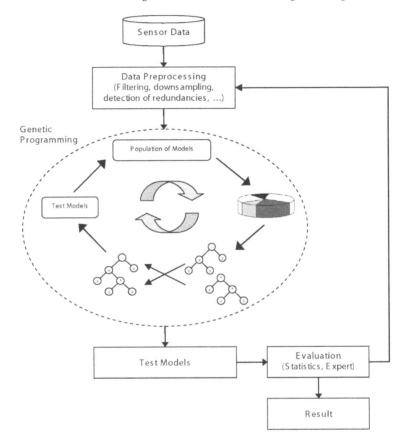

FIGURE 9.11: The basic steps of GP-based system identification.

9.2 GP-Based System Identification in HeuristicLab

9.2.1 Introduction

The HeuristicLab (HL) is a framework for developing and testing optimization methods, parameters and applying these on a multitude of problems. The project was started in 2002 and has evolved to a stable and productive optimization platform; it is continuously enhanced and topic of several publications ([WA04c], [WA04a], [WA04b], [WA05a], and [WWB+07]). On the HeuristicLab website[5] the interested reader can find installable software, in-

---

formation, documentation, and publications in the context of HeuristicLab and the research group HEAL[6].

This extensible and flexible framework enables us to combine the advanced GA concepts with genetic operators for GP; operators for analyzing dynamics in GP populations can be integrated as well as evaluators that compare training, validation, and test qualities.

Here we want to summarize how system identification problems are represented in HeuristicLab, how we have designed an appropriate solution encoding and respective operators, and finally show how we have defined a similarity measure for these solution candidates.

9.2.2 Problem Representation

A system identification problem instance has to include all data which are needed by genetic programming for generating models describing the underlying system's behavior.

The most important part of the representation of a system identification problem, that is to be tackled with genetic programming, is the *data collection* storing all available measurement data; the index of the *target variable* also has to be known and available for the modeling algorithm.

Furthermore, there also has to be an indication which data samples are to be used as *training*, *validation*, and *test* data (in our case given as start and end sample indices). The use of these data segments is different for each particular partition:

- *Training data* are the real basis for the algorithm; the modeling algorithm is able to use these training examples of the input / output behavior of the system at hand (or rather of the model that is to be learnt) for determining the quality of solution candidates (which in our case here are models / formulas).

- *Validation data* are available for the training algorithm, but normally not used for the real evolutionary optimization process. These data can for example be used for detecting overfitting phenomena, for pruning, or other additional model manipulation operations.

- *Test data*, finally, may not be considered by any part of the training algorithm. Still, these data shall be used for testing the created models on new data, i.e., data not included in the algorithm's data base, so that we can determine whether the algorithm was able to generate appropriate models or not.

Additionally, there also has to be a possibility to state which variables of the data base are really available for the modeling algorithm. For example,

---

[6]Heuristic and Evolutionary Algorithms Laboratory, Linz / Hagenberg, Austria.

this becomes relevant when sensor data are included in the data base and used for statistical analysis (correlation analysis, automated fault detection, etc.), but the models that are to be generated for a certain target variable are still not supposed to contain these variables.

Pure availability of a variable is still not sufficient information; what we also need is whether or which time offsets are allowed when referencing a variable. For example, let y be the target variable and u, v, and w possible input variables for a model for y; as we want to model y for time (sample) t we search for a model for y_t. The first crucial decision to be made is whether we want to generate static or dynamic models: In static models, only inputs at time t are considered for describing the target variable at time t; our target y_t would be described as a function $f : y_t = f(u_t, v_t, w_t)$. In dynamic modeling, on the contrary, input variables can also be referenced with a certain time lag meaning that not only values of time t are used but also "historic" data. For example, f could then be a function modeling y_t using u_{t-4}, v_{t-1}, v_{t-2}, and w_t.

In several application scenarios one also explicitly excludes input values of time t; what we get by excluding contemporary input data is a prediction model that can also be used for modeling future values on the basis of previously measures / recorded data.

Furthermore, the generation of autoregressive models also becomes possible: Autoregressive models are formulas that model an output y_t incorporating previous outputs $y_{t-1}, y_{t-2}, \ldots, y_{t-t_{max}}$; an exemplary autoregressive model for our example could be $f_{AR} : y_t = u_t + y_{t-2} + w_{t-1}$.

So, as the target variable can also be used with certain time offsets, GP is also able to generate autoregressive models.

Lots of additional information for system identification problem instances can also be very useful in the modeling process:

- Complexity limits for the models that are to be created can be given as maximum values for the height as well as the size of the models. Height hereby is equal to the height of the respective model structure tree as is to be described in Section 9.2.4; size refers to the number of nodes of the structure tree.

- Meta-information such as descriptions of the data and the underlying system, or descriptions and names of the variables in the data base, e.g.

- A collection of function and terminal definitions that can be used for compiling and evaluating models – a detailed description about the management of function bases is about to come in Section 9.2.3.

- The best solutions found so far - this of course also has to include at least information about

  - the data partitions used as training and validation data,

- the evaluation operator and respective parameter settings applied for evaluating solution candidates,

- which variables were used in the modeling process applying which minimum and maximum offsets, and

- the function and terminal definitions that were available for compiling and evaluating models.

Specific parameters for classification problems shall be described in Section 11.2 on learning classifiers using GP.

9.2.3 The Functions and Terminals Basis

9.2.3.1 Motivation, Introduction

The correct design of the functions and terminals basis used for compiling and evaluating formulas is one of the most crucial issues in the design of a GP-based system identification approach; for the sake of simplicity we will in the following refer to this pool of definitions of functions and terminals as *functions basis*. In fact, this is not wrong since terminal definitions are also functions that take several inputs such as a reference to the data basis, the variable and sample indices, a (time) offset, and a concrete coefficient for calculating the returned value. Still, as the handling of terminals differs a lot from the handling of functions, we will also treat them separately whenever necessary.

Regarding the implementation, HeuristicLab and all plugins (at least until now) are implemented in C# using the .NET framework, so the most obvious approach would be to use the functions of the .NET framework for building models; essentially, this was done in our GP implementation for the versions 1 and 1.1 of HL ([Win04], [WAW05a], [WAW05b], [WAW06a], [WAW06e], [WAW06c], [WEA+06]).

During own research activities and in the course of discussion with research partners in academics as well as industries we became more and more convinced that it would be a great benefit for GP-based modeling if the users were able to program and parameterize the functions and terminals by themselves. So, starting from the implementation in HL 2.0, a flexible and user-programmable functions basis has been used.

The definition of the evaluation of functions and terminals surely is the core of any functions and terminals management unit. So, for each function as well as for every terminal definition we have to be able to manage the source code that represents the definition of its evaluation, compile it, and provide the compiled functions to the GP process.

In detail, these definitions are designed and implemented in HeuristicLab as is explained in the following sections.

9.2.3.2 Definition of the Evaluation of Terminals

The definition of the evaluation of a terminal is given by a function that requires a reference to the data basis, variable and sample indices, a sample offset, and a coefficient as inputs; depending on the selected terminal definition, this information is processed and the return value calculated. So, a terminal definition t is a function of the data collection D, the variable index v, the sample (time) index s, a sample offset o, and a coefficient c.

Let us consider the following examples t_{var}, t_{diff}, and t_{const} representing standard variable, differential, and constant definitions:

$$t_{var}(D, v, s, o, c) = c * D[v, s - o]$$
$$t_{diff}(D, v, s, o, c) = c * (D[v, s - o] - D[v, s - o - 1])$$
$$t_{const}(D, v, s, o, c) = c$$

t_{var} calculates the product of the given coefficient multiplied with the value of variable v at sample index s shifted by o indices, thus taking the value $D[v, s-o]$. t_{diff} calculates the difference of the referenced values at $D[v, s-o]$ and its predecessor, $D[v, s-o-1]$, and returns it multiplied with the coefficient c. t_{const}, finally, simply returns the given coefficient and thus represents a constant terminal definition.

The definition of such a terminal can of course become arbitrarily simple or complex, depending on the user's intention. Anyway, in HL the definition of the evaluation functions has to be done in C# notation using the following interface:

```
public double TerminalEvaluation(double[][] Data,
    int Var, int Sample, int Offset, double Coeff);
```

The implementation of a terminal definition thus is a method following the interface given above. The respective source codes for the exemplary terminals t_{var}, t_{diff}, and t_{const} could be defined in the following way:

t_{var} : `return Coeff * Data[Var][Sample-Offset];`

t_{diff} : `return Coeff * ( Data[Var][Sample-Offset] -`
`Data[Var][Sample-Offset-1] ) ;`

t_{const} : `return Coeff;`

9.2.3.3 Definition of the Evaluation of Functions

The interface for function evaluation definitions is a lot simpler than the evaluation interface for terminals as described above: A function is simply defined by the way it calculates a value given a set of input values. Additionally, we also use a variant index so that it is possible to define several variants of functions within one function definition. So, a function definition f is a function of the input data vector *input* and the variant v.

Let us consider the following examples f_{add}, f_{div}, and t_{trig} representing addition, division, and trigonometric functions:

$$f_{add}(input, v) = sum(input)$$
$$f_{div}(input, v) = input[1]/input[2]$$
$$f_{trig}(input, v) = \begin{cases} \sin(input[1]) & : & v = 1 \\ \cos(input[1]) & : & v = 2 \\ \tan(input[1]) & : & v = 3 \\ error & : & otherwise \end{cases}$$

f_{add} calculates the sum of all input values, f_{div} divides the first argument by the second one, and f_{trig} returns the sine, the cosine, or the tangent of the first input, depending on the value of the variant index passed.

In HL the definition of the evaluation functions has to be done using the following interface:

```
public double FunctionEvaluation(double[] Args, int Var);
```

The implementation of a function definition thus is a method following the interface given above; the respective source codes for the exemplary terminals f_{add}, f_{div}, and f_{trig} could be defined in the following way:

f_{add} :
```
double d = 0;
for (int i=0; i<Args.Length; i++)
    d += Args[i];
return d;
```

f_{div} :
```
if(Args[1]==0) return double.NaN;
return (Args[0] Args[1]);
```

f_{trig} :
```
if (Var==0) return Math.Sin(Args[0]);
if (Var==1) return Math.Cos(Args[0]);
if (Var==2) return Math.Tan(Args[0]);
throw new Exception("Unknown function variant");
```

Of course, logical functions can so be integrated into the functions pool as well as boolean functions connecting logical and boolean functions.

Please note that the functions interface definition implemented in HL is a bit more sophisticated. In fact, what is also handed over to the function is an array storing a certain number of previously calculated values, i.e., a history of exactly this function:

```
public double FunctionEvaluation(double[] Args, int Var,
    double[] History);
```

If the history array is used in the evaluation function, then a number of pre-defined calculated values are automatically saved in an appropriate array, stored, and given to the function at its next evaluation. Thus, it is for example possible to implement an integral function f_{int} using the history *hist*:

$$f_{int}(input, v, hist) = hist[1] + input[1]$$

which in HL / C# notation could be implemented as

$$f_{int} : \texttt{return = History[0] + Args[0];}$$

9.2.3.4 String Representations of Terminals and Functions

Even though the evaluation on a given data basis is the most important task for a model, appropriate string representations are also necessary for representing formulas in prefix notation, standard notation (as a mixture of infix and prefix notations), or in such a way that they can be immediately incorporated in MATLAB®, *Mathematica*®, LATEX, or C/C++/C# program code.

For each representation variant there are specific interfaces for terminals and functions; in all cases character strings are returned, but the input parameters vary significantly. The terminals' string representations are given the same parameters as the evaluation functions (except for a reference to the data basis) and, in some cases, the variable name; string representation methods for functions use the string representations of the function's inputs and return composed strings representing the function and its inputs.

In the following we will pick the standard (infix/prefix) notation for demonstrating the mechanisms used. For terminals and variables we use the interfaces

```
public string Terminal_Standard(int Var, string VarName, int Offset,
    double Coeff);
public string Function_Standard(string[] Args, int Var);
```
respectively. For standard variables and the addition function, for example, the respective method implementations could be given in the following way:

t_{var} :
```
string s = "[" + Coeff.ToString() + "*";
s = s + VarName;
if (Offset==0) s = s+"(t)";
else s = s+"(t-" + Offset.ToString() + ")";
return (s + ")]");
```
f_{add} :
```
string s = "(" + Args[0];
for(int i=1; i<Args.Length; i++)
    s = s + "+" + Args[i];
s = s + ")";
return s;
```

The standard string representations of two terminals referencing variable w with time-offset 4 and coefficients 1.2 and 0.9, respectively, would so result in [1.2*w(t-4)] and [0.9*w(t-4)]; the standard string representation of the addition of these two terminals would be ([1.2*w(t-4)]+[0.9*w(t-4)]).

9.2.3.5 Parametrization of Terminals and Functions

Apart from the definitions of evaluation and string representation of terminals and functions there are several parameter settings for them that are to be summarized in this section.

Terminal definitions can be parameterized in the following ways:

- The data type and the distribution function of the coefficients allowed has to be defined: Coefficients can be

  - either integral values or real-valued, and

  - their distribution can be either uniform (defined by minimum and maximum values) or Gaussian (defined by average μ and standard deviation σ).

- The set of possible parent types can be defined, i.e., the user is able to declare which functions are allowed to use the respective terminal as input and which ones are not allowed to do so.
  This selection of possible parent functions can be done either explicitly by selecting a set of functions that are allowed as direct parents, or implicitly by defining which functions are not allowed as parents of the respective terminal type.

The parametrization possibilities for function definitions are even more than those for terminals:

- An arbitrary number of variants can be defined. Apart from considering these variants in the method code (as can be seen in the code for the trigonometric function definition in Section 9.2.3.3), each variant can be activated and de-activated independent of the other variants.

- Additionally, for each variant the function's arity (its number of input parameters) has to be defined. The arity can be either fixed or given as a range defined by minimum and maximum values.

- Each function has to define its neutral element(s), also called identity element(s). In binary operations working on elements of the set X, an element e of X is called left identity with respect to the operation $\circ$ if $e \circ a = a$ for all elements a in X; in analogy to this, e is called right identity with respect to $\circ$ if $a \circ e = a$ for all elements a in X.
  This concept of elements that leaves other elements unchanged when combined with them is here used in a slightly generalized way as we

define neutral (identity) elements for each possible input index of a function:

- There can be one neutral element that is used for all input indices, or

- neutral elements can be defined for each possible input index independently.

For the addition or subtraction functions, e.g., the neutral element for all possible indices is 0, for the multiplication function it is 1 for all inputs. But when it comes to the division, then the identity elements have to be defined separately for each input: As we divide the first input by the second one, the neutral element for the first index is 0, whereas for the second input it is 1 (because $0/a = 0$ and $a/1 = a$ for all $a \in \mathbb{R}$).

- Similar to the parent type restrictions that can be set for terminals, functions can also define a set of valid parent function definitions. Again, this can be done either directly or indirectly by selecting functions that are not allowed as parent function types.

- Finally, functions can also define child type restrictions. This can also be done directly by selecting certain function or terminal definitions as valid child types (i.e., types that are allowed as inputs for the function), or by explicitly excluding certain types from the set of possible input definitions.

  In order to maximize the flexibility of this child type management concept, these selections can be done either for all input indices uniformly or for each input index separately.

Function and terminal definitions and their respective parametrizations are collected in function and terminal management units which we here, as already mentioned before, call "functions bases". In each functions basis we not only store function and terminal definitions, but also which ones are activated and which ones are not, and an initial weighting factor is also given for each definition denoting its relative probability to be chosen when it comes to selecting a randomly chosen function or terminal.

9.2.4 Solution Representation

9.2.4.1 Representing Formulas by Structure Trees

As we have now described how function and terminal definitions are managed, we shall take a look at the representation of solution candidates for GP-based system identification. The most intuitive way to represent models is modeling them as structure trees; starting with Koza's first GP attempts using LISP structure trees, the concept of trees representing formulas has had a long tradition in GP (see [Koz92b], [KKS⁺03b], [LP02], [Kei02], or [PMR04], e.g.).

Structure trees consist of nodes and references from parent nodes to their children. Thus, for representing formulas we have to create node structures that are able to store all parameters needed as well as references to the function and terminal definitions used; this concept is visualized in Figure 9.12.

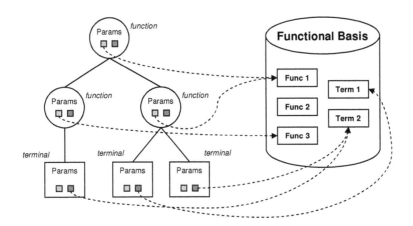

FIGURE 9.12: Structure tree representation of a formula.

The following parameters have to be stored by structure nodes in addition to references to their function or terminal definition:

- Each terminal node has to store the index of the variable it references, the sample (time) offset, and the value of the coefficient that is to be used as a multiplicative factor.

  Thus, when it comes to evaluating a terminal node for a given data base and a certain sample index, the referenced terminal definition is called using the given data and sample index as well as the parameters stored in the node; the value returned by the terminal definition function is returned as the result of the node's evaluation or representation method.

- A function node has to store not only references to its child nodes and a function definition, but also the index of the function's variant. So, when it comes to evaluating a function node or compiling its string representation for a given data base and a specific sample index, the children nodes are first evaluated with the given data and then the referenced function is called with the children's returned values and the variant index stored. The result of this function call is then returned as the result of the node's evaluation.

As we have described in Section 9.2.3.3, some functions also consider previously calculated values. So, function nodes additionally have to manage history arrays in which the calculated values are stored and which are also given to the function definition at the next sample's evaluation.

9.2.4.2 Initialization, Crossover, and Mutation

The initialization of structure trees is essentially the compilation of random tree structures referencing to randomly chosen function and terminal definitions. Of course, all constraints given by the functions basis have to be considered:

- The number of children of each function node has to fulfill the arity constraints given by the function definition parametrization; in the case of fixed arity the number of children has to be exactly this value, and in the case of variable arities the number of children may not fall below the minimum or rise above the maximum arity limit.

- Of course, parent and child constraints also have to be considered.

- The structure complexity given in the problem representation (regarding height and size of structure trees) may not be exceeded.

- Variable indices are chosen according to variable availabilities; sample offsets are initialized according to minimum and maximum sample offsets defined by the problem instance.

- Coefficients of terminal nodes are initialized according to parameter settings defined by the terminal definition.

The most frequently used crossover operator is the single-point subtree exchanging crossover already described in detail in Section 2.2.1.3. Subtrees are exchanged and new formulas are formed; the references to the function and terminal definitions are copied into the new solution candidate. Figure 9.13 illustrates this mechanism.

Of course, all constraints defined by the functions basis have to be satisfied here, too. Especially child and parent relations of the new combinations have to be checked and invalid constellations avoided. The complexity limitation requirements given by the problem instance also have to be fulfilled.

In fact, we have implemented and use three different types of crossover operators:

- The *standard* crossover variant chooses subtrees without considering their size.

- The *low level* crossover variant tries to exchange rather small subtrees of height 1 or 2, e.g.

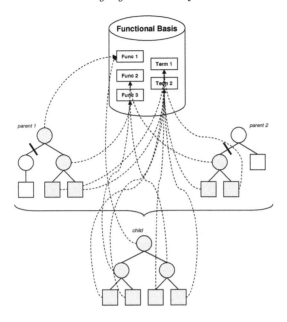

FIGURE 9.13: Structure tree crossover and the functions basis.

- The *high level* crossover variant tries to exchange rather big subtrees as for example the roots' children.

Finally, mutating a structure tree can be done in several different ways. Some structural as well as parametric mutation variants are as follows:

- A subtree could be deleted or replaced by a randomly re-initialized subtree.

- A function node could for example change its function type or turn into a terminal node.

- A terminal node representing a variable could for example change its index and thus in the following refer to another variable.

- A terminal node representing a constant could be multiplied with a factor. A good choice for the distribution of these multiplicative mutation factors could be a Gaussian distribution with average 1.0 so that the probability of smaller changes is greater than the probability of larger modification.

Up to now we have always stressed the fact that complexity limitations are given in the problem representation of the concrete system identification problem at hand. In fact, complexity limitations can also be defined by crossover and mutation operators; these operators can be parameterized so that they produce models by crossing parents or mutating formulas that fulfill size or height restrictions independently of the settings given in the problem. These limitations could for example also be modified during the execution of the GP process.

9.2.5   Solution Evaluation

9.2.5.1   Standard Solution Evaluation Operators

The primary task of an evaluation operator estimating the fitness of a system identification solution candidate is surely to measure how well the values calculated using the model fit the original target values. Numerous different evaluation functions are possible and have been reported on in the literature; in principle, the estimated values e (calculated by evaluating the model on the given data basis) are compared to the original target values o. In this context it is irrelevant for the function whether the model is evaluated on training, validation, test, or any other data partition.

Here we describe three rather simple functions that have also been implemented as evaluation operators for HeuristicLab:

- The *mean squared errors function* (*MSE*) has already been described in Sections 2.4.3 and 9.1: The function returns the average value of the squared residuals of e and o:

$$MSE(e, o) = \frac{1}{N} \sum_{i=1}^{N} (e_i - o_i)^2; N = |e| = |o| \qquad (9.5)$$

- The *coefficient of determination* (R^2) function can be used for measuring the proportion of a variable's variability that is accounted for by the model that tries to explain it; it can also be seen as the ratio of the variability of the modeled target values to the variability of the original target values. R^2 of original and modeled target values, o and e, respectively, is defined as

$$R^2(e, o) = 1 - \frac{SS_E}{SS_T}; \qquad (9.6)$$

$$SS_E = \sum_{i=1}^{N} (o_i - e_i)^2, SS_T = \sum_{i=1}^{N} (o_i - \bar{o})^2, \qquad (9.7)$$

$$\bar{o} = \frac{1}{N} \sum_{i=1}^{N} o_i, N = |e| = |o| \qquad (9.8)$$

where SS_E stands for the explained sum of squares and SS_T for the total sum of squares of the original values. The better a model is, the more the R^2 value converges to 1.

- The *variance accounted for* (*VAF*) function is defined as the fraction of the variances of the residuals and the original target values:

$$VAF(e,o) = 1 - \frac{var(o-e)}{var(o)};$$ (9.9)

$$var(x) = \frac{1}{N}\sum_{i=1}^{N}(x_i - \bar{x}), \bar{x} = \frac{1}{N}\sum_{i=1}^{N}x_i, N = |x|$$ (9.10)

The variance of the residuals, i.e., the differences between the original and modeled values, is so divided by the original values' variance; the smaller the residuals' variance is, the nearer the calculated value converges towards 1.

This main difference of this evaluation function compared to other ones as for example *mse* or R^2 is that it does not punish constant residuals; only the variance of the residuals is taken into account and might decrease a model's quality.

In the implementations of these evaluation functions we have introduced a parameter for limiting the maximum contribution of a single sample's error to the total evaluation. The residual of each specific sample can so be limited in relation to the original target values' range; this is supposed to help to cope with outliers and invalid values calculated by division by 0, for example.

9.2.5.2 Combined Solution Evaluation

Several advanced evaluation concepts are also realized in an advanced evaluation operator for HeuristicLab. Again, for the explanations given in this section let o be the original and e the estimated target values, and N the number of samples analyzed; furthermore, let $range(o)$ be the range of the original target values:

$$range(o) = max(o) - min(o)$$ (9.11)

First, instead of mean squared errors we use the *mean exponentiated error* function; the residuals are raised to the power of n, a parameter of this particular evaluation function, and the mean value of these exponentiated errors is calculated:

$$MEE(o,e,n) = \frac{1}{N}\sum_{i=1}^{N}|e_i - o_i|^n; N = |e| = |o|$$ (9.12)

Additionally, this operator is able to combine the evaluation functions given in the previous section; a combined fitness value is calculated as a linear combination of the three separate fitness values.

First, the fitness values $MEE(o,e,n)$, $R^2(o,e)$, and $VAF(o,e)$ have to be scaled so that they have comparable ranges. The exponentiated errors are scaled by dividing them by a fourth of the target values' range, so for calculating the scaled fitness value $MSEE'(o, e, n)$, $MSE(o, e, n)$ is divided by a fourth of the target data's range raised to the power of n since

$$MEE'(o, e, n) = \frac{1}{N} \sum_{i=1}^{N} \left(\frac{|e_i - o_i|}{\frac{range(o)}{4}} \right)^n ; N = |e| = |o| \tag{9.13}$$

$$MEE'(o, e, n) = \frac{1}{N} \left(\frac{1}{\frac{range(o)}{4}} \right)^n \sum_{i=1}^{N} |e_i - o_i|^n \tag{9.14}$$

$$MEE'(o, e, n) = MEE(o, e, n) * \left(\frac{1}{\frac{range(o)}{4}} \right)^n \tag{9.15}$$

where n is the exponent chosen for raising the errors to the power of n. The scaled values $R^{2'}(o, e)$ and $VAF'(o, e)$ are calculated as simply as

$$R^{2'}(o, e) = 1 - R^2(o, e), VAF'(o, e) = 1 - VAF(o, e) \tag{9.16}$$

since the range of the R^2 and VAF functions is $[0,1]$, anyway.

The minimum and maximum residuals $r_{min}(o, e)$ and $r_{max}(o, e)$ can also be considered; before using them in the combined fitness function, they are scaled in the same way as the MEE values:

$$r = e - o; r_{min}(o, e) = min(r), r_{max}(o, e) = max(r) \tag{9.17}$$

$$r_{min}'(o, e) = \frac{r_{min}(o,e)}{\frac{range(o)}{4}}, r_{max}'(o, e) = \frac{r_{max}(o,e)}{\frac{range(o)}{4}} \tag{9.18}$$

All these scaled partial fitness contribution values are multiplied with coefficients c_1, c_2, c_3, c_4, c_5, summed, and the result divided by the sum of coefficients; the result is returned as the combined fitness value $COMB(o, e, n, c)$:

$$a_1 = c_1 * MEE'(o, e, n) \tag{9.19}$$

$$a_2 = c_2 * R^{2'}(o, e, n) \tag{9.20}$$

$$a_3 = c_3 * VAF'(o, e, n) \tag{9.21}$$

$$a_4 = c_4 * r_{min}'(o, e) \tag{9.22}$$

$$a_5 = c_5 * r_{max}'(o, e) \tag{9.23}$$

$$COMB(o, e, n, c) = \frac{a_1 + a_2 + a_3 + a_4 + a_5}{c_1 + c_2 + c_3 + c_4 + c_5} \tag{9.24}$$

There are, in fact, even more sophisticated evaluation operators to be described, namely a time series analysis specific one as well as a classification specific one. These are about to be discussed in Sections 11.1 and 11.2.

9.2.5.3   Adjusted Solution Evaluation

A modification of the coefficient of determination function R^2 is the so-called *adjusted* R^2; when evaluating a model m, then this extension of the R^2

function described above also takes into account the number of explanatory terms of the model. Let N be the sample size, t the number of terms in m, and o and e again the original and estimated target values, so $R^2_{adj}(o, e)$ is calculated as

$$R^2_{adj}(o, e) = 1 - (1 - R^2(o, e)) \frac{N - 1}{N - t - 1} \qquad (9.25)$$

This add-on[7] increases the calculated quality value only if the addition of a new term to the model improves the model's performance more than what would be expected by chance; unlike R^2 it can even become a negative value.

We have adapted this concept in a slightly modified manner so that it is applicable to the partial R^2 and VAF evaluations of the combined evaluator $COMB$ described in Section 9.2.5.2. These partial evaluation results can be optionally corrected using the factor $\frac{N-1}{N-s-1}$ where s is the model's size, i.e., the number of nodes of the structure tree solution representing the model which is to be evaluated. So, the adjusted evaluation results R^2_{adj} and VAF_{adj} are calculated as

$$q = \frac{N - 1}{N - s - 1} \qquad (9.26)$$

$$R^2_{adj}(o, e) = 1 - (1 - R^2(o, e)) * q \qquad (9.27)$$

$$VAF_{adj}(o, e) = 1 - (1 - VAF(o, e)) * q \qquad (9.28)$$

9.2.5.4 Runtime Consumption Considerations

As we have now described all basic genetic operators for data-based system identification using genetic programming, we can try to estimate their relative runtime consumption.

The initialization of structure trees is not just called only once, it is also relatively cheap in terms of runtime consumption. This is because nodes, which are relatively small entities, are created according to the rules and limitations given in the problem instance and the functions basis; the connection between nodes is established by references (pointers) from parent to child nodes.

Crossover and mutation are in our case also very inexpensive with respect to runtime and memory consumption. Nodes and references are copied and parameters are modified; only in case of the creation of invalid structure trees it could be that repair routines have to be used which could, if implemented in a suboptimal way, cost significant runtime.

Anyway, it boils down to the fact that most of the runtime of a GP based system identification process is consumed by the evaluation of solution candidates. This is because models have to be evaluated on the training (and

---

[7]Of course, calling this modification an "add-on" may sound a bit misleading as it is no additive but rather a multiplicative one. The reader is asked to be so kind as to forgive this slight rhetorical incorrectness.

maybe also validation) data, i.e., on possibly hundreds or thousands of samples. Collecting these values and then calculating the fitness value can be again relatively cheap (with respect to runtime consumption) when using rather simple evaluation functions as those summarized in the Sections 9.2.5.1, 9.2.5.2, and 9.2.5.3; still, especially when using more complex functions as for example time series analysis or classification specific ones given in Sections 11.1 and 11.2, then this part of the evaluation also might cause noticeable runtime consumption.

In HeuristicLab, for instance, we have measured that even when using a graphical user interface with results display and solution protocolling, more than 99.5% of the algorithm's runtime are consumed by evaluation operators.

9.2.5.5   Early Stopping of Model Evaluation

So, what can we do to fight this problem of high computational costs of GP-based structure identification? The simplest answer would be to decrease the size of the training (and validation) data partitions. Of course this is not a generally applicable way to do this; training data should include as much information as possible in a preferably efficient way - it should be as small as possible, but at the same time also as extensive as necessary.

Sampling, i.e., evaluating the models not on the total training / validation data sets but only on certain selected samples seems to be a better idea: By only evaluating the models for a number of (at best randomly) selected sample indices, the total quality is estimated. This on the one hand surely decreases runtime consumption and on the other hand also might help to avoid overfitting as the models are evaluated on different samples at each evaluation step (so that they cannot be fit too closely to a set of samples). Still, the quality measurement might so become somehow instable; a model might be assigned completely different quality values each time it is evaluated because the samples chosen are likely to differ.

When using offspring selection as described in Chapter 4 there is even a possibility of how to speed up the evaluation without decreasing the quality of the fitness estimation method:

During the offspring selection phase, solution candidates are compared to their parents, i.e., their quality values are compared to their parents' fitness values. In the case of applying most restrictive settings, i.e., when the success ratio is set to 1.0, then models are inserted into the next generation's population only if they fulfill the quality requirements given by the parent's quality values and the comparison factor; there is no pool of possible lucky losers, solution candidates that do not fulfill the given fitness criterion are discarded. In this case the evaluation of a model can be aborted as soon as it is clear that the fitness value will surely not satisfy the fitness criterion even if the rest of the evaluation produces no additional errors.

The issue, then, is how to detect when the evaluation of a model can be aborted without decreasing the quality estimation's accuracy with respect to the total GP process. We introduce a *relative calculation interval size rcis* which is a value in the interval [0,1] (normally a value as for example 0.1, 0.2, or 0.5) used in the following way:

Let m be a model which is to be evaluated for a system identification problem p; furthermore let $p1$ and $p2$ be the parents of m, and q_{p1} and q_{p2} their respective quality values. The given comparison factor cf is then used for calculating the comparison value cv depending on whether p is a maximization or a minimization problem:

$$q_{min} = min(q_{p1}, q_{p2}); q_{max} = max(q_{p1}, q_{p2}) \qquad (9.29)$$

$$q_{range} = |q_{p1} - q_{p2}| \qquad (9.30)$$

$$cv = \begin{cases} q_{min} + q_{range} * cf & : \quad isMaximizationProblem(p) \\ q_{max} - q_{range} * cf & : \quad isMinimizationProblem(p) \end{cases} \qquad (9.31)$$

In system identification we normally deal with minimization problems when using the MSE, MEE, or $COMB$ evaluation operator as smaller fitness values are favored; when using the R^2 or VAF operator, p can be considered a maximization model since better models are assigned higher fitness values.

Let us now assume that N samples are to be evaluated; $mathbfo$ are then the N original target values, and the calculation samples interval csi is calculated as

$$csi = N * rcis \qquad (9.32)$$

The vector of estimated target values $mathbfe$ is initialized as a copy of $mathbfo$; the model's quality q_m is initially set to the worst possible fitness value (-$maxVal$ for maximization, $maxVal$ for minimization problems), and the indices i_1 and i_2 are set to 1.[8]

As long as q_m is "better" than cv (i.e., smaller if p is a minimization and greater if p is a maximization problem), the following evaluation steps are executed:

1. The index i_2 is set to $i_1 + csi - 1$; if $i_2 > N$, then $i_2 := N$.

2. The estimated values e_j are calculated for $j = [i \ldots i_2]$; these replace the values at the respective indices in e so that

$$e = [e_1, \ldots, e_{i_2}, o_{i_2+1}, \ldots, o_N] \qquad (9.33)$$

3. q_m is calculated using the given fitness function f:

$$q_m = f(o, e) \qquad (9.34)$$

---

[8]In this description we again use one-based indexing; in most modern programming languages as C, C++, Java, or C#, zero-based indexing would be used instead.

4. Now there are several ways how the evaluation is continued:

   (a) If i_2 is equal to N, i.e., if all samples have been considered, then m is assigned the fitness value q_m.

   (b) Otherwise, if q_m is no more "better" than cf, then the evaluation of m can be aborted and m can be assigned the worst possible fitness value ($maxVal$ for maximization, $-maxVal$ for minimization problems).

   As an alternative, we can also assign m an extrapolated fitness value: If p is a minimization problem and the optimal possible fitness value 0, as is the case if we use the MSE, MEE, or $COMB$ operator, then we can assign m the extrapolated fitness value $q_m * \frac{N}{i_2}$.

   (c) Otherwise, go back to step 1 and continue the evaluation of m.

By rearranging the evaluation as described above we guarantee that the quality of models that fulfill the given offspring selection criterion is accurate and calculated in the same way as when using the standard procedure. For models that perform worse than demanded and are therefore not about to fulfill the offspring selection criterion, the evaluation is aborted as soon as it is clear that the evaluation will result in such a "bad" fitness value.

Thus, a lot of runtime can be saved. For the sake of completeness we of course have to admit that the runtime consumption is increased slightly for models that are evaluated on all samples since intermediate fitness values are calculated; still, this minor drawback is accepted as the advantages outweigh by far.

9.3 Local Adaption Embedded in Global Optimization

Genetic algorithms and genetic programming are in general *global optimization methods*, i.e., their aim is to search the whole search space in an intelligent way in order to find the (or an) optimal solution. In contrast to this, *local optimization methods* are local search algorithms, which means that they move from solution to solution and so search the search space until a solution considered optimal is found (or a time-out condition is fulfilled). Well known examples for local search algorithms are the hill climbing algorithm and tabu search; please see [RN03] and [GL97] for respective explanations and discussions.

In biology, an organism's positive characteristic that has been favored by natural selection is called adaption [SG99]. This is, in fact, the central concept in evolutionary biology and of course also in evolutionary computation.

In this section we shall summarize local adaptation concepts we have introduced into the genetic programming process, namely parameter optimization as well as model structure pruning.

9.3.1   Parameter Optimization

Parameter estimation has already been mentioned in connection with classical system identification: After determining and fixing the structure of the model, appropriate parameters have to be estimated on the basis of empirical data.

In GP, the genetic process is supposed to identify the set of relevant variables, the formula structure, and appropriate parameters automatically; there are no explicit parameter estimation phases planned in the standard GP process. Furthermore, GP is very flexible regarding function and terminal definitions as well as formula structures; it is not easy to formulate general parameter optimization methods for arbitrary nonlinear model structures.

Still, in GP we have to face the problem that often models with good structures are assigned bad fitness values due to disadvantageous parameters such as coefficients or time lags. This is the reason why we have implemented a parameter optimization method based on evolution strategy (ES) concepts.

Evolution strategy is an optimization technique whose ideas are based on the natural concepts of evolution and adaption; it was created and developed since the 1960s, primarily by a German research community around Rechenberg and Schwefel ([Rec73], [Sch75], [Sch94]). As it is an evolutionary algorithm, the optimization process based on ES is executed by applying operators in a loop, i.e., main operations are applied on the solution candidates repeatedly until a given termination criterion is met. A comprehensive overview of the theory of ES can for example be found in [Bey01].

There are several similarities of evolution strategies and genetic algorithms or genetic programming; as they are all optimization methods based on evolution, they are also considered the main representatives of evolutionary algorithms (EAs). Still, there are some important differences of ESs and GAs, the most important being as follows:

- Solution candidates are in ESs represented as vectors of real-valued parameters.

- The main factors that drive evolution in ESs are mutation and selection. Whereas GAs use mutation only for avoiding stagnation, mutation is the main reproduction operator in evolution strategies: Each component of the parameter vector is mutated individually in each generation. An additive mutation is carried out, and small mutations are more likely than big ones.

- In addition to mutation, recombination can be used to create new individuals out of two parents, too.

- In contrast to nature and GAs, the selection of ESs works in a totally deterministic way: In each generation only the best individuals survive, whereas in GAs better individuals (normally) just have higher likelihood to be considered for producing new solution candidates.

In each generation of the execution of an ES, λ individuals (children) are (by mutation and optimal recombination) created out of μ individuals of the current population. Depending on the chosen strategy, the μ members of the new generation's population are selected from all $\mu + \lambda$ candidates (which is referred to as the $(\mu + \lambda)$-ES) or only from the λ children for $\lambda \gg \mu$ (which is also called the (μ, λ)-ES model). This procedure is repeated until termination criterion is reached, normally a maximum number of iterations or a state in which no more improvement can be reached.

In Section 9.1.2 we have shown the general form of a polynomial model which is characterized by its order and coefficients:

$$y = a_0 + a_1 x + a_2 x^2 + \ldots + a_n x^n \tag{9.35}$$

In this case, the optimization of the model's parameters is the task of finding appropriate coefficients $a_0 \ldots a_n$. In the much more general point of view in our GP-based approach, the parameters of a model contain a lot more; in fact, all parameter settings of the terminal nodes included in the model are also parameters for the formula which can be optimized without changing the model's structure.

For each terminal used in our GP approach, the following parameters are to be considered:

- The variable index, i.e., the number of the variable which is referenced.

- The coefficient, a value which can be used for multiplying the referenced variable's value with a given constant; this constant can be either real-valued or integral, and its distribution either uniform (defined by minimum and maximum) or Gaussian (defined by mean and variance).

- The time offset, a value which can be used for referencing to the variable's values shifted by a certain number of samples.

Thus, when it comes to optimizing a model m containing t terminal nodes, we have to consider $3 * t$ that could be manipulated by the optimization method.

As mutation is (besides selection) the most important factor in ES, we shall now discuss how mutation with respect to a model's parameters can be applied. A parameter σ is used for controlling the strength of mutation; we here see σ simply as the standard deviation of the modification added to the model's parameter values. Thus, each parameter of the model's parameters is modified, where again smaller modifications are more likely than bigger ones; variable index changes are also to be applied rather seldom (for 20% of the terminals, e.g.).

So, the whole parameter optimization procedure we have implemented for optimizing a given model m using the parameters λ, σ, it_{max} and cf_{max} is executed in the following way:

1. Collect all terminals of m in t.

2. Create λ copies of m, in the following called mutants.

3. Mutate all λ mutants individually; for each terminal of the mutant models

   - mutate the coefficient,
   - mutate the time offset, and
   - with a rather small probability mutate the variable index.

4. Evaluate all λ mutants.

5. Optionally adjust σ, a parameter steering the mutation's variance, according to Rechenberg's success rule.

6. If any of the mutants is assigned a better quality value than m, then m is replaced by the best mutant, and

   - If the number of iterations has reached a given limit (it_{max}), the algorithm is terminated and m is returned as the optimized version of the originally given formula.
   - Otherwise, the procedure is repeated starting again at step 1.

7. Otherwise, we consider this iteration a failure. If a predefined number of consecutive failures cf_{max} is reached by performing unsuccessfully for cf_{max} times in a row, the algorithm is terminated; otherwise the procedure is repeated starting again at step 1.

As we here always work on one particular model which is to be optimized and create λ mutants, this algorithm can be seen as a variant of the $(1+\lambda)-ES$ algorithm.

Obviously, the main advantage of this algorithm is that it can be applied to any kind of model without any restrictions regarding its structure or the given data basis. But, of course the major drawback of this procedure is its immense runtime consumption due to the high number of models that have to be evaluated for improving the parameters of one single model of the GP population. The use of a smaller data set (or the validation set which is normally also smaller than the training data sample) for evaluating the models can help to fight this problem, but still the use of this parameter optimization concept has to be thought out well and the parameters (σ, λ, it_{max}, and cf_{max}) set so that the runtime consumption does not get completely out of hand. As we will show in the test series analysis in later sections, this parameter optimization method does not have to be applied in every round of the GP process, and also not to all models in the population; partial use can help to control the additional runtime consumption and still use the significant benefits of this procedure.

9.3.2 Pruning

9.3.2.1 Basics and Method Parameters

Whenever gardeners and orchardists talk about pruning, then they most probably refer to the act of cutting out dead, diseased, or for any other reason unwanted branches of trees or shrubs. Even though this might harm the natural form of plants, pruning is supposed to improve the plants' health in the long run.

In informatics and especially machine learning, this term is used in analogy to describe the act of systematically removing parts of decision trees; regression or classification accuracy is decreased intentionally and thus traded for simplicity and better generalization of the model evolved. Approaches and benefits of the techniques used can be found for example in [Min89], [BB94], [HS95], or [Man97].

Obviously, the concept of removing branches of a tree can be easily transferred to GP, especially when we deal with tree-based genetic programming. Several pruning operators have already been presented for GP, see for example [ZM96] [FP98], [MK00], or more recent publications such as [dN06], [EKK04], [DH02], [FPS06], [GAT06]. In GP, pruning is often considered valuable because it helps to find more general and not over-parameterized programs; it is also referred to as an appropriate anti-bloat technique as described in Section 2.6 or [LP02], e.g.

In the case of fixed functions bases, pruning can also include the detection of really ineffective code or introns, i.e., code fragments that do not contribute to the program's (or, as in our case, model's) evaluation. For example, simply by using basic algebraic analysis, a simplification mechanism for formulas would be able to detect that -(+(x;4);4) is equal to +(+(x;4);-4) due to basic knowledge about subtraction and addition, and that this is again equal to +(x;4;-4). This then can be easily simplified to x as it is easy to implement a simplification program "knowing" that the addition of any value x and its negative counterpart $-x$ is always 0, and that 0 is the neutral element of the addition function.

But, as soon as such a fixed functions basis is not available anymore, things start to become a lot more complicated. We shall here describe pruning methods suitable for use in combination with a flexible and parameterizable set of function and terminal definitions as described in Section 9.2.3. We hereby try to consider the gain of simplicity as well as the deterioration of the model's quality caused by pruning it:

- The gain of simplicity with respect to the pruning of a model can be calculated by comparing its original tree complexity and the complexity of the pruned structure tree. The complexity of a model m, $c(m)$, can hereby be equal to the size or the height of the tree structure representing m.

  So, we calculate the model complexity decrease, $mcd(m, m_p)$, of a model m and a pruned version of m, m_p, as

  $$mcd(m, m_p) = \frac{c(m)}{c(m_p)} \qquad (9.36)$$

  Pruning a model by deleting subtrees will therefore always result in a mcd value equal to or greater than 1 as the original model's complexity (in terms of size or height of the tree structure) will always be greater than or equal to the pruned model's complexity.

- The deterioration of model caused by pruning, $deter(m, m_p)$, can be measured by calculating the ratio of the pruned model's quality $q(m_p)$ and the quality of the original formula $q(m)$ as

  $$deter(m, m_p) = \frac{q(m_p)}{q(m)} \qquad (9.37)$$

  Thus, if for example the pruned model's fitness value is 10% higher, i.e., worse than the original model's quality with respect to a given evaluation operator, then the resulting deterioration coefficient will be equal to 1.1.

  Please note that this approach yields reasonable results only when using a minimization approach, i.e., if better models are assigned smaller quality values as is the case with the MSE function, for example. If the evaluation operator applied behaves reciprocally, i.e., if for example the R^2 or VAF function is used, then the reciprocal value of $deter(m, m_p)$, $\frac{1}{deter(m,m_p)}$, is to be used instead.

These measures for the effect of pruning, namely the complexity reduction as well as the quality deterioration, are now used for parameterizing the effective pruning of models:

As we have stated above, accuracy is traded for simplicity, and now we are able to quantify this trading aspect. By giving an upper bound for the relation between the coefficients expressing the complexity deterioration and the simplification effects, the pruning mechanism can be limited; we call this composed coefficient $cp(m, m_p)$ and limit it with the upper bound cp_{max} demanding that

$$\frac{deter(m, m_p)}{mcd(m, m_p)} = cp(m, m_p) \leq cp_{max} \qquad (9.38)$$

Thus, we demand that decrease with respect to the model's quality shall not be worse than the simplicity gain multiplied with a certain factor cp_{max}.

Still, there is one major problem with this approach as tremendous loss of quality, as for example an increase of the mean squared error by a factor of 50, might be compensated by replacing a formula m_1 consisting of 60 nodes by one single constant, i.e., a model m_2 with only one node:

$$cp(m_1, m_2) = \frac{deter(m_1, m_2)}{mcd(m_1, m_2)} = \frac{\frac{q(m_2)}{q(m_1)}}{\frac{c(m_1)}{c(m_2)}} = \frac{50}{\frac{60}{1}} = \frac{50}{60} < 1 \qquad (9.39)$$

So, in order to cope with this potential problem – it is in fact really a problem since we do not want to replace all models with constant terminals – we give a second parameter for the pruning method which limits the quality deterioration, det_{max}, and so demand that

$$deter(m, m_p) = \frac{q(m_p)}{q(m)} \leq det_{max} \Leftrightarrow q(m_p) \leq det_{max} * q(m) \qquad (9.40)$$

9.3.2.2 Pruning a Structure Tree

The actual pruning of a model (with respect to one particular part of the model) in GP is rather easy as it simply consists of removing a sub-tree from the tree structure representing the formula. In the case of pruning the root node the model thereafter is simply a terminal representing the constant 0; otherwise the resected subtree is to be replaced by a constant representing the respective parent's neutral element for the respective input index. For example, pruning inputs of an addition results in the replacement of these branches by zeros, whereas children of multiplication functions have to be replaced by constants representing 1.0.

Furthermore, pruning could also include the excision of certain parts of the model, i.e., a part of a tree could be simply cut out and replaced by one its descendants.

Simple examples are shown in Figure 9.14: In the left part (a) we schematically show the replacement of the second input of an addition resulting in the insertion of the constant 0, in the middle (b) we see the replacement of a multiplication's first input by the constant 1, and in the right part (c) we see possible effects of excising two nodes and replacing them by either of their two descendants.

So, as we now know how models are pruned in general as well as what we want a pruning method to achieve, we will describe two pruning methods we have designed and implemented as operators for HeuristicLab: The first one is an exhaustive implementation that systematically tries to prune the model as much as possible, whereas the second one is inspired by evolution strategy for reducing runtime.

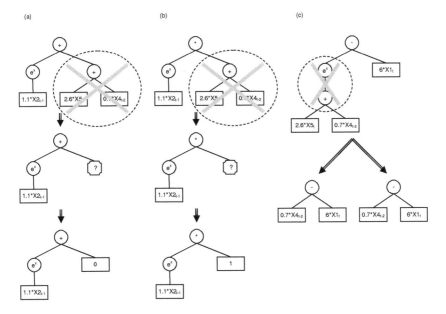

FIGURE 9.14: Simple examples for pruning in GP.

9.3.2.3 Exhaustive Pruning

When applying exhaustive pruning to a given model m we have to proceed in the following way: For each possible subtree up to a given height h_1 we create a copy of m and remove the respective branch. Furthermore, for each internal model fragment (tree) up to a given height h_2 we create a copy of m and cut out the respective fragment. After doing so, the resulting pruned models' qualities are calculated and their complexities are checked; if a pruned model meets the requirements regarding maximum deterioration and maximum coefficient of simplification and deterioration, then we go on with the procedure using this pruned formula. This routine is repeated until no more pruned model that meets the given requirements can be produced by deleting branches.

Finally, the algorithm's result is either the minimal model meeting the given requirements, or that model for which the minimal cp coefficient is calculated. This decision is controlled by the parameter $minimizeModel$ denoting whether the minimal formula is to be returned or, if this flag is set $false$, the model with the minimal cp value is to be considered the result of pruning m.

In a bit more formal way we can describe this exhaustive pruning algorithm as is given in Algorithm 9.1.

Exhaustive pruning is of course an extremely expensive method with respect

---

Algorithm 9.1 Exhaustive pruning of a model m using the parameters h_1, h_2, $minimizeModel$, cp_{max}, and det_{max}.

Initialize m_{curr} as clone of m,
Evaluate m, store calculated fitness in f
Calculate complexity of m, store result in c
Initialize $abort = false$
while $not(abort)$ **do**
　Initialize set of pruned models M
　Initialize structure tree t as tree representation of m_{curr}
　for each branch b of t with $height(b) < h_1$ **do**
　　Initialize m_{tmp} as clone of m_{curr}
　　Remove b', the corresponding branch to b in m_{tmp}
　　Evaluate m_{tmp}, store calculated fitness in f_{tmp}
　　Calculate complexity of m_{tmp}, store result in c_{tmp}
　　Calculate model complexity decrease $mcd = c/c_{tmp}$
　　Calculate quality deterioration $det = f/f_{tmp}$
　　if $det \leq det_{max} \land mcd \leq cp_{max}$ **then**
　　　Insert m_{tmp} to M
　　end if
　end for
　for each internal sub-tree st of t with $height(st) < h_2$ **do**
　　for each descendant d of st **do**
　　　Initialize m_{tmp} as clone of m_{curr}
　　　Replace st', the corresponding part to st in m_{tmp}, by d
　　　Evaluate m_{tmp}, store calculated fitness in f_{tmp}
　　　Calculate complexity of m_{tmp}, store result in c_{tmp}
　　　Calculate model complexity decrease $mcd = c/c_{tmp}$
　　　Calculate quality deterioration $det = f/f_{tmp}$
　　　if $det \leq det_{max} \land mcd \leq cp_{max}$ **then**
　　　　Insert m_{tmp} to M
　　　end if
　　end for
　end for
　if M is empty **then**
　　return m_{curr}
　else
　　if $minimizeModel$ **then**
　　　Set m_{curr} to that model in M with minimum complexity value c
　　else
　　　Set m_{curr} to that model in M with minimum mcd coefficient
　　end if
　end if
end while

---

to runtime consumption. As an alternative, a general pruning method inspired by evolution strategies is described in the following section.

9.3.2.4 ES-Inspired Pruning

As a less runtime consuming pruning method we have designed an ES-inspired pruning method: For pruning a model m, we create λ clones of m and prune those randomly; again, we use parameters h_1 and h_2 that limit the size of the branches and internal subtrees that are excised. All of the so created λ pruned mutants are checked and those that fulfill the given requirements regarding maximum deterioration and maximum coefficient of simplification and deterioration are collected. This procedure is then repeated with the best pruned mutant, whereas the best pruned model is again selected as the minimal model or the one showing the best coefficient of simplification and deterioration. As soon as this procedure is executed without any success for a given number in a row, the algorithm is terminated.

Algorithm 9.2 describes this ES-inspired pruning method in a more formal way.

9.4 Similarity Measures for Solution Candidates

Genetic diversity and population dynamics are very interesting aspects when it comes to analyzing GP processes. Measuring the entropy of a population of trees can be done for example by considering the programs' scores (as explained in [Ros95b], e.g.); entropy is there calculated as $-\sum_k p_k \cdot log(p_k)$ (where p_k is the proportion of the population P occupied by population partition k). In [McK00] the traditional fitness sharing concept from the work described in [DG89] is applied to test its feasibility in GP.

In this section we present more sophisticated measures which we have used for estimating the genetic diversity in GP populations as well as among populations of multi-population GP applications. What we use as basic measures for this are the following two functions that calculate the similarity of GP solution candidates or, a bit more specific, in our case formulas represented as structure trees:

- *Evaluation-based* similarity estimation compares the subtrees of two GP formulas with respect to their evaluation on the given training or validation data. The more similar these evaluations are with respect to the squared errors or linear correlation, the higher is the similarity for these two formulas.

- *Structural* similarity estimation directly compares the genetic material of two solution candidates: All possible pairs of ancestor and descendant

---

Algorithm 9.2 Evolution strategy inspired pruning of a model m using the parameters λ, $maxUnsuccRounds$, h_1, h_2, $minimizeModel$, cp_{max}, and det_{max}.

Initialize m_{curr} as clone of m,

Evaluate m, store calculated fitness in f

Calculate complexity of m, store result in c

Initialize $UnsuccessfulRounds := 0$

Initialize $abort := false$

while $not(abort)$ **do**

    Initialize set of pruned models M

    Initialize structure tree t as tree representation of m_{curr}

    for $i = 1 : \lambda$ **do**

        Set r to random number in $[0; 1[$

        Initialize m_{tmp} as clone of m_{curr}

        if $r < 0.5$ **then**

            Remove b, a branch of m_{tmp} with $height(b) < h_1$

        else

            Select st, an internal subtree of t with $height(st) < h_2$,

            replace st by a randomly chosen descendant of d

        end if

        Evaluate m_{tmp}, store calculated fitness in f_{tmp}

        Calculate complexity of m_{tmp}, store result in c_{tmp}

        Calculate model complexity decrease $mcd = c/c_{tmp}$

        Calculate quality deterioration $det = f/f_{tmp}$

        if $det \leq det_{max} \wedge mcd \leq cp_{max}$ **then**

            Insert m_{tmp} to M

        end if

    end for

    if M is empty **then**

        Increase $UnsuccessfulRounds$

        if $UnsuccessfulRounds = maxUnsuccRounds$ **then**

            return m_{curr}

        end if

    else

        Set $UnsuccessfulRounds := 0$

        if $minimizeModel$ **then**

            Set m_{curr} to that model in M with minimum complexity value c

        else

            Set m_{curr} to that model in M with minimum mcd coefficient

        end if

    end if

end while

---

nodes in formula trees are collected and these collections compared for pairs of formulas. So we can determine how similar the genetic make-up of formulas is without considering their evaluation.

9.4.1 Evaluation-Based Similarity Measures

The main idea of our evaluation-based similarity measures is that the building blocks of GP formulas are subtrees that are exchanged by crossover and so form new formulas. So, the evaluation of these branches of all individuals in a GP population can be used for measuring the similarity of two models m_1 and m_2:

For all subtrees in the structure-tree of model m, collected in t, we collect the evaluation results by applying these subformulas to the given data collection *data* as

$$\forall(st_i \in t)\forall(j \in [1; N]) : e_{i,j} = eval(st_i, data) \tag{9.41}$$

where N is the number of samples included in the data collection, no matter if training or validation data are considered.

The evaluation-based similarity of models m_1 and m_2, $es(m_1, m_2)$, is calculated by iterating over all subtrees of m_1 (collected in t_1) and, for each branch, picking that subtree of t_2 (containing all subtrees of m_2) whose evaluation is most "similar" to the evaluation of that respective branch. So, for each branch b_a in t_1 we compare its evaluation e_a with the evaluation e_b of all branches b_b in t_2, and the "similarity" can be calculated using the sum of squared errors (*sse*) or the linear correlation coefficient:

- When using the *sse* function, the sample-wise differences of the evaluations of the two given branches are calculated and their sum of squared differences is divided by the total sum of squares *tss* of the first branch's evaluation. This results in the similarity measure s for the given branches.

$$\overline{e_1} = \frac{1}{N} \sum_{j=1}^{N} e_a[j] \tag{9.42}$$

$$sse = \sum_{j=1}^{N} (e_a[j] - e_b[j])^2 ; tss = \sum_{j=1}^{N} (e_a[j] - \overline{e_a})^2 \tag{9.43}$$

$$s_{sse}(b_a, b_b) = 1 - \frac{sse}{tse} \tag{9.44}$$

- Alternatively the linear correlation coefficient can be used:

$$\overline{e_a} = \frac{1}{N} \sum_{j=1}^{N} e_a[j] ; \overline{e_b} = \frac{1}{N} \sum_{j=1}^{N} e_b[j] \tag{9.45}$$

$$s_{lc}(b_a, b_b) = \left| \frac{\frac{1}{n-1} \sum_{j=1}^{N} (e_a[j] - \overline{e_a})(e_b[j] - \overline{e_b})}{\sqrt{\frac{1}{n-1} \sum_{j=1}^{N} (e_a[j] - \overline{e_a})^2} \sqrt{\frac{1}{n-1} \sum_{j=1}^{N} (e_b[j] - \overline{e_b})^2}} \right|$$

(9.46)

No matter which approach is chosen, the calculated similarity measure for the branches b_a and b_b, $s(b_a, b_b)$, will always be in the interval $[0; 1]$; the higher this value becomes, the smaller is the difference between the evaluation results.

As we can now quantify the similarity of evaluations of two given subtrees, we can for each branch b_a in t_a elicit that branch b_x in t_b with the highest similarity to b_a; the similarity values **s** are collected for all branches in t_a and their mean value finally gives us a measure for the evaluation-based similarity of the models m_a and m_b, $es(m_a, m_b)$.

Optionally we can force the algorithm to select each branch in t_b not more than once as best match for a branch in t_a for preventing multiple contributions of certain parts of the models.

Finally, this similarity function can be parameterized by giving minimum and maximum bounds for the height and / or the level of the branches investigated. This is important since we can so control which branches are to be compared, be it the rather small ones, rather big ones, or all of them.

Algorithm 9.3 summarizes this evaluation-based similarity measure approach.

Algorithm 9.3 Calculation of the evaluation-based similarity of two models m_1 and m_2 with respect to data base *data*

> Collect all subtrees of the tree structure of m_1 in B_1
> Collect all subtrees of the tree structure of m_2 in B_2
> Initialize $s := 0$
> **for each** branch b_j in B_1 **do** evaluate b_j on *data*, store results in $e_{1,j}$
> **for each** branch b_k in B_2 **do** evaluate b_k on *data*, store results in $e_{2,k}$
> **for each** branch b_j in B_1 **do**
> > Initialize $s_{max} := 0$, *index* $:= -1$
> > **if** $|B_2| > 0$ **then**
> > > **for each** branch b_k in B_2 **do**
> > > > Calculate similarity s_{tmp} as similarity of b_j and b_k using $e_{1,j}$, $e_{2,k}$ and similarity function s_{sse} or s_{lc}
> > > > **if** $s_{tmp} > s_{max}$ **do** $s_{max} := s_{tmp}$; *index* $= k$
> > > **end for**
> > > **if** *PreventMultipleContribution* **do** remove b_{index} from B_2
> > **end if**
> > $s := s + s_{max}$
> **end for**
> **return** $s/|B_1|$

9.4.2   Structural Similarity Measures

Structural similarity estimation is, unlike the evaluation-based described before, independent of data; it is calculated on the basis of the genetic make-up of the models which are to be compared.

Koza [Koz92b] used the term variety to indicate the number of different programs in populations by comparing programs structurally and looking for exact matches. The Levenshtein distance [Lev66] can be used for calculating the distance between trees, but it is considered rather far from ideal ([Kei96], [O'R97], [LP02]); in [EN00] an edit distance specific to genetic programming parse trees was presented which considered the cost of substituting between different node types.

A very comprehensive overview of program tree similarity and diversity measures has been given for instance in [BGK04]. The standard tree structures representation in GP makes it possible to use more fine-grained structural measures that consider nodes, subtrees, and other graph theoretic properties (rather than just entire trees). In [Kei96], for example, subtree variety is measured as the ratio of unique subtrees over total subtrees and program variety as a ratio of the number of unique individuals over the size of the population; [MH99] investigated diversity at the genetic level by assigning numerical tags to each node in the population.

When analyzing the structure of models we have to be aware of the fact that often structurally different models can be equivalent. Let us for example consider the formulas $*(+(2,X2),+(X3)$ and $+(*(X2,X3),*(X3,2))$: As we know about distributivity we know that these formulas can be considered equivalent, but any structure analysis approach taking into account size, shape or parent / child relationships in the structure tree would assign these models a rather low similarity value. This is why we have designed and implemented a method that systematically collects all pairs of ancestor and descendant nodes and information about the properties of these nodes. Additionally, for each pair we also document the distance (with respect to the level in the model tree) and the index of the ancestor's child tree containing the descendant node. The similarity of two models is then, in analogy to the method described in the previous section, calculated by comparing all pairs of ancestors and descendants in one model to all pairs of the other model and averaging the similarity of the respective best matches.

Figure 9.15 shows a simple formula and all pairs of ancestors and descendants included in the structure tree representing it; the input indices as well as the level differences ("level delta") are also given. Please note: The pairs given on the right side of Figure 9.15 are shown intentionally as they symbolize the pairs of nodes with level difference 0, i.e., nodes combined with themselves.

We define a *genetic item* as a 6-tuple storing the following information about the ancestor node a and descendant node d:

- $type_a$, the type of the ancestor a

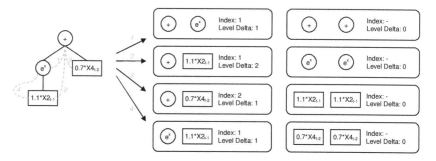

FIGURE 9.15: Simple formula structure and all included pairs of ancestors and descendants (genetic information items).

- $type_d$, the type of the descendant d

- δl, the level delta

- $index$, the index of the child branch of a that includes d

- np_a, the node parameters characterizing a

- np_d, the node parameters characterizing d

where the parameters characterizing nodes are represented by tuples containing the following information:

- var, the variant (of functions)

- $coeff$, the coefficient (of terminals)

- to, the time offset (of terminals)

- vi, the variable index (of terminals)

Now we can define the similarity of two *genetic items* gi_1 and gi_2, $s(gi_1, gi_2)$, as follows:

Most important are the types of the definitions referenced by the nodes; if these are not equal, then the similarity is 0 regardless of all other parameters:

$$\forall(gi_1, gi_2) : gi_1.type_a \neq gi_2.type_a \Rightarrow s(gi_1, gi_2) = 0 \qquad (9.47)$$

$$\forall(gi_1, gi_2) : gi_1.type_d \neq gi_2.type_d \Rightarrow s(gi_1, gi_2) = 0 \qquad (9.48)$$

If the types of the nodes correspond correctly, then the similarity of gi_1 and gi_2 is calculated using the similarity contributors $s_1 \dots s_{10}$ of the parameters of gi_1 and gi_2 weighted with coefficients $c_1 \dots c_{10}$.

The differences regarding input index, variant, and variable index are not in any way scaled or relativized; their similarity contribution is 1 in the case of equal parameters for both genetic items and 0 otherwise. The differences regarding level difference, coefficient, and time offset, on the contrary, are indeed scaled:

- The level difference is divided by the maximum tree height $height_{max}$,

- the difference of coefficients is divided by the range of the referenced terminal definition (in case of uniformly distributed coefficients) or divided by the standard deviation σ (in case coefficients are normally distributed), and

- the difference of the time offsets is divided by the maximum time offset allowed $offset_{max}$.

$$\forall(gi_1, gi_2 : gi_1.type_a = gi_2.type_a \& gi_1.type_d = gi_2.type_d) : \tag{9.49}$$

$$s_1 = 1 - \frac{|gi_1.\delta l - gi_2.\delta l|}{height_{max}} \tag{9.50}$$

$$s_2 = \begin{cases} gi_1.index \neq gi_2.index & : & 0 \\ gi_1.index = gi_2.index & : & 1 \end{cases} \tag{9.51}$$

$$s_3 = \begin{cases} gi_1.np_a.var \neq gi_2.np_a.var & : & 0 \\ gi_1.np_a.var = gi_2.np_a.var & : & 1 \end{cases} \tag{9.52}$$

$$s_4 = \begin{cases} gi_1.np_d.var \neq gi_2.np_d.var & : & 0 \\ gi_1.np_d.var = gi_2.np_d.var & : & 1 \end{cases} \tag{9.53}$$

$$\delta c_a = |gi_1.np_a.coeff - gi_2.np_a.coeff| \tag{9.54}$$

$$s_5 = 1 - \begin{cases} isUniformTerminal(gi_1.type_a) & : & \frac{\delta c_a}{gi_1.type_a.max - gi_1.type_a.min} \\ isGaussianTerminal(gi_1.type_a) & : & \frac{\delta c_a}{gi_1.type_a.\sigma*4} \end{cases}$$

$$\tag{9.55}$$

$$\delta c_d = |gi_1.np_d.coeff - gi_2.np_d.coeff| \tag{9.56}$$

$$s_6 = 1 - \begin{cases} isUniformTerminal(gi_1.type_d) & : & \frac{\delta c_d}{gi_1.type_d.max - gi_1.type_d.min} \\ isGaussianTerminal(gi_1.type_d) & : & \frac{\delta c_d}{gi_1.type_d.\sigma*4} \end{cases}$$

$$\tag{9.57}$$

$$s_7 = 1 - \frac{|gi_1.np_a.to - gi_2.np_a.to|}{offset_{max}} \tag{9.58}$$

$$s_8 = 1 - \frac{|gi_1.np_d.to - gi_2.np_d.to|}{offset_{max}} \tag{9.59}$$

$$s_9 = \begin{cases} gi_1.np_a.vi \neq gi_2.np_a.vi & : & 0 \\ gi_1.np_a.vi = gi_2.np_a.vi & : & 1 \end{cases} \tag{9.60}$$

$$s_{10} = \begin{cases} gi_1.np_d.vi \neq gi_2.np_d.vi & : & 0 \\ gi_1.np_d.vi = gi_2.np_d.vi & : & 1 \end{cases} \tag{9.61}$$

Finally, there are two possibilities how to calculate the structural similarity of gi_1 and gi_2, $sim(gi_1, gi_2)$: On the one hand this can be done in an *additive* way, on the other hand in a *multiplicative* way.

- When using the *additive* calculation, which is the obviously more simple way, $sim(gi_1, gi_2)$ is calculated as the sum of these similarity contributions $s_{1...10}$ weighted using the factors $c_{1...10}$ and, for the sake of normalization of results, divided by the sum of the weighting factors:

$$sim(gi_1, gi_2) = \frac{\sum_{i=1}^{10} s_i \cdot c_i}{\sum_{i=1}^{10} c_i} \tag{9.62}$$

- Otherwise, when using the *multiplicative* calculation method, we first calculate a punishment factor p_i for each s_i (again using weighting factors c_i, $0 \leq c_i \leq$ for all $i \in [1; 10]$) as

$$\forall (i \in [1; 10]) : p_i = (1 - s_i) \cdot c_i \tag{9.63}$$

and then get the temporary similarity result as

$$sim_{tmp}(gi_1, gi_2) = \prod_{i=1}^{10} (1 - p_i). \tag{9.64}$$

In the worst case scenario we get $d_i = 0$ for all $i \in [1; 10]$ and therefore the worst possible sim_{tmp} is

$$sim_{worst} = \prod_{i=1}^{10} (1 - ((1 - d_i) \cdot c_i)) = \prod_{i=1}^{10} (1 - c_i). \tag{9.65}$$

As sim_{worst} is surely greater than 0 we linearly scale the results to the interval $[0; 1]$:

$$sim(gi_1, gi_2) = \frac{sim_{tmp}(gi_1, gi_2) - sim_{worst}}{1 - sim_{worst}}. \tag{9.66}$$

In fact, we prefer this *multiplicative* similarity calculation method since it allows more specific analysis: By setting a weighting coefficient c_j to a rather high value (i.e., near or even equal to 1.0) the total similarity will become very small for pairs of genetic items that do not correspond with respect to this specific aspect, even if all other aspects would lead to a high similarity result.

Based on this similarity measure it is easy to formulate a similarity function that measures the similarity of two model structures. In analogy to the approach presented in the previous section, for comparing models m_1 and m_2 we collect all pairs of ancestors and descendants (up to a given maximum level difference) in m_1 and m_2 and look for the best matches in the respective opposite model's pool of genetic items, i.e., pairs of ancestor and descendant nodes. As we are able to quantify the similarity of genetic items, we can elicit for each genetic item gi_1 in the structure tree of m_1 exactly that genetic item gi_x in the model structure m_2 with the highest similarity to gi_1; the similarity values **s** are collected for all genetic items contained in m_1 and their mean value finally gives us a measure for the structure-based similarity of the models m_1 and m_2, $sim(m_1, m_2)$.

Optionally we can force the algorithm to select each genetic item of m_2 not more than once as best match for an item in m_1 for preventing multiple contributions of certain components of the models.

This function is defined in a more formal way using pseudo-code in Algorithm 9.4.

Algorithm 9.4 Calculation of the structural similarity of two models m_1 and m_2

Collect all genetic items m_1 in GI_1
Collect all genetic items m_2 in GI_2
Initialize $s := 0$
for each branch gi_j in GI_1 **do**
    Initialize $s_{max} := 0$, $index := -1$
    if $|B_2| > 0$ **then**
        for each genetic item gi_k in GI_2 **do**
            Calculate similarity s_{tmp} as similarity of gi_j and gi_k
            if $s_{tmp} > s_{max}$ **do** $s_{max} := s_{tmp}$; $index = k$
        end for
        if $PreventMultipleContribution$ **do** remove gi_{index} from GI_2
    end if
    $s := s + s_{max}$
end for
return $s/|GI_1|$

Obviously, it is possible that some model contains all pairs of genetic items that are also incorporated in another model, but not vice versa. Thus, this similarity measure $sim(m_1, m_2)$ is not symmetric, i.e., $sim(m_1, m_2)$ does not necessarily return the same result as $sim(m_2, m_1)$ for any pair of models m_1 and m_2.

Of course, this similarity concept for GP individuals cannot be the basis of theoretical concepts comparable to those based on GP (hyper)schemata, for example; we do here not want to give any statements about the probability of certain parts of formulas to occur in a given generation. In the presence of mutation or other structure modifying operations (as for example pruning) we are interested in measuring the structural diversity in GP populations; using this structural similarity measure we are able to do so.

Chapter 10

Applications of Genetic Algorithms: Combinatorial Optimization

DOI: 10.1201/9781420011326-11

Within Chapter 7 the knowledge about the global optimum has been used in order to analyze and highlight certain properties of the considered algorithms. In case of practical applications of considerable dimension this information is not available.

The analyses described in this chapter do not consider information about the genotypes of global optima and are therefore limited to the observation of the dynamics of genetic diversity in populations and in subpopulations of parallel GA concepts. The main conclusions of Chapter 7 were that it is most beneficial for the evolutionary process of genetic algorithms if the essential genetic information (the alleles of the globally optimal solution) establishes slowly in the population, which is important for gaining high quality results. As already indicated in previous chapters, this can be achieved by offspring selection. In this chapter results for several benchmark problem instances will be reported on in terms of achievable solution qualities, i.e., best and average solution qualities. The results for the TSP benchmark problem instances obtained using standard GAs, GAs with offspring selection, and the SASEGASA have been taken from [Aff05] and [AW04b]. Additionally, some characteristic aspects of certain algorithm variants are analyzed in greater detail by observing the genetic diversity over time similar to the genetic diversity analyses reported on in Section 6.2. For the CVRP we have also compared the performance of standard GAs to the performance of GAs with offspring selection. By doing so, the observation of genetic diversity over time has again been used to point out selected aspects that are representative for the respective algorithms when applied to the CVRP.

Beside the increased robustness of offspring selection described in Chapter 7 we here also consider the effects of a greater number of subpopulations for the SASEGASA. The most important fact is that we can in this context observe the scalability of achievable global solution qualities by applying greater numbers of subpopulations.

As is shown in this chapter, in this context we can observe that a slow decrease of genetic diversity caused by the evolutionary forces supports the GA in producing high quality results.

10.1 The Traveling Salesman Problem

All TSP benchmark problems used here have been taken from the TSPLIB [Rei91] using updated information[1] about the best or at least the best known solutions. The results for the TSP are represented as the relative difference to the best known solution defined as

$$relativeDifference = \left(\frac{ResultQuality}{OptimalQuality} - 1 \right) \cdot 100 \, [\%] \qquad (10.1)$$

All values presented in the following tables are the best and average relative differences of five independent test runs executed for each test case. The average number of evaluated solutions gives a quite objective measure of the computational effort.

10.1.1 Performance Increase of Results of Different Crossover Operators by Means of Offspring Selection

The first aspect to be considered is the effect of the offspring selection model on the quality improvement using different crossover operators. In order to visualize the positive effects of the new methods in a more obvious way, we also present results that were generated by a standard GA with proportional selection, generational replacement, and 1-elitism.

In Table 10.2 the results achieved with the conventional GA are listed. The fixed parameter values that were used for all algorithms in the different test runs are given in Table 10.1.

As we want to see how the algorithmic concepts presented in the first part of this book influence the ability of GAs to produce high quality results, the effects of offspring selection are here given on the basis of a number of experiments which were performed on a single population in order to not dilute the effects of offspring selection principles with the effects of the segregation and reunification strategies. Table 10.3 recapitulates the results for a selection of commonly applied crossover operators suggested for the path representation of the TSP ([Mic92], [LKM+99]) each on its own, as well as one combination of more effective crossover operators.

Remarkable in this context is that also the use of crossover operators, that are commonly considered rather unsuitable for the TSP [LKM+99], leads to quite good results in combination with offspring selection. The reason for this behavior is that in our selection principle only children that have emerged as a good combination of their parents' attributes are considered for the further evolutionary process, if the success ratio is set to a higher range.

---

[1]Updates for the best (known) solutions can for example be found on http://www.iwr.uni-heidelberg.de/groups/comopt/software/TSPLIB95/.

Table 10.1: Overview of algorithm parameters.

Parameters for the standard GA (Results presented in Tab. 10.2)	
Generations	100,000
Population Size	120
Elitism Rate	1
Mutation Rate	0.05
Selection Operator	Roulette
Mutation Operator	Simple Inversion
Parameters for the offspring selection GA (Results presented in Tab. 10.3)	
Population Size	500
Elitism Rate	1
Mutation Rate	0.05
Selection Operator	Roulette
Mutation Operator	Simple Inversion
Success Ratio	0.7
Maximum Selection Pressure	250

Table 10.2: Experimental results achieved using a standard GA.

Problem	Crossover	Best	Average	Evaluated Solutions
berlin52	OX	0.00	3.76	12,000,000
berlin52	ERX	5.32	7.73	12,000,000
berlin52	MPX	21.74	26.52	12,000,000
ch130	OX	3.90	5.41	12,000,000
ch130	ERX	142.57	142.62	12,000,000
ch130	MPX	83.57	85.07	12,000,000
kroa200	OX	3.14	4.69	12,000,000
kroa200	ERX	325.92	336.19	12,000,000
kroa200	MPX	146.94	148.08	12,000,000

Table 10.3: Experimental results achieved using a GA with offspring selection.

Problem	Crossover	Best	Average	Evaluated Solutions	Change to standard GA
berlin52	OX	0.00	1.90	14,250,516	-1.86
berlin52	ERX	0.00	1.97	6,784,626	-5.76
berlin52	MPX	0.00	0.76	6,825,199	-25.76
berlin52	OX,ERX,MPX	0.00	0.90	7,457,451	–
ch130	OX	1.54	2.26	13,022,207	-3.15
ch130	ERX	0.57	2.11	4,674,485	-140.51
ch130	MPX	1.11	3.18	9,282,291	-81.89
ch130	OX, ERX, MPX	0.68	1.18	5,758,022	–
kroa200	OX	2.73	3.51	15,653,950	-1.18
kroa200	ERX	3.21	5.40	19,410,458	-330.79
kroa200	MPX	3.28	4.65	13,626,348	-143.43
kroa200	OX, ERX, MPX	2.34	3.04	9,404,241	–

In combination with higher values for the maximum selection pressure, genetic search can be guided advantageously also for poor crossover operators as the larger amount of handicapped offspring are simply not considered for the further evolutionary process. Figure 10.1 shows this effect in detail for the *berlin52* TSP instance. Quite good results in terms of global convergence could also be achieved using a combination of different crossover operators, as additional genetic diversity is so brought into the population and inferior crossover results are not considered due to the enhanced offspring selection model.

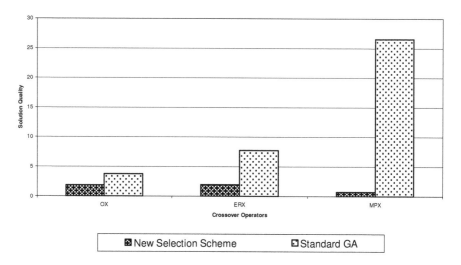

FIGURE 10.1: Quality improvement using offspring selection and various crossover operators (taken from [AW04b]. This figure is displayed with kind permission of Springer Science and Business Media.

10.1.2   Scalability of Global Solution Quality by SASEGASA

In this part of the experimental section we present the main effects of SASEGASA when applied to a practical implementation in a distributed environment: A higher number of subpopulations at the beginning of the evolutionary process allows to achieve scalable improvements in terms of global convergence.

Table 10.4: Parameter values used in the test runs of the SASEGASA algorithms with single crossover operators as well as with a combination of the operators.

Parameters for SASEGASA with 1 crossover operator (OX) (Results presented in Tab. 10.5)	
Subpopulation Size	100
Elitism Rate	1
Mutation Rate	0.05 resp. 0.00
Selection Operator	Roulette
Crossover Operators	OX
Mutation Operator	Simple Inversion
Success Ratio	0.8
Maximum Selection Pressure	30
Parameters for SASEGASA with 1 crossover operator (ERX) (Results presented in Tab. 10.6)	
Subpopulation Size	100
Elitism Rate	1
Mutation Rate	0.05 resp. 0.00
Selection Operator	Roulette
Crossover Operators	ERX
Mutation Operator	Simple Inversion
Success Ratio	0.8
Maximum Selection Pressure	30
Parameters for SASEGASA with 1 crossover operator (MPX) (Results presented in Tab. 10.7)	
Subpopulation Size	100
Elitism Rate	1
Mutation Rate	0.05 resp. 0.00
Selection Operator	Roulette
Crossover Operators	MPX
Mutation Operator	Simple Inversion
Success Ratio	0.8
Maximum Selection Pressure	15
Parameters for SASEGASA with a combination of crossover operators (OX, ERX, MPX) (Results presented in Tab. 10.8)	
Subpopulation Size	100
Elitism Rate	1
Mutation Rate	0.05 resp. 0.00
Selection Operator	Roulette
Crossover Operators	OX, ERX, MPX
Mutation Operator	Simple Inversion
Success Ratio	0.8
Maximum Selection Pressure	15

Table 10.5: Results showing the scaling properties of SASEGASA with one crossover operator (OX), with and without mutation.

		Results with mutation			Results without mutation		
Problem	Sub-populations	Best	Average	Evaluated Solutions	Best	Average	Evaluated Solutions
berlin52	1	12.13	20.86	41280	29.63	44.99	22,577
berlin52	5	4.92	8.67	731,191	18.58	27.20	242,195
berlin52	10	2.29	5.80	1,007,320	7.54	12.30	751,379
berlin52	20	0.72	3.21	2,802,620	5.48	7.36	2,368,694
berlin52	40	0.00	1.20	8,407,988	1.12	3.59	7,117,442
berlin52	80	0.00	0.72	25,154,907	0.00	2.21	25,045,133
berlin52	160	0.00	0.00	90,775,916	0.00	1.36	87,850,762
ch130	1	59.21	89.27	88,326	207.94	217.57	28,809
ch130	5	21.72	24.89	834,049	136.10	190.53	240,916
ch130	10	7.45	12.88	2,210,398	114.65	120.26	914,765
ch130	20	4.50	6.25	5,410,587	63.68	85.23	2,743,967
ch130	40	4.36	5.19	13,912,314	50.44	60.91	9,104,041
ch130	80	1.80	3.33	40,283,441	24.73	32.92	30,082,798
ch130	160	1.34	2.58	117,919,398	15.04	25.95	102,551,323
kroa200	1	90.87	136.71	139,629	371.87	412.20	34,315
kroa200	5	18.29	53.60	1,299,129	259.92	372.42	253,757
kroa200	10	16.55	21.18	3,155,000	199.69	227.65	1,066,148
kroa200	20	7.07	9.26	7,689,795	141.47	179.60	3,189,587
kroa200	40	4.02	5.30	21,251,916	120.60	148.41	9,688,113
kroa200	80	3.05	4.23	58,042,978	72.75	95.34	32,909,364
kroa200	160	2.59	2.85	175,599,138	45.77	75.30	116,522,803

Table 10.6: Results showing the scaling properties of SASEGASA with one crossover operator (ERX), with and without mutation.

Problem	Sub-populations	Results with mutation			Results without mutation		
		Best	Average	Evaluated Solutions	Best	Average	Evaluated Solutions
berlin52	1	4.77	7.11	34,578	1.40	5.39	30,031
berlin52	5	0.00	0.20	310,088	0.00	2.75	239,678
berlin52	10	0.00	0.00	809,083	0.00	0.00	692,015,
berlin52	20	0.00	0.00	2,229,713	0.00	0.00	1,962,213
berlin52	40	0.00	0.00	6,753,499	0.00	0.00	6,358,343
berlin52	80	0.00	0.00	23,020,154	0.00	0.00	22,299,205
berlin52	160	0.00	0.00	84,402,610	0.00	0.00	82,851,322
ch130	1	14.33	24.88	141,314	22.56	24.57	127,335
ch130	5	6.84	10.86	911,371	18.59	20.99	702,917
ch130	10	6.31	10.56	´1,820,004	16.87	18.87	1,572,299
ch130	20	4.22	5.41	4,831,614	11.98	17.16	3,779,535
ch130	40	1.06	1.93	13,271,120	4.09	7.79	10,354,983
ch130	80	0.23	0.72	36,602,158	2.34	6.69	32,090,886
ch130	160	0.00	0.52	111,218,379	1.00	2.10	104,042,226
kroA200	1	29.82	37.11	453,954	40.42	46.01	441,940
kroA200	5	14.22	25.52	2,458,083	38.68	41.63	2,300,084
kroA200	10	8.89	18.08	´5,462,657	32.51	36.57	4,624,114
kroA200	20	6.27	8.04	12,076,655	29.16	34.53	9,923,258
kroA200	40	3.86	5.03	28,810,360	27.23	32.10	22,506,282
kroA200	80	2.50	2.95	73,702,312	23.79	27.45	59,028,450
kroA200	160	0.36	1.82	171,391,466	21.26	26.26	146,796,110

Table 10.7: Results showing the scaling properties of SASEGASA with one crossover operator (MPX), with and without mutation.

Problem	Sub-populations	Results with mutation			Results without mutation		
		Best	Average	Evaluated Solutions	Best	Average	Evaluated Solutions
berlin52	1	9.15	18.98	80,635	8.46	20.72	58,985
berlin52	5	0.00	3.16	497,211	6.89	11.05	418,175
berlin52	10	0.00	1.08	1,216,238	1.75	4.93	1,153,493
berlin52	20	0.00	0.00	3,302,870	2.66	4.05	2,445,796
berlin52	40	0.00	0.00	10,875,130	0.00	1.00	9,227,596
berlin52	80	0.00	0.00	18,414,626	0.00	0.00	19,769,438
berlin52	160	0.00	0.00	92,662,669	0.00	0.00	56,682,137
ch130	1	141.16	160.39	59,847	140.38	158.13	63,547
ch130	5	22.05	101.46	504,065	36.73	79.08	585,371
ch130	10	14.27	32.64	1,867,440	26.40	35.14	1,602,154
ch130	20	5.48	10.27	4,665,532	10.85	22.84	3,875,043
ch130	40	1.83	6.11	11,096,130	14.62	18.81	8,837,255
ch130	80	2.24	3.90	27,379,806	4.70	10.07	26,085,696
ch130	160	0.44	1.72	75,905,160	4.11	5.98	74,759,771
kroA200	1	198.47	243.65	94,830	180.91	249.89	89,746
kroA200	5	26.52	102.53	1,461,829	61.80	135.68	942,102
kroA200	10	13.22	30.33	4,096,990	30.99	76.45	2,932,813
kroA200	20	4.56	24.94	9,154,913	17.59	37.14	7,907,319
kroA200	40	4.82	15.69	19,573,066	11.56	27.96	18,798,917
kroA200	80	4.41	7.43	44,363,179	4.62	14.08	57,105,958
kroA200	160	2.57	4.12	116,759,298	5.33	8.31	115,013,599

Indeed, as the Tables 10.5–10.8 show for the different crossover operators and their combinations, respectively, achievable solution quality can be pushed to highest quality regions also for higher dimensional problems with only linearly increasing computational effort by simply increasing the initial number of subpopulations. Especially when using a combination of considered crossover operators (see Table 10.8), which becomes possible due to self-adaptive offspring selection, the global optimum could be detected for all considered benchmark problems when using a combination of different crossover operators. The SASEGASA parameter settings used for the results given before have been chosen in such a way in order to point out certain aspects; the parameters used for achieving the results given in Table 10.8 are quite advantageous for applying SASEGASA to the TSP. Therefore, also higher dimensional test cases are shown in Table 10.8 for which the global optimum

Table 10.8: Results showing the scaling properties of SASEGASA with a combination of crossover operators (OX, ERX, MPX), with and without mutation.

	Sub-populations	Results with mutation			Results without mutation		
Problem		Best	Average	Evaluated Solutions	Best	Average	Evaluated Solutions
berlin52	1	0.72	6.12	76,920	3.79	11.20	49,866
berlin52	5	0.00	0.46	774,770	0.00	0.00	548,284
berlin52	10	0.00	0.00	1,670,107	0.00	0.00	1,296,130
berlin52	20	0.00	0.00	4,247,216	0.00	0.00	3,275,660
berlin52	40	0.00	0.00	11,240,451	0.00	0.00	7,394,200
berlin52	80	0.00	0.00	53,844,262	0.00	0.00	21,129,365
berlin52	160	0.00	0.00	246,725,814	0.00	0.00	61,538,007
ch130	1	20.32	34.57	131,650	35.24	43.22	117,720
ch130	5	6.30	7.12	1,243,637	4.86	6.74	961,172
ch130	10	4.12	4.45	3,275,072	3.01	4.24	2,578,121
ch130	20	0.98	1.78	9,092,937	1.60	2.58	6,475,903
ch130	40	0.23	0.98	32,446,649	0.62	1.13	15,027,715
ch130	80	0.23	0.32	77,406,460	0.00	0.24	41,921,823
ch130	160	0.00	0.08	170,273,008	0.00	0.17	96,545,540
kroA200	1	30.80	40.91	311,334	70.50	74.09	208,108
kroA200	5	6.91	10.34	2,094,110	17.03	22.90	1,555,680
kroA200	10	4.27	6.42	5,165,175	10.23	12.54	3,778,171
kroA200	20	1.88	2.25	18,477,477	3.74	4.44	9,321,037
kroA200	40	0.33	1.68	68,132,626	1.16	2.63	29,112,958
kroA200	80	0.33	1.16	134,467,940	1.06	1.53	68,299,249
kroA200	160	0.00	0.58	201,322,654	0.11	0.25	131,669,520
lin318	1	64.78	77.56	403,431	128.55	143.94	242,534
lin318	5	17.60	20.53	3,292,861	34.08	42.41	2,338,523
lin318	10	7.95	10.93	8,093,264	21.04	22.46	5,680,243
lin318	20	2.38	4.35	26,534,811	10.25	11.36	13,394,560
lin318	40	1.97	3.19	200,885,952	4.67	6.36	33,267,177
lin318	80	1.54	2.34	624,986,088	2.01	2.46	93,879,278
lin318	160	0.80	1.12	959,258,717	1.48	2.55	256,372,204
lin318	320	0.00	0.52	2,116,724,528	0.88	1.02	632,882,394
fl417	1	60.31	89.11	585,102	120.83	142.99	408,664
fl417	5	9.86	13.78	5,104,971	32.43	39.92	3,615,174
fl417	10	2.70	5.14	26,586,653	21.49	22.76	8,451,114
fl417	20	1.50	2.35	106,892,925	4.80	8.35	22,441,329
fl417	40	0.29	0.91	664,674,431	0.56	2.29	236,658,335
fl417	80	0.21	0.27	1,310,193,035	0.19	0.79	519,000,908
fl417	160	0.00	0.11	2,122,856,932	0.11	1.55	802,368,224
fl417	320	0.00	0.00	4,367,217,575	0.00	0.34	2,231,720,072

could also be found by simply increasing the number of demes. Apart from the computational effort which becomes higher and higher in a single processor environment, the degree of difficulty may be increased by increasing the problem dimension.

The scalability of achievable solution qualities, that comes along with a linearly increasing number of generated solutions, is a real advancement to classical serial and parallel GA concepts, where a greater number of evaluated solutions cannot improve global solution quality anymore after the GA has prematurely converged. As theoretically considered in the previous chapters, the reasons for this beneficial behavior are given by the interplay between genetic drift and migration embedded in the self-adaptive selection pressure steering concept. Even if the achieved results without mutation are not quite as good as those achieved by the SASEGASA with a standard mutation rate, it is remarkable that the scaling property still holds. We have also executed experiments with smaller numbers of larger subpopulations as well as with greater numbers of smaller subpopulations. Still, these results are not documented here, as these test series showed basically the same results with a comparable total effort of evaluated solutions. This is an interesting aspect for an efficient practical adaptation to a concrete parallel environment.

Even if the achieved results are clearly superior to most of the results reported for applications of evolutionary algorithms to the TSP [LKM+99], it has to be pointed out again that all introduced and applied additions to a standard evolutionary algorithm are generic and absolutely no problem-specific local pre- or post-optimization techniques have been used in our experiments.

10.1.3   Comparison of the SASEGASA to the Island-Model Coarse-Grained Parallel GA

The island model is the most frequently applied parallel GA model. Moreover, the island model is closer related to the newly introduced concepts of the present work than other coarse- and fine-grained parallel GA models. Therefore, this part of the empirical studies discusses the results that are achievable with a conventional island GA compared to the results of SASEGASA.

A main difference between an island GA and a SASEGASA is the self-adaptive selection pressure steering concept which as a side effect allows the detection of premature convergence in the algorithm's subpopulations. It therefore becomes possible to select the dates of migration phases dynamically and the SASEGASA algorithm is no more dependent on static migration intervals as the island model is. Furthermore, especially in the migration phases, the self-adaptive selection pressure steering concept of the SASEGASA enables the algorithm to join the genetic information of individuals descending from different subpopulations in a more directed way than within the island model. Less fit offspring, that may especially emerge in the migration phases as children from parents descending from very different regions of the solution space, are simply not considered for the ongoing evolutionary process due to offspring selection. In addition to this, it is also not useful to apply a combination of crossover operators within the standard island model, as each crossover result would become part of the ongoing evolutionary process since no offspring selection steps are performed. In contrast to this, the SASEGASA maintains only those crossover results that represent a successful combination of their parents' attributes, which makes a combination of more operators reasonable.

The Tables 10.10–10.12 show the results for the island GA using the same TSP benchmarks as those that we have also used for testing the SASEGASA applying either OX (see Table 10.10), ERX (see Table 10.11), or MPX (see Table 10.12) as crossover mechanisms, each with and without mutation.

Table 10.9: Parameter values used in the test runs of a island model GA with various operators and various numbers of demes.

Parameters for the Island GA	
(Results presented in Tab. 10.10–Tab. 10.12)	
Deme Population Size	100
Elitism Rate	1
Mutation Rate	0.05 resp. 0.00
Selection Operator	Roulette
Crossover Operators	OX, ERX, resp. MPX
Mutation Operator	Simple Inversion
Migration Interval	20 Rounds
Migration Rate	0.15 (of deme)
Migration Topology	unidirectional ring
Migration Selection	Best
Migration Insertion	Random

Table 10.10: Results showing the scaling properties of an island GA with one crossover operator (OX) using roulette-wheel selection, with and without mutation.

	Number of	Results with mutation			Results without mutation		
Problem	demes	Best	Average	Evaluated Solutions	Best	Average	Evaluated Solutions
berlin52	1	2.29	4.06	1,500,000	63.02	78.05	1,500,000
berlin52	5	2.29	5.44	7,500,000	20.67	28.90	7,500,000
berlin52	10	2.69	4.66	15,000,000	14.37	18.35	15,000,000
berlin52	20	2.29	3.68	30,000,000	9.67	19.64	30,000,000
berlin52	40	0.00	0.46	60,000,000	4.61	12.95	60,000,000
berlin52	80	0.00	0.61	120,000,000	4.55	7.73	120,000,000
berlin52	160	0.00	0.00	240,000,000	5.17	8.49	240,000,000
ch130	1	5.52	7.81	1,500,000	283.50	330.73	1,500,000
ch130	5	2.80	5.60	7,500,000	158.18	184.96	7,500,000
ch130	10	1.42	2.84	15,000,000	139.17	168.70	15,000,000
ch130	20	2.47	3.45	30,000,000	91.93	122.74	30,000,000
ch130	40	1.21	2.52	60,000,000	112.95	120.46	60,000,000
ch130	80	0.65	1.71	120,000,000	91.83	108.93	120,000,000
ch130	160	0.55	1.32	240,000,000	88.02	107.12	240,000,000
kroa200	1	14.19	17.00	1,500,000	479.92	526.06	1,500,000
kroa200	5	5.66	7.41	7,500,000	323.76	340.63	7,500,000
kroa200	10	5.79	6.39	15,000,000	245.37	308.34	15,000,000
kroa200	20	3.11	5.10	30,000,000	227.65	234.68	30,000,000
kroa200	40	2.35	4.05	60,000,000	189.73	226.60	60,000,000
kroa200	80	0.61	2.51	120,000,000	178.99	204.01	120,000,000
kroa200	160	1.30	2.20	240,000,000	167.42	194.77	240,000,000

As already noticed for the conventional GA (see Table 10.2), the results of the island GA are also quite good when using the OX crossover operator (see Table 10.10) and (in terms of solution quality) comparable to the SASEGASA results obtained using the OX crossover. Still, the computational effort (i.e., the number of evaluated solutions) is comparatively high in order to achieve the results as migration is in the island GA applied in a less goal-oriented way. As only mutation and migration are qualified to regain alleles that are lost due to genetic drift, there is further empirical evidence that migration works less effectively in the island model when considering the island GA results; these, in fact, are really bad when deactivating mutation. The SASEGASA is, in contrast to this, still able to scale up solution quality to high quality regions even without mutation (which can be seen in Tables 10.5– 10.8).

The results returned by the island GA using ERX and MPX crossovers are rather weak regardless of mutation, and are significantly outperformed by the SASEGASA results. As we have already seen for the conventional GA, the

Table 10.11: Results showing the scaling properties of an island GA with one crossover operator (ERX) using roulette-wheel selection, with and without mutation.

Problem	Number of demes	Results with mutation			Results without mutation		
		Best	Average	Evaluated Solutions	Best	Average	Evaluated Solutions
berlin52	1	10.90	13.78	1,500,000	6.30	10.56	1,500,000
berlin52	5	0.00	5.20	7,500,000	2.77	4.57	7,500,000
berlin52	10	0.72	4.76	15,000,000	3.30	5.00	15,000,000
berlin52	20	0.00	2.42	30,000,000	0.00	0.40	30,000,000
berlin52	40	0.00	0.96	60,000,000	0.00	0.00	60,000,000
berlin52	80	0.00	0.00	120,000,000	0.00	0.20	120,000,000
berlin52	160	0.00	0.00	240,000,000	0.00	0.00	240,000,000
ch130	1	154.45	176.38	1,500,000	152.96	165.12	1,500,000
ch130	5	109.18	116.27	7,500,000	118.59	125.68	7,500,000
ch130	10	82.24	91.16	15,000,000	93.21	103.03	15,000,000
ch130	20	48.17	62.94	30,000,000	72.93	77.12	30,000,000
ch130	40	53.18	55.63	60,000,000	61.10	68.79	60,000,000
ch130	80	41.65	48.34	120,000,000	42.32	51.34	120,000,000
ch130	160	35.22	39.51	240,000,000	38.81	54.72	240,000,000
kroa200	1	401.47	421.13	1,500,000	405.27	420.05	1,500,000
kroa200	5	302.29	316.97	7,500,000	324.59	352.06	7,500,000
kroa200	10	247.51	268.50	15,000,000	280.62	309.44	15,000,000
kroa200	20	232.59	241.82	30,000,000	271.53	282.68	30,000,000
kroa200	40	209.18	220.53	60,000,000	237.78	253.12	60,000,000
kroa200	80	178.87	199.01	120,000,000	216.09	233.08	120,000,000
kroa200	160	180.91	190.64	240,000,000	220.42	224.12	240,000,000

Table 10.12: Results showing the scaling properties of an island GA with one crossover operator (MPX) using roulette-wheel selection, with and without mutation.

Problem	Number of demes	Results with mutation			Results without mutation		
		Best	Average	Evaluated Solutions	Best	Average	Evaluated Solutions
berlin52	1	33.63	36.52	1,500,000	29.20	32.34	1,500,000
berlin52	5	11.32	18.77	7,500,000	12.83	16.53	7,500,000
berlin52	10	8.71	12.10	15,000,000	8.83	11.67	15,000,000
berlin52	20	3.74	7.27	30,000,000	9.06	12.41	30,000,000
berlin52	40	0.00	5.90	60,000,000	3.92	6.63	60,000,000
berlin52	80	0.00	4.66	120,000,000	1.31	4.44	120,000,000
berlin52	160	0.00	1.54	240,000,000	1.31	3.28	240,000,000
ch130	1	129.00	131.82	1,500,000	117.77	125.84	1,500,000
ch130	5	71.34	75.19	7,500,000	54.73	67.72	7,500,000
ch130	10	48.90	52.51	15,000,000	42.31	48.11	15,000,000
ch130	20	44.34	47.52	30,000,000	32.57	37.07	30,000,000
ch130	40	36.24	40.77	60,000,000	30.95	39.48	60,000,000
ch130	80	31.47	38.93	120,000,000	30.51	34.16	120,000,000
ch130	160	34.91	39.03	240,000,000	29.31	32.84	240,000,000
kroa200	1	208.52	219.21	1,500,000	201.85	209.58	1,500,000
kroa200	5	122.68	132.46	7,500,000	111.61	125.41	7,500,000
kroa200	10	100.13	108.64	15,000,000	87.75	102.55	15,000,000
kroa200	20	91.77	98.36	30,000,000	91.14	94.76	30,000,000
kroa200	40	83.34	91.31	60,000,000	84.85	88.23	60,000,000
kroa200	80	88.37	90.04	120,000,000	77.87	81.14	120,000,000
kroa200	160	77.34	80.94	240,000,000	80.64	83.29	240,000,000

island model also does not offer concepts for dropping out disadvantageous crossover results.

Thus, in contrast to most of the enhanced GA concepts discussed in literature which are in most cases tuned for some specific purpose, it appears that the SASEGASA algorithm acts very stabilizing under various conditions. It is also quite impressive that the generic applicability and transference of the positive attributes of SASEGASA appear unimpaired when considering a completely different optimization problem - namely the optimization of hard real-valued benchmark test functions in high dimensions [Aff05]. It has been reported in [Aff05] that the SASEGASA algorithm without any problem-specific adaptations is able to find the global optimal solution for all considered benchmark test functions (Rosenbrock, Rastrigin, Griewangk, Ackley, and Schwefel's sine root function) in dimensions that have hardly been discussed in GA literature (up to $n = 2000$).

10.1.4   Genetic Diversity Analysis for the Different GA Types

As already mentioned in the introductory part of this chapter, we do not confine ourselves to report the results in table form and try to compare the internal functioning of the certain algorithmic variants also here. In contrast to Chapter 7 we here consciously abandon information about globally optimal solutions which are unknown also in practical applications.

Results that are as interesting as those achieved by observing the dynamics of the global optimal alleles can be obtained by analyzing the genetic diversity distribution during the run of a GA. For this purpose it is necessary to define an appropriate distance measure between two solution candidates for the problem representation at hand. In contrast to GP-based structure identification diversity analyses such a distance measure is quite intuitive and easy to describe for the TSP.

The similarity measure between two TSP-solutions t_1 and t_2 used here is defined as a similarity value sim between 0 and 1:

$$sim(t_1, t_2) = \frac{\mid e : e \in E(t_1) \wedge e \in E(t_2) \mid}{\mid E(t_1) \mid} \in [0, 1] \tag{10.2}$$

giving the quotient of the number of common edges in the TSP solutions t_1 and t_2 and the total number of edges. E here denotes the set of edges in a tour. The according distance measure can then be defined as

$$d(t_1, t_2) = 1 - sim(t_1, t_2) \in [0, 1] \tag{10.3}$$

Thus, the similarity or the distance of two concrete TSP solutions can be measured on a linear scale between the values 0 and 1.

A very detailed representation of genetic diversity in a population is the graphical display of pairwise similarities or distances for all members of a population. An appropriate measure, which is provided in the HeuristicLab framework, is to illustrate the similarity as a $n \times n$ matrix where each entry indicates the similarity in form of a grey scaled value. Figure 10.2 shows an example: The darker the $(i, j) - th$ entry in the $n \times n$ grid is, the more similar are the two solutions i and j. Not surprisingly, the diagonal entries, which stand for the similarity of solution candidates with themselves, are black indicating maximum similarity.

Unfortunately, this representation is not very well suited for a static monochrome figure. Therefore, the dynamics of this $n \times n$ color grid over

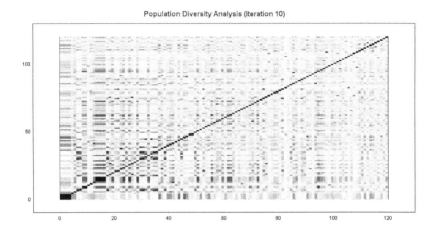

FIGURE 10.2: Degree of similarity/distance for all pairs of solutions in a
SGA's population of 120 solution candidates after 10 generations.

the generations is shown in numerous colored animations available at the
website of this book[2].

For a meaningful figure representation of genetic diversity over time it is
necessary to summarize the similarity/distance information of the entire population in a single value. An average value of all $\binom{n}{2}$ combinations of solution
pairs in form of a mean/max similarity value of the entire population as a
value between 0 and 1 can be calculated according to the Formulas 6.5 to 6.8
stated in Chapter 6. This form of representation allows to display genetic
diversity over the generations in a single curve. Small values around 0 indicate low average similarity, i.e., high genetic diversity and vice versa high
similarity values of almost 1 indicate little genetic diversity in the population.
In the following we show results of exemplary test runs of GAs applied to the
kroA200 200 city TSP instance taken from the TSPLib using the parameter
settings given in Table 10.1 and OX crossover.

Figures 10.3 and 10.4 show the genetic diversity curves over the generations
for a conventional standard genetic algorithm as well as for a typical offspring
selection GA. The gray scaled values in the Figures 10.3, 10.4, and 10.5 show
the progress of mean similarity values of each individual (compared to all
others in the population); average similarity values are represented by solid
black lines.

For the standard GA it is observable that the similarity among the solution candidates of a population increases very rapidly causing little genetic

---

[2]http://gagp2009.heuristiclab.com

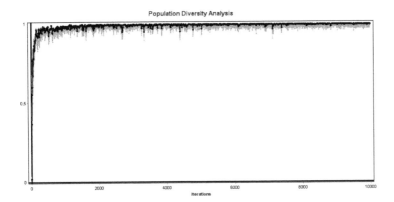

FIGURE 10.3: Genetic diversity in the population of a conventional GA over time.

diversity already after a couple of generations; it is only mutation which is responsible for reintroducing some new diversity keeping the evolutionary process going. As already explained in Chapter 7, without mutation the standard GA tends to prematurely converge very rapidly.

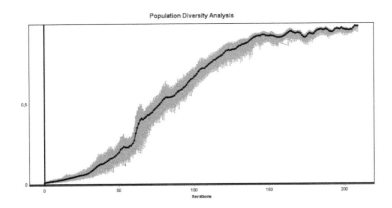

FIGURE 10.4: Genetic diversity of the population of a GA with offspring selection over time.

Equipped with offspring selection the results turn out to be completely different: The average similarity in the population increases slowly and steadily from 0 to 1. This means that the high degree of genetic diversity, which is available initially, is decreased very slowly and carefully yielding to a state of convergence when no more diversity is available in the population which is detected by the algorithm by reaching a maximum selection pressure value (see Chapter 4). As already discussed in Chapter 7 by analyzing the dynamics of the alleles of the global optimal solution, also the dynamic of genetic diversity shows that an offspring selection GA is rarely dependent on mutation and operates much closer to the general assumptions stated in the building block hypotheses and the according schema theorem as a comparable conventional GA is able to do for general problem formulations.

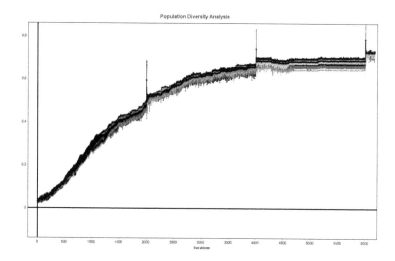

FIGURE 10.5: Genetic diversity of the entire population over time for a SASEGASA with 5 subpopulations.

The analysis of genetic diversity over time for the SASEGASA is shown in Figure 10.5. Similar to the diversity analyses for a conventional GA and for the offspring selection GA, the genetic diversity analysis has also been done for the SASEGASA applied to the *kroA200* 200-city benchmark TSP instance using the parameters given in Table 10.4 and the combination of crossover operators OX, ERX, and MPX. As we can see in Figure 10.5, the genetic diversity is still rather high at the first reunification phase around iteration 2000 where the genetic diversity in each single subpopulation is already lost. This means that even if there is no more genetic diversity in each of the subpopulations itself, there is still quite a lot of genetic diversity in the entire population. This means that the certain subpopulations must have drifted to quite different regions of the search space which is consistent with the theoretical considerations of Chapter 5. After each reunification step (the next one from 4 to 3 subpopulations is around iteration 4000) the average similarity (which is inversely proportional to the average genetic diversity) stabilizes at a higher level indicating lower average genetic diversity after each reunification. The iterations between the certain migration phases are responsible for getting essential alleles (which are part of the global optimum or at least of a high quality solution) fixed in order to let the SASGEASA operate beneficially; this is in fact the case in our concrete example, as we also see in the results stated in Table 10.8).

Summarizing these results it can be stated for the TSP experiments that the analysis of genetic diversity in the population confirmed the results of Chapter 7 without using any information about the concrete search space topology. The illustration in form of a static figure is certainly some kind of restriction when the dynamics of a system should be observed. For that reason the book's website contains some additional material showing the dynamics of pairwise similarities for all members of the population (as indicated in Figure 10.2) in the form of short motion pictures.

10.2   Capacitated Vehicle Routing

Similar to the practical study on the TSP problem we have also applied several algorithms on the capacitated vehicle routing problem, to several instances of the Taillard benchmark set [Tai93]. This set consists of 14 instances from 75 to 385 cities of which we picked the first two instances of those with 75 cities, one with 100 cities, and one with 150 cities. There is no proven globally optimal solution to these instances. Several authors, including Taillard himself, have published best known solutions; a new best known solution in one instance with 75 cities was discovered recently [AD06].

The instances were interp reted according to the definition of a CVRPTW as presented in Chapter 8.2. Since the CVRP does not include time windows, but only demands, artificial ready times, service times, and due dates have been added such that the size of the time window is 2^{16}. This is high enough so that the time windows do not constrain the solution. Additionally, there is no maximum number of vehicles given; thus the number of possible vehicles was predefined by the number of customers, so that in the worst case every customer was serviced by a separate vehicle. Since any additional vehicle will always remain unused, our constraint on the maximum number of vehicles did not also constrain the solution space.

The representation that is chosen is a path encoding with trip delimiters, similar to the approach given in [PB96]. The number 0 represents the depot; all numbers greater than 0 represent customers which are visited in the order

they appear in the array when reading it from left to right. There are as many
0s in any array as there are vehicles plus an additional 0 which terminates a
representation. Since all customers have to be visited and no customer can
be visited more than one time, the size of the encoding is of fixed length
throughout the optimization process. Unused vehicles are marked as empty
routes and are represented by two subsequent 0s. During crossover, each
string is sorted so that empty routes are listed after the last active vehicle has
completed its tour. There is, however, no specific order for active vehicles.

10.2.1 Results Achieved Using Standard Genetic Algorithms

The genetic algorithm which we applied uses some of the operators described in Chapter 8 and was applied in six different configurations shown
in Table 10.13. The algorithm has been applied five times per instance. By
experimentation we want to analyze the GA on the one hand by using different mutation operators and on the other hand by choosing different selection
operators.

The two following main test scenarios are set up, the first one with lower selection pressure using roulette wheel selection and the second one with higher
selection pressure using 3-tournament selection with a group size of three.
Both of these scenarios are tested with different settings for mutation operators, among them the previously described M1, M2, and LSM as optimizing mutation operators which aim to improve solutions with some knowledge
about the fitness function (the distance between the cities) as well as nonoptimizing mutation operators which do not know about the fitness function
and therefore make unbiased choices. We group the mutation operators within
a single genetic algorithm and give them the same probability by dividing the
mutation rate through the number of mutation operators in the particular
test. The only exception is the LSM which only has a 0.0001% chance of
being selected due to its computational complexity.

The fitness function is described in Chapter 8; it simply calculates the total
traveled Euclidean distance.

10.2.1.1 Quality Progress of the Genetic Algorithm

The GA is barely able to thrive its population towards a high quality region.
Optimization mainly depends on the presence of 1-elitism, which preserves the
best found solution from generation to generation. Given this behavior it is
not completely puzzling that local search methods like tabu search achieve
good performances on these problems, as the GA in this form is not able
to exploit much of the genetic information in the population; this happens
especially when selection pressure is lower, as we see for example in the results

Table 10.13: Parameter values used in the CVRP test runs applying a standard GA.

| **Parameters for the SGA** | |
(Results presented in Tab. 10.14)	
Generations	2000
Population Size	200
Elitism Rate	1
Mutation Rate	0.06
Selection Operator	Roulette, 3-Tournament
Crossover Operators	{SBX, RBX}
	{M1, M2}
Mutation Operators	{M1, M2, LSM}
	{Relocate, Exchange, 2-Opt}

shown for the SGA using roulette wheel selection. In Figure 10.6 we compare the results according to the used parent selection operators. Using higher selection pressure, the average and worst qualities are maintained at a slightly better level when picking the best performing test for each. Still with both selection strategies the average and worst qualities are not improving over the course of 2,000 generations. From the quality charts we are able to see that the diversity is very high, which will become obvious again when we take a look at the diversity progress.

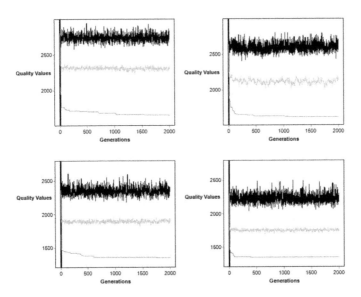

FIGURE 10.6: Quality progress of a standard GA using roulette wheel selection on the left and 3-tournament selection the right side, applied to instances of the Taillard CVRP benchmark: tai75a (top) and tai75b (bottom).

From our observations it seems that the RBX is better suited to further optimize the best solution. Analyzing one of the tests in a more detailed way, we see that RBX is responsible in approximately 75% of the evolutionary cycles of selection, crossover, and mutation in which a new best solution was found whereas SBX is responsible for only 25% of the cases. We will find similar behavior for the genetic algorithm using offspring selection where the SBX operator is working better when the population is diverse and of worse quality than when the population has converged and is of better quality. Both operators also benefit from a local search like behavior in the repair procedure. As already described in Chapter 8, many different approaches enhance the GA with local search to treat the VRP.

10.2.1.2 Diversity Progress of the Genetic Algorithm

In the diversity progress charts shown in Figure 10.7 we see how lower selection pressures leave individuals in the population which do not have many common edges, while there is higher mutual similarity when using 3-tournament selection for example. Similar individuals have several edges in common, and when these are crossed, the common edges will remain and other edges will be taken from either one of the two parents. Through selection by fitness the common edges amounting in the population are those of good quality. So, ideally the mutual similarity of the GA's individuals should increase slowly in order to go from exploration to exploitation. Thus, the algorithm should start with a diverse population and end up with a highly similar population with each solution being either the optimal solution or with a quality close to it.

The similarity measure for two VRP solutions t_1 and t_2 is calculated in analogy to the TSP similarity using edgewise comparisons. However, as big routes in the VRP are subdivided into smaller routes, a maximum similarity sim_{max} is calculated for each route $r \in t_1$ to all routes $s \in t_2$. These values are summed for all routes r_i and finally divided by the number of routes.

As we have seen already in the quality chart the GA is in this example not able to decrease the diversity over the course of the optimization. This could also result in good solutions as is shown when examining the final achieved solution qualities in Table 10.14. Overall, the GA shows a behavior closer to that of a trajectory-based approach than a population-based approach. In a trajectory-based approach, there is only a single solution which is slightly modified by mutation and accepted as new solution if some criteria are met. One characteristic of trajectory-based approaches is their ability to exploit the search space in local regions by finding local optima. As is the case with the GA here, the best individual of a generation is saved to the next generation and maintains a strong line of good quality genes.

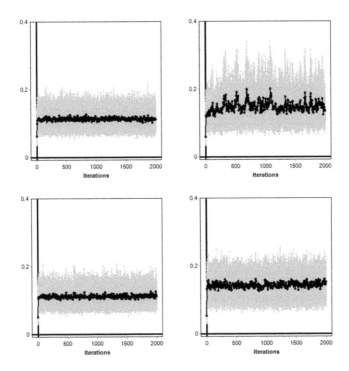

FIGURE 10.7: Genetic diversity in the population of a GA with roulette wheel selection (shown on the left side) and 3-tournament selection (shown on the right side).

10.2.1.3   Empirical Results

Results are listed in Table 10.14 showing the average of the best qualities found and the standard deviation from the mean in percent as well as the quality of the best found solution. These results show that the selection pressure applied by roulette wheel selection was not enough to guarantee good solution qualities; the average best quality is worse than 20-30% worse than the best known solution. Additionally, the quality values vary to a greater degree, which also suggests that the results are not close to a potential optimum.

When using higher selection pressure, for example by applying tournament selection with higher group sizes, the GA is able to achieve formidable average best qualities on the two benchmark instances used here. The results are around 1% worse than the best known solution in most of the cases, but still the GA was not able to find the best known solution. Interesting in this context is that the choice of mutation operators matters more when the GA performs worse, as is the case with roulette-wheel selection, than when it performs better. In our example using 3-tournament selection the choice of mutation operators is less important and good results can be achieved using both optimizing as well as nonoptimizing mutation operators.

Table 10.14: Results of a GA using roulette-wheel selection, 3-tournament selection and various mutation operators.

	Roulette Wheel Selection			Best Known
Problem	Relocate/ Exchange/2-Opt	M1/M2	M1/M2/LSM	
tai75a	1729.57±1.65%	**1665.86**±1.48%	1670.18±0.98%	1618.36
Best found	1713.00	1641.36	1654.62	
tai75b	1396.37±0.80%	**1361.75**±0.74%	1365.54±0.36%	1344.62
Best found	1387.64	1352.05	1360.36	
	3-Tournament Selection			
Problem	Relocate/ Exchange/2-Opt	M1/M2	M1/M2/LSM	
tai75a	1635.23±0.61%	1637.16±0.67%	**1634.29**±0.54%	1618.36
Best found	1622.66	1619.22	1623.57	
tai75b	**1353.51**±0.17%	1358.12±1.15%	1355.35±0.23%	1344.62
Best found	1350.85	1347.05	1352.02	

A statistical comparison on the results between the GA with roulette wheel selection and 3-tournament selection shows the advantageous performance of the GA with 3-tournament selection for these benchmark instances. A box plot of the results is shown in Figure 10.8. We have compared these results pairwise, on the one hand roulette-wheel selection and on the other hand 3-tournament selection each time with the same mutation operators using a two sided t-test. The hypothesis that the mean values of the results are equal is rejected at a significance level of 0.05 in four out of the six comparisons. As the means of the results achieved using 3-tournament selection are lower than those achieved using roulette-wheel selection, we conclude that a higher selection pressure is responsible for better performance.

10.2.2 Results Achieved Using Genetic Algorithms with Offspring Selection

A genetic algorithm with offspring selection is quite successful insofar as it can direct the whole population towards the global optimum or a solution with a quality close to it. In this test the GA with OS does not make use of parental selection operators, but randomly selects parents, crosses them, and mutates the children with a certain probability. Accepted offspring individuals must have a quality better than the best parent (the comparison factor is set

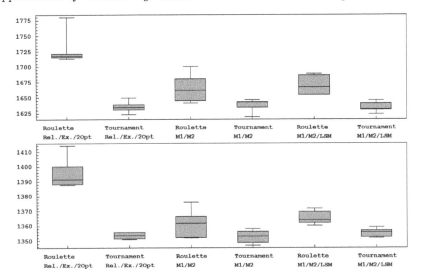

FIGURE 10.8: Box plots of the qualities produced by a GA with roulette and 3-tournament selection, applied to the problem instances tai75a (top) and tai75b (bottom).

to 1). Infeasible solutions are penalized using high punishment factors; thus, the algorithm, starting with a randomly created but feasible population, will remain in the feasible region at any time. The parameters for these tests are listed in Table 10.15.

Similar to the tests with the standard GA several scenarios have been selected and compared. The GA with offspring selection is applied to 75 customer CVRP instances using population sizes of 200 as well as 400, and also to higher instances with population sizes of 500 and 1000. For each test scenario the selection operators, crossover operators, and mutation rate are fixed, but the mutation operators vary between nonoptimizing operators for which we have chosen relocate, exchange, and 2-Opt similar to the GA tests as well as optimizing operators such as M1, M2, and LSM with the same considerations as above; one test was done without mutation.

We have used two termination criteria that will stop the execution as soon as one of them is satisfied. The first one is based on reaching the maximum selection pressure barrier and the other one limits the number of evaluated solutions to 400,000 which is the same number of evaluations the standard GA has been given on these instances. The GA with OS and a population size of 200 always terminates before the maximum number of evaluated solutions has been reached and lists around 250,000 to 300,000 evaluated solutions at the end. Using a population size of 400 it has always terminated because it reached the upper limit of 400,000 evaluated solutions prior to reaching maximum selection pressure. The GA with OS and a population size of 500 is given a maximum amount of 1,500,000 evaluations for the 100 customer problem and 2,000,000 for the 150 customer problem due to the bigger complexity of the instances it has to solve. A GA with offspring selection and a population size of 1000 was also applied to the *tai*385 problem instance with a maximum of 10,000,000 evaluations.

Table 10.15: Parameter values used in CVRP test runs applying a GA with OS.

Parameters for the GA with OS	
(Results presented in Table 10.16–Table 10.17)	
Population Size	200, 400, 500
Elitism Rate	1
Mutation Rate	0.06, 0.0
Selection Operator	Random
Crossover Operators	{SBX, RBX}
	{M1, M2}
Mutation Operators	{M1, M2, LSM}
	{Relocate, Exchange, 2-Opt}
Success Ratio	1
Comparison Factor Bounds	1–1
Maximum Selection Pressure	200

10.2.2.1   Improvement in Quality Progress with Offspring Selection

A benefit of using offspring selection is the automatic adaption of the necessary selection pressure. Instead of choosing between roulette wheel, linear rank or tournament selection, and an appropriate group size, it is feasible to simply use random parent selection. The selection and reproduction phases will be repeated as long as the necessary amount of individuals fulfilling the success criterion can been generated. Thus, the algorithm will use less selection pressure when the criterion is easily satisfied and will apply more selection pressure when the criterion is harder to be satisfied. Random selection allows the worst individual to be selected as often as the best individual.

The GA with OS is quite successful even without mutation; this is what we had expected given the analyses in Chapter 7. Figure 10.9 shows the quality progresses of the offspring selection GA. The number of generations is fairly low compared to a conventional genetic algorithm, but more work is done per generation. The curves show the typical behavior of a GA with OS where worst, average, and best quality values converge at the end of the evolutionary process. The result is a highly similar population of good quality; at this point genetic diversity, as we will see below, has mostly been lost. So, the algorithm cannot proceed further to create new better solutions and terminates. Using a higher population size such as 400 in our case, but with the same number of evaluated solutions the algorithm terminates before it reduces the genetic diversity to the point where no further better solution can be created. Nevertheless, the GA with OS finds better solutions as the higher population size can hold more genetic information as well as it uses about 100,000 evaluations more than with a population size of 200.

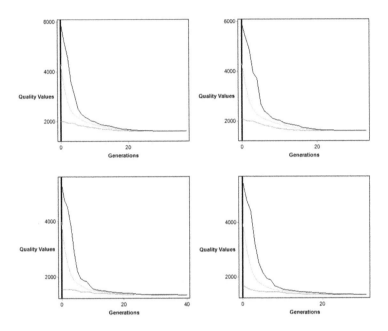

FIGURE 10.9: Quality progress of the offspring selection GA for the instances (from top to bottom) tai75a and tai75b. The left column shows the progress with a population size of 200, while in the right column the GA with offspring selection uses a population size of 400.

In Figure 10.10 the influence of the crossover operators in each generation is shown. It shows how many offspring in each generation are created by crossing them with SBX or RBX in percent. The higher the values, the more frequently one operator was able to create better offspring which exceeds the quality of the best parent in the GA with OS here. It can be seen that SBX initially is able to produce slightly more successful children as the RBX, but as the population converges and improves in quality RBX produces better offspring to a higher degree.

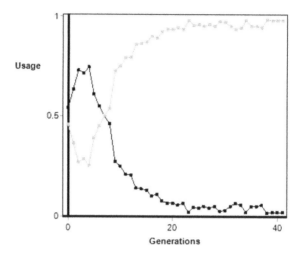

FIGURE 10.10: Influence of the crossover operators SBX and RBX on each generation of an offspring selection algorithm. The lighter line represents the RBX; the darker line represents the SBX.

10.2.2.2 Improved Diversity Progress with Offspring Selection

The diversity progress shown in Figure 10.11 is similar to what has been observed for the TSP. The GA with OS starts with the same diverse initial population that the GA starts with, but is able to slowly spread the good genetic information among the population so that in the end the similarity of the individuals rises and the algorithm progresses from exploration to exploitation. At the end of the search, genetic diversity is close to 1, so almost all the individuals in the population share the same edges. This behavior has already been analyzed in Chapter 7. The results are slightly different when using a higher population size. The algorithm finishes before it can reduce the genetic diversity in the population and thus the diversity progress looks cut off. Nevertheless, as we will see in the next section, the results are improved. Since the GA with OS and a population size of 400 has room for further optimization as there is still enough diversity, allowing more evaluations could result in even better results.

10.2.2.3 Empirical Results

Results show a very sound performance of the GA using offspring selection: It is able to get very close to the optimum and to find it in even much more cases than the GA without offspring selection. Increasing the population size to 400 individuals allowed offspring selection to find the best known solution much more often. The only exception is the *tai75b* instance where the best

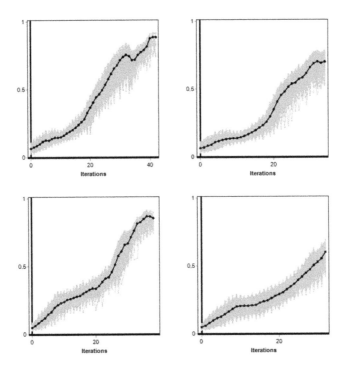

FIGURE 10.11: Genetic diversity in the population of an GA with offspring selection and a population size of 200 on the left and 400 on the right for the problem instances tai75a and tai75b (from top to bottom).

known solution is not found that easily; it seems that it also requires a bit of luck. Finding the "2nd best known solution", which had been the best-known solution for a while, is considerably easier: In 14 out of 20 runs the GA with OS and a population size of 400 was able to find it, but only in a single run out of 20 it could find the currently best known solution. It may be possible that the best known solution does not lie within an attracting region for the GA, which is probably also the reason for its late discovery in [AD06]. Regarding solution quality, the currently best known solution quality is 1344.618 while the "2nd best known solution" has a quality of 1344.637.

Analyzing the results reported in Table 10.15 we see that the choice of the mutation operator in our genetic algorithm using offspring selection is again of less importance. The best results are computed using nonoptimizing mutation operators as well as a combination of M1 and M2 with local search. Omitting mutation leads to good results in general with average best solution qualities close to the best known solution qualities.

Table 10.16: Results of a GA with offspring selection and population sizes of 200 and 400 and various mutation operators. The configuration is listed in Table 10.15.

	Population Size 200				Best Known
Problem	Relocate/ Exchange/2-Opt	M1/M2	M1/M2/LSM	No Mutation	
tai75a Best found	**1620.26**±0.16% 1618.36	1622.03±0.15% 1618.36	1622.48±0.06% 1621.96	1622.72±0.07% 1621.95	1618.36
tai75b Best found	1346.26±0.24% 1344.64	1345.78±0.06% 1344.64	1345.85±0.12% 1344.64	**1345.71**±0.16% 1344.64	1344.62

	Population Size 400				
Problem	Relocate/ Exchange/2-Opt	M1/M2	M1/M2/LSM	No Mutation	
tai75a Best found	1620.68±0.11% 1618.71	1620.52±0.18% 1618.36	**1618.73**±0.03% 1618.36	1621.02±0.13% 1618.36	1618.36
tai75b Best found	1344.67±0.00% 1344.64	1344.64±0.00% 1344.64	**1344.63**±0.00% 1344.62	1344.83±0.03% 1344.64	1344.62

From the results we can also see that the GA with OS benefits from a higher population size insofar as it is able to get closer to the best known solution on average and finding it more often. Given the small number of replications, however, no statistical significance can be drawn; still, as the box plots in Figure 10.12 show, using a higher population size results in more robust tests with smaller standard deviations of the results' qualities as well as quality values closer to that of the best known solutions. This is not surprising as it has been discussed that a larger initial population is more likely to hold all the relevant alleles which are to be identified and assembled in a single solution during the optimization process. A larger population can hold more diverse solutions which prevents important alleles from disappearing. Naturally, it takes more effort for a larger population to converge and thus the number of evaluated solutions increases with the population size. Population size in an offspring selection genetic algorithm is a tradeoff between achievable quality and effort; in a traditional GA, increasing the population size has a similar effect only when the parent selection pressure is increased accordingly. This may for example be achieved by using tournament selection with an increased tournament group size.

The results returned by the standard GA and the GA with OS are compared in Figure 10.13 which shows the box plots of the results' qualities of these two GA variants. Here we see that the results of the GA using offspring selection are generally more robust insofar as they are of good quality and do not spread as much as the results returned by the standard GA using 3-tournament selection. Again, a pairwise two sided t-test of the results computed with the standard GA compared to the offspring selection GA rejected the hypothesis that the means of these results are equal at a significance level of 0.05. As the means of the offspring selection GA are lower than standard GA, it is thus feasible to assume that the offspring selection GA performs indeed better than the standard GA.

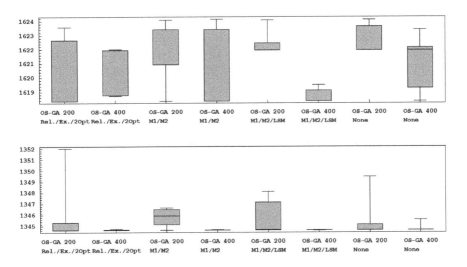

FIGURE 10.12: Box plots of the offspring selection GA with a population size of 200 and 400 for the instances tai75a and tai75b.

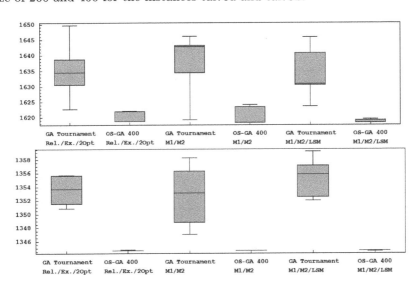

FIGURE 10.13: Box plots of the GA with 3-tournament selection against the offspring selection GA for the instances tai75a (shown in the upper part) and tai75b (shown in the lower part).

We have also applied a GA with offspring selection for solving more complex problem instances, specifically one with 100 and one with 150 customers as well as on the most complex instance with 385 customers. The algorithm is suited well to get very close to the best known solution in this configuration, though it is likely that population size still needs to be increased as the best known solution could not be reached in any of the test runs. The best solution found for the *tai*100*a* instance has a quality of 2062.25 and is about 1% worse than the currently best known solution; the offspring selection GA achieved average best qualities 1-2% worse than the currently best known solution. In all cases except two it could finish before reaching the maximum amount of evaluated solutions, having evaluated on average 1.2 to 1.3 million solutions. For the *tai*150*a* instance the algorithm finished on average having evaluated 1.9 million solutions; some runs, however, ran into the maximum of 2 million solution evaluations. The best solution in 5 test runs has a quality of 3068.04 and is 0.42% worse than the best known one; it also achieves average best qualities approximately 1% worse than the best known solution. The results are given in Table 10.17.

For the *tai*385 problem instance, which is the largest instance in Taillard's benchmark set, a good result could be achieved as well. Here the customers are modeled according to the locations of the most important towns or villages in the smallest political entities in the Swiss canton of Vaud. The demand is modeled proportional to the number of inhabitants living there [Tai93]. Using an offspring selection GA with a population size of 1000 without mutation, the final tour length found is 25,498.40 after 10 million evaluated solutions. This is 4.37% higher than the currently best known solution with a tour length of 24,431.44. It is likely that better results can be achieved with even higher population sizes, a point where parallelization becomes more and more important in order to achieve results in adequate time.

Table 10.17: Showing results of a GA with offspring and a population size of 500 and various mutation operators. The configuration is listed in Table 10.15.

Problem	Relocate/ Exchange/2-Opt	Population Size 500 M1/M2	M1/M2/LSM	No Mutation	Best Known
tai100a Best found	2081.30±0.22% 2074.56	**2077.89**±0.48% 2062.25	2079.99±0.21% 2073.22	2078.60±0.26% 2073.55	2041.34
tai150a Best found	3082.48±0.37% 3068.04	**3078.79**±0.23% 3071.54	3087.86±0.39% 3074.44	3086.44±0.48% 3068.54	3055.23

Chapter 11

Data-Based Modeling with Genetic Programming

DOI: 10.1201/9781420011326-12

11.1 Time Series Analysis

Whenever (input or output) data of any kind of system are recorded over time and compiled in data collections as sequences of data points, then these sequences are called *time series*; typically, these data points are recorded at time intervals which are often, but not always uniform.

The collection of methods and approaches which are used for trying to understand the underlying mechanisms that are documented in time series is called *time series analysis*; but not only do we want to know what produced the data, but what we are also interested in is to predict future values, i.e., we want to develop models that can be used as predictors for the system at hand.

There is a lot of literature on theory and different approaches to time series analysis. One of the most famous approaches is the so-called Box-Jenkins approach as described in [BJ76] and [And76], e.g., which includes separate model identification, parameter estimation, and model checking steps. Detailed discussions of other methods and their mathematical and statistic background can be found for example in [And71], [Ken73], [Pan83], [KO90], [Pan91], [BD91], [Ham94], and [BD96]; more recent research and applications are for example given in [PTT01], [Cha01], [Dei04], [Wei06], and [MJK07].

The main principle can be formulated in the following way: For a given target time series T storing the values $T_{(1)}, \ldots, T_{(n)}$ and a given set of variables $X_1, \ldots, X_N$ we search for a model f that describes T as

$$T_{(t)} = f(X_{1(t)}, X_{1(t-1)}, \ldots, X_{1(t-t_{max})},$$

$$\ldots,$$

$$X_{N(t)}, X_{N(t-1)}, \ldots, X_{N(t-t_{max})}) + \epsilon_t$$

where t_{max} is the maximum number of past values, and ϵ_t is an error term. If the target variable's values are also allowed to be considered, then a so-called autoregressive part is added so that we search for a model f so that

$$T_{(t)} = f(X_{1(t)}, X_{1(t-1)}, \ldots, X_{1(t-t_{max})},$$

$$\ldots,$$

$$X_{N(t)}, X_{N(t-1)}, \ldots, X_{N(t-t_{max})},$$

$$T_{(t-1)}, \ldots, T_{(t-t_{max})}) + \epsilon_t$$

Of course, the field of applications of time series analysis is huge and includes for example astronomy, sociology, economics, or, which is what we are going to do in the course of the application examples given in this section, the analysis of physical systems. Of course it is not at all natural that any physical system, may it be technical or not, can be represented by a simple and easily understandable model. In this context the authors strongly recommend reading Eugene P. Wigner's article "The Unreasonable Effectiveness of Mathematics in the Natural Sciences" [Wig60]. In this article Wigner points out that, although so many natural phenomena such as, e.g., gravitation or planetary motion can be described by astoundingly simple equations, it is not at all natural that "laws of nature" exist and even much less that man is able to discover them.

Especially in the context of analyzing physical systems, the models which are to be created for describing a system can be seen as so-called *virtual sensors*: The goal is to develop models of sufficient quality so that these models (functions) can be used instead of real sensors, i.e., they are virtual sensors. Of course, these virtual sensors can be used in various ways, for example also in addition to real sensors enabling fault detection.

In this section we will concentrate on time series analysis with genetic programming: GP is used for evolving models that describe target time series using other data time series collections. Of course we in principle use the GP methods for structure identification described in the previous sections, but some time series specific details are to be described here, especially a time series specific evaluation operator described in Section 11.1.1. Test results are given and discussed in Section 11.1.2.

11.1.1 Time Series Specific Evaluation

In principle there is no reason why one should not use means squared errors or any other of the evaluation functions presented in Section 9.2.3.3 for evaluating time series models produced by GP. Still, in time series we do not only want to produce models that approximate the given target values, but also the dynamics of the underlying system that are represented in the measured data. Thus, we also want to estimate a model's quality with respect to the local changes in the data as well as the accumulated values.

This can be done by calculating the differential and integral values. For a given time series $\mathbf{x}$, the differential of order o is defined as $diff(\mathbf{x}, o)$ and the integral as $int(\mathbf{x})$:

$$diff(\mathbf{x}, o)_i = \mathbf{x}_i - \mathbf{x}_{i-o} \tag{11.1}$$

$$int(\mathbf{x})_i = \sum_{i=1}^{i} \mathbf{x}_i \tag{11.2}$$

for each index $i \in [1; |\mathbf{x}|]$.

For evaluating a time series model m on the basis of target values $\mathbf{o}$ we calculate all respective values $\mathbf{e}$ by evaluating m and then calculate the combined fitness values (as described in Section 9.2.5.2) for the plain values, the differential (of a predefined order o), and the integral values. These partial results are weighted using the coefficients c_1, c_2, and c_3, and the final result is calculated in the following way:

$$TS(\mathbf{o}, \mathbf{e}, o, c_{plain}, c_{diff}, c_{int}, c_1, c_2, c_3) :$$

$$a_1 = COMB(\mathbf{o}, \mathbf{e}, n, c_{plain}) \tag{11.3}$$

$$a_2 = COMB(diff(\mathbf{o}, o), diff(\mathbf{e}, o), n, c_{diff}) \tag{11.4}$$

$$a_3 = COMB(int(\mathbf{o}), int(\mathbf{e}), n, c_{int}) \tag{11.5}$$

$$TS(\mathbf{o}, \mathbf{e}, o, n, c_{plain}, c_{diff}, c_{int}, c_1, c_2, c_3) = \frac{\sum_{i=1}^{3} a_i \cdot c_i}{\sum_{i=1}^{3} c_i} \tag{11.6}$$

with c_{plain}, c_{diff}, and c_{int} being the coefficients needed by the combined evaluation function for weighting the partial MEE, VAF, and R^2 results as well as the maximum negative and positive errors.

Of course, early stopping of model evaluations as described in Section 9.2.5.5 is also possible for this time series evaluation function.

11.1.2 Application Example: Design of Virtual Sensors for Emissions of Diesel Engines

The first research work of members of the Heuristic and Evolutionary Algorithms Laboratory (HEAL) in the area of system identification using GP was done in cooperation with the Institute for Design and Control of Mechatronical Systems (DesCon) at JKU Linz, Austria. The framework and the main infrastructure was given by DesCon who maintain a dynamical motor test bench (manufactured by AVL, Graz, Austria) shown in Figure 11.1. A BMW diesel motor is installed on this test bench, and a lot of parameters of the ECU (engine control unit) as well as engine parameters and emissions are measured; for example, air mass flows, temperatures, and boost pressure values are measured, nitric oxides (NO_x, to be described later) are measured using a Horiba Mexa 7000 combustion analyzer, and an opacimeter is used for estimating the opacity of the engine's emissions (in order to measure the emission of particulate matters, i.e., soot).

FIGURE 11.1: Dynamic diesel engine test bench at the Institute for Design and Control of Mechatronical Systems, JKU Linz.

During several years of research on the identification of NO_x and soot emissions, members of DesCon have tried several modeling approaches, some of them being purely data-based as for example those using artificial neural networks (ANNs). Due to rather unsatisfactory results obtained using ANNs, the ability of GP to produce reasonable models was investigated in pilot studies; we are here once again thankful to Prof. del Re for initiating these studies.

In this context, our goal is to use system identification approaches in order to create models that are designed to replace or support physical sensors; we want to have models that can be potentially used instead of these physical sensors (which can be damageable or simply expensive). This is why we are here dealing with the design of so-called *virtual sensors*.

11.1.2.1 Designing Virtual Sensors for Nitric Oxides (NO_x)

In general, being able to predict NO_x emissions on-line (i.e., during engine operation) would be very helpful for low emissions engine control. While NO_x formation is widely understood (see for example [dRLF+05] and the references given therein), the computation of NO_x turns out to be too complex and - at the moment - not easy to be used for control. The reason for this is that in theory it would be possible to calculate the engine's NO_x emissions if all relevant parameters (pressures, temperatures, ...) of the combustion chambers were known, but (at least at the moment) we are not able to measure all these values.

As already mentioned above, ANNs have been used for data-based modeling of NO_x emissions of a BMW diesel engine. These results were not very satisfying, as is for example documented in [dRLF$^+$05]: Even though modeling quality on training data was very good, the model's ability to predict correct values for operating points not included in the training data was very poor.

We therefore designed and implemented a first GP approach based on the HeuristicLab 1.0; preliminary results were published in [WAW04a] and [WAW04b]. In [WAW04b] we documented the ability of GP using offspring selection to produce reasonable models for NO_x, including lots of statistics showing that the results obtained applying rigid offspring selection were significantly better than those obtained without using OS or even OS with less strict parameter settings, i.e., lower success ratio and comparison factor parameters.

NO_x values were recorded by DesCon members following the standard procedure defined by the Federal Test Procedure (FTP); a whole standardized test run is therefore called a FTP cycle. FTP tests were executed on the DesCon test bench in two different ways as it is possible to activate or to deactivate exhaust gas recirculation (EGR). In principle, recirculating a portion of an engine's exhaust gas back to the engine cylinders is called EGR; the incoming air is intermixing with recirculated exhaust gas, which lowers the adiabatic flame temperature and reduces the amount of excess oxygen (at least in diesel engines). Furthermore, the peak combustion temperature is decreased; since the formation of NO_x progresses much faster at high temperatures, EGR can also be used for decreasing the generation of NO_x. Further information about EGR and its effects on the formation of NO_x can for example be found in [Hey88] and [vBS04].

We shall therefore here take a closer look at the following two modeling tasks:

- Situation (1): Use data recorded with deactivated EGR;

- Situation (2): Use data recorded with activated EGR.

In both cases the data were recorded at 20 Hz; the execution of the cycles took approximately 23 minutes. In total, 33 variables are recorded; here we do not give a total overview of the statistic parameters of these variables but rather restrict ourselves to the linear correlation of the input variables to the target variable: All linear correlations[1] of the potential input variables and the target variable NO_x are summarized in Table 11.1; all variables were filtered using a median filter of order five[2] before calculating the correlation coefficients.

---

[1] We here use the same standard formula for calculating linear correlation coefficients of time series as described in Section 9.4.1.

[2] Applying a median filter means that a moving window is shifted over the data and all samples are replaced by the median value of their respective data environment. For calculating the filtered value y_i using median filtering of order 5 we collect the original values

Table 11.1: Linear correlation of input variables and the target values (NO_x) in the NO_x data set I.

Variable	Correlation Coefficient		Variable	Correlation Coefficient	
	Situation (1)	Situation (2)		Situation (1)	Situation (2)
time	-0.141	-0.129	*alpha*	0.437	0.462
CO_2	0.477	0.941	*COH*	0.099	0.259
COL	0.222	0.390	*KW_VAL*	0.763	0.853
M_T01F	0.414	0.515	*ME_MES1*	0.408	0.416
ME_MES2	0.460	0.488	*ME_MES3*	0.043	0.054
ME_MES4	0.000	0.000	*ME_MES5*	-0.024	-0.007
ME_MES6	0.000	0.000	*ME_MES7*	-0.092	0.015
ME_MES8	0.492	0.451	*ME_MES9*	0.660	0.592
ME_MES10	0.135	-0.133	*ME_MES11*	0.449	0.532
ME_MES12	0.253	0.321	*ME_MES13*	-0.052	0.376
ME_MES14	0.091	0.101	*ME_MES15*	0.364	0.314
ME_MES16	0.392	0.478	*ME_MES17*	-0.438	-0.470
N_MOTOR	0.404	0.413	*OPA_OPAC*	0.248	0.419
T_EXH	0.347	0.474	*T_LLNK*	0.133	0.004
T_LLVK	0.531	0.315	*T_OIL*	0.096	-0.181
THC_VK	-0.074	0.205	*TWA*	0.149	0.064

Obviously, activating EGR significantly increases the correlation of NO_x and all exhaust variables such as CO_2 or THC, for example.

So, in addition to this, the next question is whether to incorporate gas emissions as for example CO_2 in the modeling process; of course, estimating NO_x is a lot easier if CO_2 is known since there is a high correlation (especially when EGR is activated), but NO_x models that do not need CO_2 information are more useful as they can be applied without having to measure other emission values. Furthermore, we also excluded the variables *alpha*, COH, COL, THC, M_T01F, $ME_MES01 - 07$, ME_MES10, ME_MES14, and ME_MES17 from the set of valid input variables for building models that do not incorporate exhaust information.

We applied GP using populations of 700 individuals for modeling the measured NO_x data; 1-elitism was applied, the mutation rate was set to 0.07, and rigid offspring selection was applied (maximum selection pressure: 300). The first 3,000 samples (representing 2.5 minutes) of the data sets were neglected; in strategy (1) the samples 3,001 – 10,000 were used as training data, in strategy (2) the samples 3,001 – 13,000. The rest of the data was used as validation / test samples.

Amongst other tests, we attacked modeling situation (1) without using exhaust information (hereafter called test strategy (1)), and modeling situation (2) using exhaust information (test strategy (2)); both test strategies were executed 5 times independently leading to the mean squared errors on training data summarized in Table 11.2.

---

x_{i-2}, x_{i-1}, x_i, x_{i+1}, and x_{i+2}; after sorting these values we get $x'_{i,j}$ for $j \in [1,5]$ with $x'_{i,j} < x'_{i,j+1}$ for $j \in [1,4]$. y_i is then set to the median value of x'_i, i.e., $y_i = x'_{i,3}$.

Table 11.2: Mean squared errors on training data for the NO_x data set I.

	Test Strategy (1)	Test Strategy (2)
Average	49.867	13.454
Minimum	43.408	11.259
Maximum	58.126	18.432

Let us now have a closer look at the best models (with respect to training data) produced for these test scenarios; their evaluations are both displayed in Figures 11.2 and 11.3, respectively.

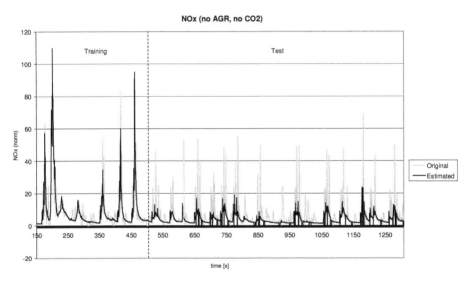

NOx_vK(t) = ((([0,950183*T_OEL(t-21)]*(([1,145998*T_LLVK(t-38)]-[1,023461*ME_MES17(t-28)])/[1,065444*ME_MES10(t-18)]+[0,951174*N_MOTOR(t-20)]))-((([0,825300*
ME_MES9(t-27)]+[0,871124*ME_MES9(t-22)])/[0,933514*ME_MES9(t-13)])/[1,060086*N_MOTOR(t-21)]/[0,991764*ME_MES9(t-26)])*((([0,691293*
ME_MES9(t-15)]/[0,705167*N_MOTOR(t-26)])/[0,947497*N_MOTOR(t-25)]/[1,136678*ME_MES11(t-8)]))/(([1,082096*N_MOTOR(t-1)]+[1,266062*T_LLVK(t-37)])/
-9,654))))+((([1,205990*T_OEL(t-30)]+[0,937356*T_OEL(t-10)])*([1,066581*T_LLVK(t-40)]/[0,928334*T_OEL(t-6)]+[1,184806*ME_MES7(t-12)]))-[0,888947*
ME_MES9(t-14)]*((([1,181614*ME_MES9(t-28)]/[0,666182*N_MOTOR(t-25)])/[0,631016*N_MOTOR(t-24)]/[1,301366*ME_MES11(t)]))/([0,867211*
N_MOTOR(t-40)]+[1,049105*T_LLVK(t-34)])/-7,628)))))

FIGURE 11.2: Evaluation of the best model produced by GP for test strategy (1).

The best model for test strategy (1) has a worse fit on test data ($mse_{test}(best_1) = 60.636$ in contrast to $mse_{training}(best_1) = 43.408$); the best model for test strategy (2) surprisingly even has a better fit on test data ($mse_{test}(best_2) = 5.809$) than on training data ($mse_{train}(best_2) = 11.259$).

We also tested standard GP without offspring selection, but with proportional as well as tournament ($k = 3$) parent selection, 1000 individuals, 2000 iterations, 7% mutation rate and the same data base as the one described previously.

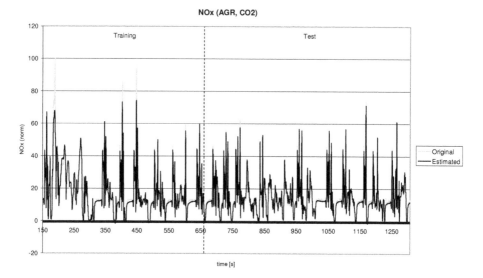

FIGURE 11.3: Evaluation of the best model produced by GP for test strategy (2).

Especially the use of proportional selection did not yield reasonable results, the evaluation of the best model for test strategy (1) returned mean squared error 110.23 on training data, and for the best for test strategy (2) the mean squared error was 21.34. The results obtained using tournament selection, which is suggested in GP literature (as for example in [KKS+03a] or [LP02]), were a lot better, but still not as good as those produced by extended GP: The best model for test strategy (1) showed mean squared error 61.92 on training data, and the best for test strategy (2) showed mean squared error 14.33. These results were no surprise, especially as we had seen on synthetic data sets that GP using rigid OS and gender specific parent selection performs a lot better than standard GP ([WAW04b], [Win04]).

Comparing these results to those achieved using other methods, we saw that they were indeed promising, but still not completely satisfactory. In fact, we then started a series of data-based tests using GP in the context of the analysis of mechatronical systems; this encouraged us to enforce research on the use of extended GP concepts in the identification of mechatronical systems.

11.1.2.2 Designing Virtual Sensors for Particulate Emissions (Soot)

A lot of research work was done by DesCon members on the identification of particulate emissions of a BMW diesel engine. The main results have been published in [AdRWL05] and [LAWR05]; we shall here only summarize these results in a rather compact way.

In short, first attempts to use GP for producing models for soot were not very successful; GP did not produce any useful solution without restriction of the search space. Therefore, a two step approach was used:

"In a first step, a statistical analysis was done on the basis of steady state measurements. Expert knowledge was combined with statistical correlations to yield an accurate steady state model. The advantage of steady state analysis is the secure validation of the model; any delay time or sensor dynamics are irrelevant. However, such a model could never meet the requirements of estimating the highly dynamical process of an IC engine. Therefore the steady state model is used as origin for the genetic programming cycle." (Taken from [AdRWL05] where this static model is given in detail.)

Using this static (steady state) model, an additional variable was calculated and inserted into the set of potential input variables; this so enhanced variables set was then used as basis for data-based identification of soot.

This extended data basis was used by two modeling approaches, namely a neural network training algorithm as well as GP; the best results for the ANN approach were achieved using a network structure with 2 hidden layers and 25 hidden nodes per layer. The parameters of the GP-based training algorithm were set to our standard GP settings (1000 individuals, 10% mutation rate, rigid OS, 1-elitism). Again, the data were measured during a standard FTP engine test lasting approximately 23 minutes; the first approximately 8 minutes were taken as training, the rest as validation / test data set.

Figure 11.4 shows a detail of the evaluation of the models produced by GP and ANN on validation data: As we see clearly, both virtual sensors do not capture the behavior completely correctly, but the GP model's fit seems to be better than the one of the ANN model. This suspicion becomes clearer by analyzing the distribution of errors which is shown in Figure 11.5: The errors caused by the evaluation of the model produced by GP are more symmetric than those of the ANN[3] which can be considered an indication for a rather good model. The cumulative errors of these models are shown in 11.6, and we here see that the model produced by GP is able to reproduce the engine's cumulated soot emissions quite well.

Again, these results were by far not completely satisfactory; of course, the ANN model could be improved by changing the network structure or the number of training iterations, and the GP process was not enhanced with local optimization or pruning operations. Still, again, these results sustained our confidence in GP's ability to produce reasonable models for mechatronical systems.

---

[3]In addition to GP and ANN, an auto-regressive moving-average with exogenous inputs (ARMAX) modeling approach was also calculated for reasons of comparison; the distribution of the errors caused by the evaluation of this model are also shown in Figure 11.5. Please see [BJ76] for explanations and application examples of ARMA(X) models.

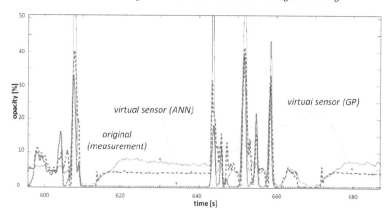

FIGURE 11.4: Evaluation of models for particulate matter emissions of a diesel engine (snapshot showing the evaluation of the model on validation / test samples), as given in [AdRWL05].

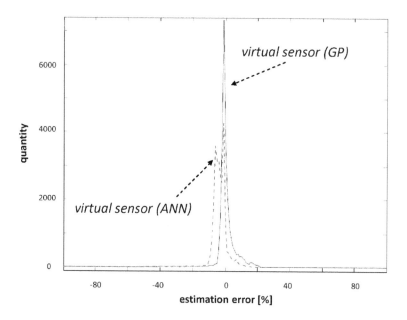

FIGURE 11.5: Errors distribution of models for particulate matter emissions, as given in [AdRWL05].

11.1.2.3 NO_x Data Sets Used for Further Empirical Studies

The NO_x data set described previously in this section was used for several research activities of DesCon members as well as in our project investigating GP for the design of virtual sensors. Nevertheless, in the course of further

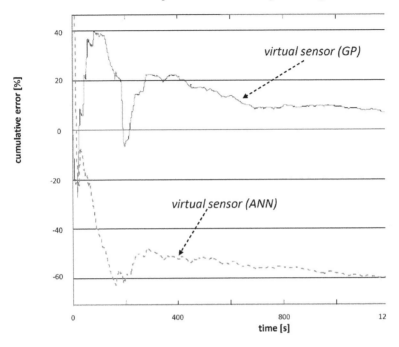

FIGURE 11.6: Cumulative errors of models for particulate matter emissions, as given in [AdRWL05].

research work several other measurements were recorded and analyzed; two of them were also used for test series that will be reported on in the following chapters. This is why we describe and characterize these data sets here.

NO_x Data Set II

Recorded in 2006 by members of the Institute for Design and Control of Mechatronical Systems at JKU Linz at the test bench already mentioned, this NO_x data set includes the variables listed in Table 11.3. The data set available in this context again contains measurements taken from a 2 liter 4 cylinder BMW diesel engine. Again, several emissions (including NO_x, CO, and CO_2) as well as several other engine parameters were recorded at 100 Hz and downsampled to 20 Hz. 22 signals were recorded over approximately 18 minutes, but only 9 variables were considered in further identification test series.

Several variables were measured over approximately 30 minutes at 100 Hz recording frequency; they have been downsampled to 20 Hz, so that the resulting data set includes ~36,000 samples. From the variables recorded several have been removed (as for example CO, CO_2, and *time*) due to irrelevance or high correlations with the target variable Nox_true; the 10 remaining variables are characterized in Table 11.3. Figure 11.7 shows a graphical representation of the target values over the whole recording time.

The variable NOx_Can represents values given by a quick, but also rather imprecise estimation for the NO_x emissions; the actual NO_x emissions were again measured using a Horiba Mexa 7000 combustion analyzer; the respective values are stored in variable Nox_true.

Table 11.3: Statistic features of the identification relevant variables in the NO_x data set II.

Variable	Minimum	Maximum	Mean	Variance
(0) Eng_nAvg	0.00	3,311.00	1,618.80	413,531.96
(1) $AFSCD\_mAirPerCyl$	-44.56	1,161.36	453.12	60,952.03
(2) $VSACD\_rOut$	5.00	96.00	33.59	1,706.83
(3) NOx_CAN	-0.30	6.72	1.52	2.87
(4) T_OEL	78.68	100.83	87.57	31.05
(5) (T) Nox_true	62,46	1,115.23	225.25	60,673.98
(6) $InjCrv\_qPil1Des$	0.00	1.40	0.88	0.10
(7) $InjCrv\_qMI1Des$	0.00	57.93	12.63	122.73
(8) $InjCrv\_phiMI1Des$	-3.86	10.61	2.80	18.70
(9) $BPSCD\_pFltVal$	986.20	2,318.00	1214.89	104,434.00

All pairwise linear correlations[4] are summarized in Table 11.4; again, all variables were filtered using a median filter of order 5 before calculating the correlation coefficients. Obviously, there is a rather high linear correlation between the target variable and the input variables $BPSCD\_pFltVal$ and NOx_CAN; the values stored in $AFSCD\_mAirPerCyl$ and $InjCrv\_qMI1Des$ are also remarkably correlated to the designated target values.

NO_x Data Set III

During the time in which we were doing the research work discussed here, maintenance work was repeatedly done at the DesCon test bench; amongst other aspects, several sensors were removed or replaced by newer ones.

---

[4]We here use the same standard formula for calculating linear correlation coefficients of time series as described in Section 9.4.1.

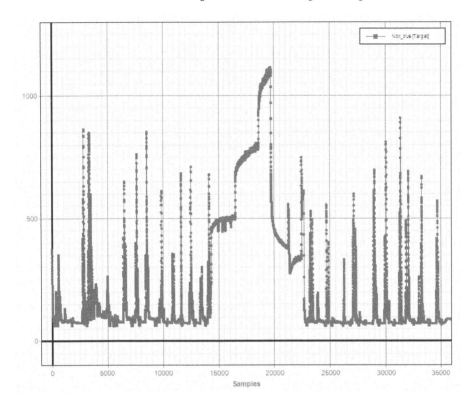

FIGURE 11.7: Target NO_x values of NO_x data set II, recorded over approximately 30 minutes at 20Hz recording frequency yielding $\sim$36,000 samples.

The third NO_x data set was recorded in 2007 by members of DesCon; again, several variables were measured at the test bench while testing a 2 liter 4 cylinder BMW diesel engine (simulated vehicle: BMW 320d Sedan). The mean engine speed was set to 2,200 revolutions per minute (rpm), and in each engine cycle 15mg fuel were injected.

Once again, several emissions (including NO_x, CO, and CO_2) as well as several other engine parameters were recorded; this time the measurements were recorded over approximately 18.3 minutes at 100 Hz and then downsampled to 10 Hz, yielding a data set containing $\sim$11,000 samples. The target values (the engine's NO_x emissions measured by a Horiba combustion analyzer) are stored in variable $HoribaNOx$.

In [Win08], tests have been documented in which we have used this data set for testing the ability of GP to incorporate physical knowledge. For this purpose we have also used a synthetic variable HFM^*:

$$HFM^* = \frac{HFM}{N} \cdot \frac{1000}{60} \qquad (11.7)$$

Table 11.4: Linear correlation coefficients of the variables relevant in the NO_x data set II.

	(0)	(1)	(2)	(3)	(4)	(5)	(6)	(7)	(8)	(9)
(0) Eng_nAvg	1.00	0.80	0.52	0.70	0.61	0.65	0.59	0.65	0.68	0.75
(1) $AFSCD\_mAirPerCyl$	0.80	1.00	0.78	0.90	0.80	0.91	0.60	0.88	0.63	0.95
(2) $VSACD\_rOut$	0.53	0.78	1.00	0.73	0.77	0.78	0.38	0.77	0.63	0.81
(3) NOx_CAN	0.70	0.90	0.73	1.00	0.74	0.93	0.63	0.86	0.58	0.94
(4) T_OEL	0.61	0.80	0.77	0.74	1.00	0.78	0.51	0.75	0.49	0.81
(5) (T) NOx_true	0.65	0.91	0.78	0.93	0.78	1.00	0.61	0.90	0.60	0.95
(6) $InjCrv\_qPil1Des$	0.59	0.60	0.38	0.63	0.51	0.61	1.00	0.70	0.03	0.62
(7) $InjCrv\_qMI1Des$	0.65	0.88	0.77	0.86	0.75	0.90	0.70	1.00	0.50	0.87
(8) $InjCrv\_phiMI1Des$	0.68	0.63	0.63	0.58	0.49	0.60	0.03	0.50	1.00	0.66
(9) $BPSCD\_pFltVal$	0.75	0.95	0.81	0.94	0.81	0.95	0.61	0.87	0.66	1.00

This synthetic variable is also included in NO_x data set III; detailed explanations regarding the meaning of this additional variable can be found in [Win08].

Figure 11.8 visualizes all target $HoribaNOx$ values available (in total approximately 11,000 samples); Figure 11.9 shows a detail of these data, namely the $HoribaNOx$ of samples 6000 – 7000.

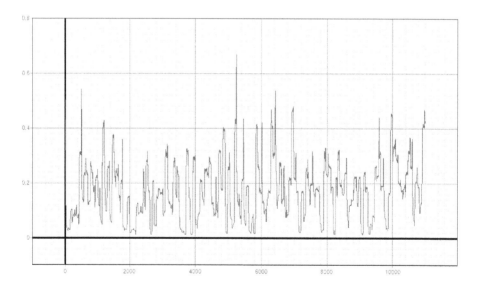

FIGURE 11.8: Target $HoribaNOx$ values of NO_x data set III.

In detail, Table 11.5 summarizes the main statistic parameters of the variables relevant in this identification task. Again, all pairwise linear correlations have also been calculated, with the results summarized in Table 11.6; all variables were again filtered using a median filter of order 5 before calculating the correlation coefficients. As we see in this table, there are no remarkably

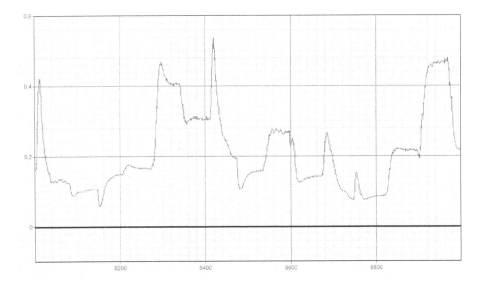

FIGURE 11.9: Target $HoribaNOx$ values of NO_x data set III, samples 6000 – 7000.

high correlations except for the obvious one between HFM and HFM^*; the correlation coefficient of HFM^* and the target, $HoribaNOx$, is above average (0.72), but not high enough to build a reasonable model only using this variable as input.

In Sections 11.3, 11.4, and 11.6 we will present research results achieved using these NO_x data sets II and III.

Table 11.5: Statistic features of the variables in the NOx data set III.

Variable	Minimum	Maximum	Mean	Variance
(0) (T) $HoribaNOx$	0.011	0.670	0.171	0.011
(1) qMI	8.010	21.960	15.232	16.992
(2) pMI	-0.727	8.016	3.424	6.525
(3) qPI	0.000	2.480	0.929	0.627
(4) $tiPI$	0.018	6.690	4.425	1.358
(5) $pRAIL$	487.900	927.400	709.355	13,334.040
(6) N	1,906.000	2,507.000	2,208.384	27,668.381
(7) $pBOOST$	981.000	1906.000	1209.841	28,618.435
(8) HFM	15.148	241.628	101.290	1,226.203
(9) HFM^*	0.105	1.627	0.763	0.062

Table 11.6: Linear correlation coefficients of the variables relevant in the NO_x data set III.

	NOx	qMI	pMI	qPI	$tiPI$	$pRAIL$	N	$pBOOST$	HFM	HFM^*
(0) (T) NOx	1.00	0.01	0.15	-0.13	0.61	-0.05	-0.14	0.25	0.59	0.67
(1) qMI	0.01	1.00	0.03	0.04	-0.39	-0.04	-0.05	0.37	0.29	0.32
(2) pMI	0,15	0.03	1.00	-0.03	-0.11	0.01	0.18	-0.05	-0.06	-0.10
(3) qPI	-0.13	0.04	-0.03	1.00	-0.14	-0.10	0.02	0.11	0.01	0.00
(4) $tiPI$	0,61	-0.39	-0.10	-0.14	1.00	-0.01	0.11	0.37	0.66	0.68
(5) $pRAIL$	-0.05	-0.04	0.01	-0.10	-0.01	1.00	-0.02	-0.05	-0.02	-0.02
(6) N	-0.14	-0.05	0.18	0.02	0.11	-0.02	1.00	0.14	0.30	0.08
(7) $pBOOST$	0.25	0.37	-0.05	0.11	0.37	-0.05	0.14	1.00	0.73	0.73
(8) HFM	0.59	0.29	-0.06	0.01	0.66	-0.02	0.30	0.73	1.00	0.97
(9) HFM^*	0.67	0.32	-0.10	0.00	0.68	-0.02	0.08	0.73	0.97	1.00

11.2 Classification

11.2.1 Introduction

Classification is understood as the act of placing an object into a set of categories, based on the object's properties. Objects are classified according to a (in most cases hierarchical) classification scheme also called taxonomy. Amongst many other possible applications, examples of taxonomic classification are biological classification (the act of categorizing and grouping living species of organisms), medical classification, and security classification (where it is often necessary to classify objects or persons for deciding whether a problem might arise from the present situation or not). A statistical classification algorithm is supposed to take feature representations of objects and map them to a special, predefined classification label. Such classification algorithms are designed to learn (i.e., to approximate the behavior of) a function which maps a vector of object features into one of several classes; this is done by analyzing a set of input-output examples ("training samples") of the function. Since statistical classification algorithms are supposed to "learn" such functions, we are dealing with a specific area of *machine learning* and, more generally, *artificial intelligence*.

In a more formal way, the classification problem can be formulated in the following way: Let the data consist of a set of samples, each containing k feature values $x_{i1}, \ldots, x_{ik}$ and a class value y_i. What we look for is a function f that maps a sample x_i to one of the c classes available:

$$f : X \to C; \tag{11.8}$$

$$\forall (x \in X) : f(x) = f(x_1, \ldots, x_k) = y; y \in \{C_1, \ldots, C_c\} \tag{11.9}$$

where X denotes the feature vector space and C the set of classes.

There are several approaches which are nowadays used for solving data mining and, more specifically, classification problems. The most common ones are (as for example described in [Mit00]) decision tree learning, instance-based learning, inductive logic programming (such as Prolog, e.g.), and reinforcement learning.

11.2.2 Real-Valued Classification with Genetic Programming

In this section we shall concentrate on GP-based classification. In fact, we will here restrict ourselves to real-valued classification tasks, i.e.,

$$X \subset \mathbb{R}^k, C \subset \mathbb{R} \tag{11.10}$$

Thus, we can apply the GP-based system identification approach described in the previous sections; the representations of the problems (Section 9.2.2), the solution candidates (9.2.4), and the genetic operators (9.2.4.2) can be used without any restrictions.

The only critical aspect is that the evaluation and the quality estimation of classifiers have to be modified: Evaluating a model m on a set of input features $(x_1, \ldots, x_k)$ will lead to a target value $y \in \mathbb{R}$, but y does not necessarily have to be exactly one certain class value, i.e., we might get $y \notin C$. The exact mapping of feature vectors and their respective target values to class values is done using sets of thresholds $t1, \ldots, t_{c-1}$ placed between the class values $C_1, \ldots, C_c$:

$$\forall (i \in [1; c-1]) : C_i < t_i < C_{i+1} \tag{11.11}$$

Based on a set of thresholds T we can classify a sample for which the target value y has been calculated as belonging to class c_t using the mapping function f':

$$f' : \{\mathbb{R}, \mathbb{R}\} \to C \ ; \ c_t = f'(c_t, T) \tag{11.12}$$

$$y < t_1 \Rightarrow f'(c_t, T) = C_1 \tag{11.13}$$

$$y > t_{c-1} \Rightarrow f'(c_t, T) = C_c \tag{11.14}$$

$$\forall (i \in [1; c-2]) : t_i < y < t_{i+1} \Rightarrow f'(c_t, T) = C_{i+1} \tag{11.15}$$

11.2.3 Analyzing Classifiers

11.2.3.1 Classification Rates and Confusion Matrices

When it comes to analyzing classifiers, the most important aspect is of course how many samples are classified correctly. For each feature vector sample x we have an original classification y, and by applying the classifier which is to be evaluated we get the predicted class y'. As described before, this classification of x is done using a classification model yielding $y = f(x)$ and an optional post-processing step using thresholds T yielding $y = f'(y, T)$.

Let us assume that we analyze n samples $x_{1 \ldots n}$ (classified into c classes $C_1 \ldots C_c$) with their respective original classifications $y_{1 \ldots n}$; by applying a classification model m we get the respective predicted classifications $y'_{1 \ldots n}$ as described above. The ratios of correctly classified samples for all classes or each class separately are calculated as cc and cc_i, respectively:

$$cc = \frac{|j : j \in [1; n] \ \wedge \ y_j = y'_j|}{n} \tag{11.16}$$

$$\forall (i \in [1; c]) : cc_i = \frac{|j : j \in [1; n] \ \wedge \ y_j = y'_j \ \wedge \ y_j = C_i|}{|j : j \in [1; n] \ \wedge \ y_j = C_i|} \tag{11.17}$$

For more detailed analysis, confusion matrices [KP98] contain information about actual and predicted classifications done by classification systems. In general, a confusion matrix cm is a table containing $c \times c$ cells that states how many samples of each given class are classified as belonging to a specific class; for example, each column of the matrix can represent the instances of a predicted class while each row represents the instances in the original (actual) class (or vice versa). So, the value $cm_{i,j}$ stores the number of samples of class i that are classified as class j.

An example is given in Table 11.7 in which each row of the matrix represents the instances in a predicted class while each column represents the instances in the original (actual) class; additionally, the numbers of samples not classified $(nc_1 \ldots nc_c)$ are also given as well as the total rate of correct classifications. Please note that the sum of all cells has to be equal to the number of samples n, i.e.,

$$\sum_{i=1}^{c}\sum_{j=1}^{c} cm_{i,j} + \sum_{i=1}^{c} nc_i = n \tag{11.18}$$

Table 11.7: Exemplary confusion matrix with three classes

		Actual Class			
		"1"	"2"	"3"	
Estimated	"1"	$cm_{1,1}$	$cm_{2,1}$	$cm_{3,1}$	
Class	"2"	$cm_{1,2}$	$cm_{2,2}$	$cm_{3,2}$	
	"3"	$cm_{1,3}$	$cm_{2,3}$	$cm_{3,3}$	
Not classified		nc_1	nc_2	nc_3	
Correct Classifications Ratio					$\frac{\sum_{i=1}^{c} cm_{i,i}}{n}$

The special case of binary classification into two classes (i.e., $c = 2$) is frequently found as it is in many applications necessary to decide for given samples whether or not some given condition is fulfilled. There are the four different possible outcomes of a single predicted (estimated) classification in the case of binary classification into classes "positive" ("yes," "1," "true") and "negative" ("no," "0," "false"):

- A false positive classification is done when a sample is incorrectly classified as "positive" which is in fact "negative,"

- a false negative classification is done when a sample is incorrectly classified as "negative" which is in fact "positive," and

- true positive as well as true negative classifications are respective correct classifications.

A typical "positive / negative" example is given in Table 11.8:

Table 11.8: Exemplary confusion matrix with two classes

		Actual Class	
		Positive	Negative
Estimated	Positive	a (true positive)	b (false positive)
Class	Negative	c (false negative)	d (true negative)

In this case,

- the *accuracy* is defined as $ACC = \frac{a+d}{a+b+c+d}$,

- the *true positive rate* (also called *sensitivity*) as $TP = \frac{a}{a+c}$,

- the *true negative rate* (also called *specificity*) as $TN = \frac{d}{b+d}$,

- the *false positive rate* as $FP = \frac{b}{b+d}$ (which is in fact the probability of classifying a sample as "positive" when it is actually "negative"),

- the *false negative rate* as $FN = \frac{c}{a+c}$ (which is in fact the probability of classifying a sample as "negative" when it is actually "positive"), and finally

- the *precision* as $P = \frac{a}{a+b}$.

11.2.3.2 Receiver Operating Characteristic (ROC) Curves

Receiver operating characteristic (ROC) analysis provides a convenient graphical display of the trade-off between true and false positive classification rates for two class (binary) problems [FE05]. Since its introduction in the medical and signal processing literatures ([HM82], [Zwe93]), ROC analysis has become a prominent method for selecting an operating point; for a recent snapshot of applications and methodologies see [FBF+03] and [HOFLe04]. ROC analysis often includes the calculation of the area under the ROC curve (AUC).

In the context of two class classification, ROC curves are calculated in the following way: For each possible threshold value discriminating two given classes (e.g., 0 and 1, "true" and "false" or "positive" and "negative"), the numbers of true and false classifications for one of the classes are calculated. For example, if the two classes "true" and "false" are to be discriminated using a given classifier, a fixed set of equidistant thresholds is tested and the true positives (TP) and the false positives (FP) are counted for each of them. Each pair of TP and FP values produces a point of the ROC curve; examples are graphically shown in Figure 11.10. Slightly different versions are also often used; for example the positive predictive value (= TP / (TP + FP)) or the negative predictive value (= TN / (FN + TN)) could be displayed instead.

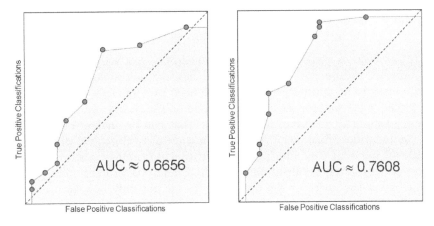

FIGURE 11.10: Two exemplary ROC curves and their area under the ROC curve (AUC).

The most common quantitative index describing an ROC curve is the area under it. The bigger the area under a ROC curve is, the better the discriminator model is; if the two classes can be ideally separated, the ROC curve goes through the upper left corner and, thus, the area under it reaches its maximal possible value which is exactly 1.0.

This method is very useful for analyzing the qualtity of two class classifiers, but unfortunately it is not directly applicable for more than two classes. When it comes to measuring or graphically illustrating the quality of multi-class classifiers, one possibility is to define symmetric areas around the original class values; for each class value C_i the corresponding area is defined as $[C_i - r, C_i + r]$. Successively increasing the parameter value r from 0 to $\frac{C_{i+1}-C_i}{2}$ and calculating the numbers of correct and incorrect classifications for each r yields a set of pairs of FP/TP values. Jiang and Motai [JM05], for example, use this technique for illustrating and analyzing the classification performance in the context of automatic motion learning.

Although this method can be used very easily, it is not generally applicable because it is restricted to symmetric areas. Emerson and Fieldsend [FE05] propose a different approach and define the ROC surface for the Q-class problem in terms of a multi-objective optimization problem in which the goal is to simultaneously minimize misclassification rates when the misclassification costs and parameters governing the classifier's behavior are unknown. The problem with this approach is that the estimated Pareto fronts presented in [FE05] can be illustrated and used for graphical interpretation for a classification problem involving not more than three classes. This is why we here in the following section propose the use of sets of ROC curves for each class separately.

11.2.3.3   Sets of Receiver Operating Characteristic Curves and their Use in the Evaluation of Multi-Class Classification

In this section we present an extension to ROC analysis making it possible to measure the quality of classifiers for multi-class problems. Unlike other multi-class-ROC approaches which have been presented recently (see [FE05] or [Sri99], e.g.) we propose a method based on the theory of ROC curves that creates sets of ROC curves for each class that can be analyzed separately or in combination. Thus, what we get is a convenient graphical display of the trade-off between true and false classifications for multi-class problems. We have developed a generalization of this AUC analysis for multi-class problems which gives the operator the possibility to see not only how accurately, but also how clearly classes can be separated from each other.

The main idea presented here is that for each given class C_i the numbers of true and false classifications are calculated for each possible pair of threshold between the classes C_{i-1} and C_i as well as between C_i and C_{i+1}. This is in fact done under the assumption that the c classes are ordered and that $C_i < C_{i+1}$ holds for every $i \in [1, (n-1)]$ (with c being the number of classes).

For a given class C_i the corresponding TP and FP values (on the basis of the N original values $\vec{o}$ and estimated values $\vec{e}$) are calculated as:

$$\forall (\langle t_a, t_b \rangle | (C_{i-1} < t_a < C_i) \wedge (C_i < t_b < c_{i+1})): \tag{11.19}$$
$$TP(t_a, t_b) = |\{e_j : (t_a < e_j < t_b) \wedge (t_a < o_j < t_b)\}| \tag{11.20}$$
$$FP(t_a, t_b) = |\{e_j : (t_a < e_j < t_b) \wedge (o_j < t_a \vee o_j > t_b)\}| \tag{11.21}$$

This approach has been published first in [WAW06d] and then described in detail (including application examples) in [WAW07].

The resulting tuples of (FP,TP) values are stored in a matrix which can be plotted as is exemplarily illustrated in Figure 11.11: On the basis of synthetic data $10^2 = 100$ ROC points for 10 thresholds between the chosen class C_i and C_{i-1} as well as between C_i and C_{i+1} were calculated. This obviously yields a set of points which can be interpreted in analogy to the interpretation of "normal" ROC curves: The closer the points are located to the upper left corner, the higher is the quality of the classifier at hand.

For getting sets of ROC curves instead of ROC points, the following change is introduced: An arbitrary threshold t_a between the classes C_{i-1} and C_i is fixed and the FP and TP values for all possible thresholds t_b between C_i and C_{i+1} are calculated. What we get is one single ROC curve; this calculation is executed for all possible values of t_a (i.e., for all possible threshold between C_{i-1} and C_i). This procedure also has to be executed the other way around, i.e., also has to choose an arbitrary threshold t_b between C_i and C_{i+1}, calculate all corresponding ROC points, and repeat this for all values for all possible values of t_a.

Finally, what we get is a set of ROC curves; an example showing 10 ROC curves is given in Figure 11.11.

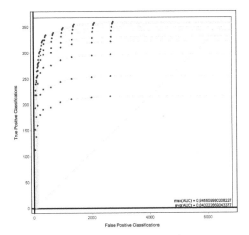

FIGURE 11.11: An exemplary graphical display of a multi-class ROC (MROC) matrix.

Of course this procedure cannot be executed in exactly this way for the classes C_1 and C_n. For c_1 it is only possible to calculate the ROC points (and therefore the ROC curve) for all possible thresholds between C_1 and C_2; for C_c this is done analogically with all possible thresholds between C_{c-1} and C_c. This is why sets of ROC curves can be calculated for the classes $C_2 \ldots C_{c-1}$ whereas only simple ROC curves can be produced for C_1 and C_c.

As already mentioned in the previous section, the area under the ROC curve (AUC) is a very common quantitative index describing the classifier's quality. In the context of multi-class ROC (MROC) curves the two following values can be calculated assuming that all m ROC curves for a given class have already been calculated:

- The maximum AUC ($MaxAUC$) is the maximum of all areas under the ROC curves calculated for a specific class. It measures how exactly this class is separated from the others using the best thresholds parameter setting.

$$MaxAUC = \max_{i=1..m} AUC(ROC_i)$$

- The average AUC ($AvgAUC$) is calculated as the mean value of all areas under the ROC curves for a specific class. It measures how clearly this class is separated from the others since it takes into account all possible thresholds parameter settings.

$$AvgAUC = \frac{\sum_{i=1..m} AUC(ROC_i)}{m}$$

We will in the following turn to a topic very much related to what we have discussed in the previous sections, namely the evaluation of classifiers evolved by GP.

11.2.4   Classification Specific Evaluation in GP

Of course, there is on the one hand no reason why standard evaluation functions such as the MSE / MEE, VAF, or R^2 functions could not be used for estimating the quality of classification model during the GP process. The reason for this is that we here want the identification algorithm to produce a model that is able to reproduce the given target data as well as possible, similar to when dealing with regression or time series analysis.

Still, on the other hand the evaluation of classification models may also include several aspects for which the standard evaluation functions are not suitable. This is why we shall describe several aspects that may contribute to a classification specific evaluation function for GP solution candidates in the context of real-valued learning of classifiers with genetic programming.

11.2.4.1   Preprocessing of Estimated Target Values

Before we compare original and estimated class values we suggest the following classification specific preprocessing step:

The errors of predicted values that are lower than the lowest class value or greater than the greatest class value should not have a quadratic or even worse, but rather partially only linear contribution to the fitness of a model. To be a bit more precise: Given n samples with original classifications o_i divided into c classes $C_1, ..., C_c$ (with C_1 being the lowest and C_c the greatest class value), the so preprocessed estimated values $preproc(e_i)$ shall be calculated as follows:

$$\forall (i \in [1, n]) :$$
$$(e_i < C_1) \Rightarrow preproc(e_i, x) = C_1 - (C_1 - e_i)^{\frac{1}{x}} \qquad (11.22)$$
$$(e_i > C_c) \Rightarrow preproc(e_i, x) = C_c + (e_i - C_c)^{\frac{1}{x}} \qquad (11.23)$$

with x being an exponential parameter which depends on the evaluation function that uses these preprocessed values. For example, when using the mean squared error or any other function that incorporates the use of squared differences between original and estimated value, x is to be set to 2, whereas when using the MEE function it has to be set to the chosen exponent.

The reason for this is that values that are greater than the greatest class value or below the lowest value are anyway classified as belonging to the class having the greatest or the lowest class number, respectively; using a standard evaluation function without preprocessing of the estimated values would punish a formula producing such values more than necessary.

11.2.4.2 Considering Standard Evaluation Functions

For quantifying the quality of classifiers we can use all functions described in Section 9.2.5; in contrast to standard applications, we can also apply these functions for each class individually.

In the standard case, all n values are evaluated using the MEE, VAF, and R^2 values as well as the minimum and maximum errors $error_{min}$ and $error_{max}$; these can optionally be calculated using the preprocessed values $preproc(e_i)$ instead of e_i for all $i \in [1; n]$. Thus, we get partial values mee, vaf and r^2, $error_{min}$ and $error_{max}$ which can be weighted using the factors w_{mee}, w_{vaf}, w_{r^2}, $w_{err_{min}}$, and $w_{err_{max}}$.

This approach of course does not consider the distribution of samples to the classes; for example, if 98% of the samples belong to class 0 and only 2% to class 1, then the evaluation of a model classifying all samples as 0 will be fairly good when using these standard evaluation functions even though this classifier is more or less useless.

In order to overcome this problem we could for example sample the data so that all classes are represented by the same number of samples; we instead here describe the application of these evaluation functions to the classes given separately:

The sets of estimated values $ec_1 \ldots ec_c$ contain the values estimated for each class $C_1 \ldots C_c$, and in analogy to this the sets $oc_1 \ldots oc_c$ are sets of the corresponding class values:

$$\forall(i \in [1; n]) : o_i = k \Rightarrow e_i \in ec_k, o_i \in oc_k \tag{11.24}$$

Additionally, we also need class weights $w_1 \ldots w_c$ (with $w = \sum_{i=1}^{c} w_i$) and can so calculate the partial fitness values as

$$mee = \frac{1}{w} \sum_{i=1}^{c} mee(oc_i, ec_i, n) \cdot w_i \tag{11.25}$$

$$r^2 = \frac{1}{w} \sum_{i=1}^{c} r^2(oc_i, ec_i) \cdot w_i \tag{11.26}$$

$$vaf = \frac{1}{w} \sum_{i=1}^{c} \left(1 - \frac{var(oc_i - ec_i)}{var(o)} \right) \cdot w_i \tag{11.27}$$

$$error_{min} = \frac{1}{w} \sum_{i=1}^{c} r_{min}(oc_i, ec_i) \cdot w_i \tag{11.28}$$

$$error_{max} = \frac{1}{w} \sum_{i=1}^{c} r_{max}(oc_i, ec_i) \cdot w_i \tag{11.29}$$

Again, these values can optionally be calculated using the preprocessed values $preproc(e_i)$ instead of e_i for all $i \in [1; n]$.

Of course, the adjusted functions described in Section 9.2.5.3 could be used instead of the standard functions.

11.2.4.3 Considering Classification Specific Aspects

We propose the consideration of the following classification specific aspects in the evaluation of classifier models:

- The range of the values estimated for each of the given classes,

- how well the classes are separated correctly from each other depending on the choice of appropriate thresholds, and

- the area under ROC curves or, in the case of multi-class classification, the area under sets of MROC curves.

Class Ranges

For calculating the class ranges $cr_1 \ldots cr_c$ we definitively need the sets of estimated values for each class, $ec_1 \ldots ec_c$:

$$\forall (i \in [1; c]) : cr_i = max(ec_i) - min(ec_i) \tag{11.30}$$

and can so calculate the class ranges' contribution cr as

$$cr = \sum_{i=1}^{c} cr_i \cdot w_i \tag{11.31}$$

Figure 11.12 exemplarily displays several samples with original class values C_1, C_2, and C_3; the class ranges result from the estimated values for each class and are indicated as cr_1, cr_2, and cr_3.

Thresholds Analysis

As is indicated in Figure 11.12 we do not only want to consider class ranges but also a more classification-like approach. Between each pair of contiguous classes we set m equally distributed temporary thresholds:

$$\forall (i \in [1; c-1]) \forall (k \in [1; m]) : t_{i,k} = C_i + k \cdot \frac{C_{i+1} - C_i}{m+1} \tag{11.32}$$

Then, for each threshold we count the numbers of samples which are classified incorrectly; here we also consider a given matrix storing misclassification punishments mcp for each pair of classes giving the misclassification punishment for classifying a sample of class a as class b as $mcp_{a,b}$ for all a and b in $[1; c]$:

$$\forall (i \in [1; c-1]) \forall (k \in [1; m]) : \forall (j \in [1; n]) :$$

$$p(i, k, j) = \begin{cases} mcp_{i,i+1} \cdot \frac{1}{freq_{o_j}} & : & o_j < t_{i,k} \wedge e_j > t_{i,k} \\ mcp_{i+1,i} \cdot \frac{1}{freq_{o_j}} & : & o_j > t_{i,k} \wedge e_j < t_{i,k} \\ 0 & : & else \end{cases} \tag{11.33}$$

$$p(i, k) = \sum_{j=1}^{n} p(i, k, j) \tag{11.34}$$

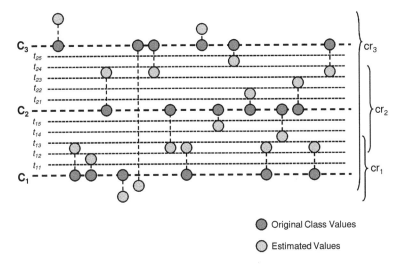

FIGURE 11.12: Classification example: Several samples with original class values C_1, C_2, and C_3 are shown; the class ranges result from the estimated values for each class and are indicated as cr_1, cr_2, and cr_3.

assuming that a sample j is (temporarily) classified as class $(i+1)$ if $e_j > t_{i,k}$ and as class i if $e_j < t_{i,k}$; $freq_a$ is the frequency of class a, i.e., the number of samples that are originally classified as belonging to class a.

The thresholds' contribution to the classifier's fitness, $thresh$, can be now calculated in two different ways: We can consider the minimum sum of punishments for each pair of contiguous classes as

$$thresh = \sum_{i=1}^{c-1} min_{k \in [1;m]} p(i, k) \tag{11.35}$$

or consider all thresholds which are weighted using threshold weights $tw_{1...m}$ as

$$thresh = \sum_{i=1}^{c-1} \frac{1}{\sum_{k=1}^{m} tw_k} \sum_{k=1}^{m} p(i, k) \cdot tw_k \tag{11.36}$$

Normally, we define the threshold weights tw using minimum and maximum weights, weighting the thresholds at near to the original class values minimally and those in the "middle" maximally:

$$tw_1 = tw_{min}, \; tw_m = tw_{min}; \; tw_{range} = tw_{max} - tw_{min} \qquad (11.37)$$

$$m \bmod 2 = 0 \Rightarrow \begin{cases} l = m/2 \\ tw_l = tw_{l+1} = tw_{max} \\ \forall (i \in [2; l-1]): \\ \quad tw_i = tw_{min} + \frac{tw_{range}}{l-1} \cdot (i-1) \\ \forall (i \in [l+1; m-1]): \; tw_{m-i+1} = tw_i \end{cases} \qquad (11.38)$$

$$m \bmod 2 = 1 \Rightarrow \begin{cases} l = (m+1)/2 \\ tw_l = tw_{max} \\ \forall (i \in [2; l-1]): \\ \quad tw_i = tw_{min} + \frac{tw_{range}}{(m-1)/2} \cdot (i-1) \\ \forall (i \in [l+1; m-1]): \; tw_{m-i+1} = tw_i \end{cases} \qquad (11.39)$$

(M)ROC Analysis

Finally, we also consider the area under the (M)ROC curves as described in Section 11.2.3.3: For each class we calculate the AUC values for ROC curves and sets of MROC curves (with a given number of thresholds checked for each class), and then we can either use the average AUC or the maximum AUC for each class weighted with the weighting factors already mentioned before:

$$auc = \begin{cases} \sum_{i=1}^{c} AvgAUC(C_i) \cdot w_i &: \texttt{consider average AUCs} \\ \sum_{i=1}^{c} MaxAUC(C_i) \cdot w_i &: \texttt{consider maximum AUCs} \end{cases} \qquad (11.40)$$

11.2.4.4   Combined Classifier Evaluation

As we have now compiled all information needed for estimating the quality of a classifier model in GP, $CLASS$, we calculate the final overall quality using respective weighting factors:

$$\begin{aligned}
a_1 &= mee \cdot c_1 & (c_1 &= w_{mee}) \\
a_2 &= vaf \cdot c_2 & (c_2 &= w_{vaf}) \\
a_3 &= r^2 \cdot c_3 & (c_3 &= w_{r^2}) \\
a_4 &= error_{min} \cdot c_4 & (c_4 &= w_{err_{min}}) \\
a_5 &= error_{max} \cdot c_5 & (c_5 &= w_{err_{max}}) \\
a_6 &= cr \cdot c_6 & (c_6 &= w_{cr}) \\
a_7 &= thresh \cdot c_7 & (c_7 &= w_{thresh}) \\
a_8 &= auc \cdot c_8 & (c_8 &= w_{auc}) \\
CLASS(o, e) &= \frac{\sum_{i=1}^{8} a_i \cdot c_i}{\sum_{i=1}^{8} c_i}
\end{aligned} \qquad (11.41)$$

11.2.5   Application Example: Medical Data Analysis

11.2.5.1   Benchmark Data Sets

For testing GP-based training of classifiers here we have picked the following data sets: The *Wisconsin Breast Cancer*, the *Melanoma*, and the *Thyroid* data sets.

- The *Wisconsin* data set is a part of the UCI machine learning repository[5]. In short, it represents medical measurements which were recorded while investigating patients potentially suffering from breast cancer. The number of features recorded is 9 (all being continuous numeric ones); the file version we have used contains 683 recorded examples (by now, 699 examples are already available since the data base is updated regularly).

- The *Thyroid* data set represents medical measurements which were recorded while investigating patients potentially suffering from hypo- or hyperthyroidism; this data set has also been taken from the UCI repository. In short, the task is to determine whether a patient is hypothyroid or not. Three classes are formed: Euthyroid (the state of having normal thyroid gland function), hyperthyroid (overactive thyroid), and hypothyroid (underactive thyroid).
  In total, the data set contains 7200 samples. The samples of the *Thyroid* data set are not equally distributed to the three given classes; in fact, 166 samples belong to class "1" ("subnormal functioning"), 368 samples are classified as "2" ("hyperfunction"), and the remaining 6666 samples belong to class "3" ("euthyroid"); a good classifier therefore has to be able to correctly classify significantly more than 92% of the samples simply because 92 percent of the patients are not hypo- or hyperthyroid. 21 attributes (15 binary and 6 continuous ones) are stored in this data set.

- The *Melanoma* data set represents medical measurements which were recorded while investigating patients potentially suffering from skin cancer. It contains 1311 examples for which 30 features have been recorded; each of the 1311 samples represents a pigmented skin lesion which has to be classified as a melanoma or a nonhazardous nevus. This data set has been provided to us by Prof. Dr. Michael Binder from the Department of Dermatology at the Medical University Vienna, Austria.
  A comparison of machine learning methods for the diagnosis of pigmented skin lesions (i.e., detecting skin cancer based on the analysis of visual data) can be found in [DOMK+01]; in this paper the authors describe the quality of classifiers produced for a comparable data

---

[5]http://www.ics.uci.edu/~mlearn/.

Table 11.9: Set of function and terminal definitions for enhanced GP-based classification.

Functions		
Name	**Arity**	**Description**
+	2	Addition
*	2	Multiplication
-	2	Subtraction
/	2	Division
e^x	1	Exponential Function
IF	3	If [Arg0] then return [Then] branch ([Arg1]), otherwise return [Else] branch ([Arg2])
$\leq, \geq$	2	Less or equal, greater or equal
&&, ||	2	Logical AND, logical OR
Terminals		
Name	**Parameters**	**Description**
var	x, c	Value of attribute x multiplied with coefficient c
const	d	A constant double value d

collection using k-NN classification, ANNs, decision trees, and SVMs. The difference is that in the data collection used in [DOMK$^+$01] all lesions were separated into three classes (common nevi, dysplastic nevi, or melanoma); here we use data representing lesions that have been classified as benign or malign, i.e., we are facing a binary classification problem.

All three data sets were investigated via 10-fold cross-validation. This means that each original data set was divided into 10 disjoint sets of (approximately) equal size. Thus, 10 different pairs of training (90% of the data) and test data sets (10% of the data) can be formed and used for testing the classification algorithm.

11.2.5.2 Solution Candidate Representation Using Hybrid Tree Structures

The selection of the functions library is an important part of any GP modeling process because this library should be able to represent a wide range of systems; Table 11.9 gives an overview of the function set as well as the terminal nodes used for the classification experiments documented here. As we can see in Table 11.9, mathematical functions and terminal nodes are used as well as Boolean operators for building complex arithmetic expressions. Thus, the concept of decision trees is included in this approach together with the standard structure identification concept that tries to evolve nonlinear mathematical expressions. An example showing the structure tree representation of a combined formula including arithmetic as well as logical functions is displayed in Figure 11.13.

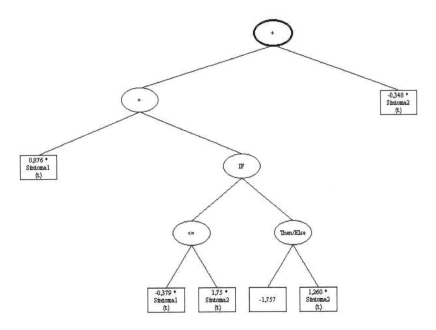

FIGURE 11.13: An exemplary hybrid structure tree of a combined formula including arithmetic as well as logical functions.

11.2.5.3 Evaluation of Classification Models

There are several possible functions that can serve as fitness functions within the GP process. For example, the ratio of misclassifications (using optimal thresholds) or the area under the corresponding ROC curves ([Zwe93], [Bra97]) could be used. Another function frequently used for quantifying the quality of models is the R^2 function that takes into account the sum of squared errors as well as the sum of squared target values; an alternative, the so-called *adjusted* R^2 function, is also utilized in many applications.

We have decided to use a *variant of the squared errors function* for estimating the quality of a classification model. There is one major difference of this modified mean squared errors function to the standard implementation of this function: The errors of predicted values that are lower than the lowest class value or greater than the greatest class value do not have a totally quadratic, but partially only linear contribution to the fitness value. To be a bit more precise: Given N samples with original classifications o_i divided into n classes

$c_1, ..., c_n$ (with c_1 being the lowest and c_n the greatest class value), the fitness value F of a classification model producing the estimated classification values e_i is evaluated as follows:

$$\forall (i \in [1, N]) :$$
$$(e_i < c_1) \Rightarrow f_i = (o_i - c_1)^2 + \mid c_1 - e_i \mid,$$
$$(c_1 \leq e_i \leq c_n) \Rightarrow f_i = (e_i - o_i)^2, \tag{11.42}$$
$$(e_i > c_n) \Rightarrow f_i = (o_i - c_n)^2 + \mid c_n - e_i \mid$$

$$F = \frac{1}{N} \sum_{i=1}^{N} f_i \tag{11.43}$$

The reason for this is that values that are greater than the greatest class value or below the lowest value are anyway classified as belonging to the class having the greatest or the lowest class number, respectively; using a standard implementation of the squared error function would punish a formula producing such values more than necessary.

11.2.5.4 Finding Appropriate Class Thresholds: Dynamic Range Selection

Of course, a mathematical expression alone does not yet define a classification model; thresholds are used for dividing the output into multiple ranges, each corresponding to exactly one class. These regions are defined before starting the training algorithm in static range selection (SRS, see for example [LC05] for explanations), which brings along the difficulty of determining the appropriate range boundaries a priori. In the GP-based classification framework discussed here we have therefore used dynamic range selection (DRS) which attempts to overcome this problem by evolving the range thresholds along with the classification models: Thresholds are chosen so that the sum of class-wise ratios of misclassifications for all given classes is minimized (on the training data, of course).

In detail, let us consider the following: Given N (training) samples with original classifications o_i divided into n classes $c_1, \ldots, c_n$ (with c_1 being the lowest and c_n the greatest class value), models produced by GP can be in general used for calculating estimated values e_i for all N samples. Assuming thresholds $T = t_1, \ldots, t_{n-1}$ (with $c_j < t_j < c_{j+1}$ for $j \in [1; n - 1]$), each sample k is classified as ec_k:

$$e_k < t_1 \Rightarrow ec_k(T) = c_1 \tag{11.44}$$
$$t_j < e_k < t_{j+1} \Rightarrow ec_k(T) = c_{j+1} \tag{11.45}$$
$$e_k > t_{n-1} \Rightarrow ec_k(T) = c_n \tag{11.46}$$

Thus, assuming a set of thresholds T_m, for each class c_k we get the ratio of correctly classified samples cr_k as

$$total_k(T_m) = |\{a : (\forall(x \in a) : o_x = ec_k(T_m))\}| \tag{11.47}$$

$$correct_k(T_m) = |\{b : (\forall(x \in b) : o_x = ec_k(T_m) \wedge o_x = c_k)\}| \tag{11.48}$$

$$cr_k(T_m) = \frac{correct_k(T_m)}{total_k(T_m)} \tag{11.49}$$

The sum of ratios of correctly classified samples is – dependent on the set of thresholds T_m – calculated as

$$cr(T_m) = \sum_{i=1}^{n} cr_i(T_m) \tag{11.50}$$

So, finally we can define the set of thresholds applied as that set T_{opt} so that each other set of thresholds leads to equal or lower sums of classification accuracies[6]:

$$T_d \neq T_{opt} \Rightarrow cr(T_d) \leq cr(T_{opt}) \tag{11.51}$$

These thresholds, that are optimal for the training samples, are fixed and also applied on the test samples.

Please note that this sum of class-wise classification accuracies is not equal to the total ratio of correctly classified samples which is used later on in Sections 11.2.5.5 and 11.2.5.8; the total classification accuracy for a set of thresholds $acc(T_m)$ (assuming original and estimated values $\mathbf{o}$ and $\mathbf{e}$) is defined as

$$z(T_m) = |\{a|(\forall(x \in a) : o_x = ec_x(T_m))\}| \tag{11.52}$$

$$acc(T_m) = \frac{z}{N}. \tag{11.53}$$

11.2.5.5 First Results, Identification of Optimal Operators and Parameter Settings

As first reported in detail in [WAW07], during our thorough test series we have identified the following GP-relevant parameter settings as the best ones for solving classification problem instances:

- **GP-algorithm:** Enhanced GP using strict offspring selection.

- **Mutation rate:** $10\% - 15\%$.

- **Population size:** $500 - 2,000$.

---

[6]Please note here that it could happen that more than one combination of thresholds can be optimal, simply because there could be more than one optimal threshold for any given pair of class values. This is why we here give an inequation in (11.51).

- **Selection operators:** Whereas standard GA implementations use only one selection operator, the SASEGASA requires two, namely the so-called female selection operator as well as the male selection operator. Similar to our experience gained during the tests on the identification of mechatronical systems, it seems to be the best to choose the roulette-wheel selection in combination with the random selection operator. The reason for this is that apparently merging the genetic information of rather good individuals (models, formulas) with randomly chosen ones is the best strategy when using the SASEGASA for solving identification problems.

- **Success ratio and selection pressure:** As for instance described in [AW04b], there are some additional parameters of the SASEGASA regarding the selection of those individuals that are accepted to be a part of the next generation's population. These are the success ratio and the maximal selection pressure that steer the algorithm's behavior regarding offspring selection. For model structure identification tasks in general and especially in case of dealing with classification problems, the following parameter settings seem to be the best ones:

  - Success ratio = 1.0, and
  - Maximum selection pressure = 100 – 500 (this value has to be defined before starting a identification process depending on other settings of the genetic algorithm used and the problem instance which is to be solved).

As has already been explained in further detail in previous chapters, these settings have the effect that in each generation only offspring survive that are really better than their parent individuals (since the success ratio is set to 1.0, only better children are inserted into the next generation's population). This is why the selection pressure becomes very high as the algorithm is executed, and therefore the maximum selection pressure has to be set to a rather high value (as, e.g., 100 or 500) to avoid premature termination.

- **Crossover operators:** We have implemented and tested three different single-point crossover procedures for GP-based model structure identification: One that exchanges rather big subtrees, one that is designed to exchange rather small structural parts (e.g., only one or two nodes), and one that replaces randomly chosen parts of the respective structure trees. Moreover, for each crossover operator we have also implemented an extended version that additionally randomly mutates all terminal nodes (i.e., manipulates the parameters of the represented formula). The following 6 structure identification crossover operators are available: *StandardSPHigh*, *StandardSPMedium*, *StandardSPLow*, *ExtendedSPHigh*, *ExtendedSPMedium*, and *ExtendedSPLow*.

Since arbitrarily many crossover operators can be selected when applying the SASEGASA[7], the task was not to find out which operator can be used to produce the best results but rather which subset of operators is to be chosen. According to what we experienced, the following set of crossover operators should be applied: All three standard operators (*StandardSPHigh*, *StandardSPMedium*, and *StandardSPLow*) plus one of the extended ones, for instance *ExtendedSPLow*.

- **Mutation operators:** The basic mutation operator for GP structure identification we have implemented and tested, *GAStandard*, works as already described in Chapter 9: A function symbol could become another function symbol or be deleted; the value of a constant node or the index of a variable could be modified. Furthermore, we have also implemented an extended version (*GAExtended*) that additionally randomly mutates all terminal nodes (in analogy to the extended crossover operators).

  As the latest test series have shown, the choice of the crossover operators influences the decision which mutation operator to apply to the SASEGASA: If one of the extended crossover operators is selected, it seems to be best to choose the standard mutation operator. But if only standard crossover methods are selected, picking the extended mutation method yields the best results.

Selected experimental results of the standard GP implementation and the SASEGASA algorithm for the *Thyroid* data set using various parameter settings are presented in Table 11.10. For each parameter settings version the 10-fold cross validation test runs were executed, the resulting average results are listed. In all cases, the population size was 1000; furthermore, the following parameter settings were used:

(1) crossover: *ExtendedSPMedium*; mutation: *GAStandard*; selection: roulette.

(2) crossover: *StandardSPMedium*; mutation: *GAExtended*; selection: roulette.

(3) crossover: *all 6 available operators*; mutation: *GAExtended*; selection: *random* and *roulette* (maximum selection pressure: 500).

(4) crossover: *all 6 available operators*; mutation: *GAStandard*; selection: *Random* and *roulette* (maximum selection pressure: 500).

---

[7]Using more than one crossover operator within the SASEGASA does not mean using a combination of several operators for creating one new solution, but rather in the following way: Every time a new child is to be produced using two parent individuals, one of the given crossover operators is chosen randomly; the chance of being applied is equal for each operator.

Table 11.10: Experimental results for the *Thyroid* data set.

Using standard GP implementation		
Parameter	*Correct classifications*	
settings	*Evaluation*	*Prognosis*
(1)	92.80%	92.13%
(2)	93.91%	93.25%
Using the SASEGASA		
Parameter	*Correct classifications*	
settings	*Evaluation*	*Prognosis*
(3)	97.15%	96.34%
(4)	98.21%	98.07%
(5)	97.70%	97.25%
(6)	**98.93%**	**98.53%**

(5) crossover: *all 3 standard operators plus ExtendedSPLow*; mutation: *GA-Standard*; selection: *roulette* and *roulette* (maximum selection pressure: 500).

(6) crossover: *all 3 standard operators plus ExtendedSPLow*; mutation: *GA-Standard*; selection: *random* and *roulette* (maximum selection pressure: 500).

As an example, the model produced for cross validation partition 3 using the parameter settings combination (6) is shown in Figure 11.17.

These insights have been used also in the more extensive test series documented later on in this chapter.

11.2.5.6 Graphical Classifier Analysis

Graphical analysis can often help analyzing results achieved to any kind of problem; this is of course also the case in machine learning and in data-based classification.

The most common and also simplest way how to illustrate classification results is to plot the target values and the estimated values into one chart; Figure 11.14 shows a graphical representation of the best result obtained for the *Thyroid* data set, cross-validation set 9.

In Figure 11.15 we show 4 ROC chart examples that were generated for the classes '0' and '2' of the *Thyroid* data set, 10-fold cross validation set number 9:

(a) ROC curve for an unsuitable classifier for class '2', evaluated on training data;

Table 11.11: Summary of the best GP parameter settings for solving classification problems.

Parameter	Optimal Value
GP algorithm	SASEGASA (single population, i.e., GP with offspring selection)
Mutation rate	10% – 15%
Population size	1,000
Maximum selection pressure	100 – 1,000
Parent selection operators	Random, roulette
Crossover Operators	StandardSPLow, StandardSPMedium, StandardSPHigh, ExtendedSPLow
Mutation operator	GAStandard
Ratio of weighting the evaluation contributions (SumOfSquaredErrors : separability : class ranges)	4 : 1 : 1

(b) ROC curve for the best identified classifier for class '0', evaluated on training data;

(c) ROC curve for the best identified classifier for class '0', evaluated on test data;

(d) ROC curve for the best identified classifier for class '2', evaluated on test data.

In Figure 11.16 finally we show 4 MROC chart examples that were generated for the intermediate classes '1' of the *Thyroid* data set, again on the basis of 10-fold CV-set number 9:

(a) MROC curve for an unsuitable classifier for class '1', evaluated on training data;

(b) MROC curve for an unsuitable classifier for class '1', evaluated on test data;

(c) MROC curve for the best identified classifier for class '1', evaluated on training data;

(d) MROC curve for the best identified classifier for class '1', evaluated on test data.

On the webpage of this book[8] interested readers can find a collection of 10 example models (exactly one for each partition of the 10-fold cross-validation)

---

[8]http://gagp2009.heuristiclab.com/.

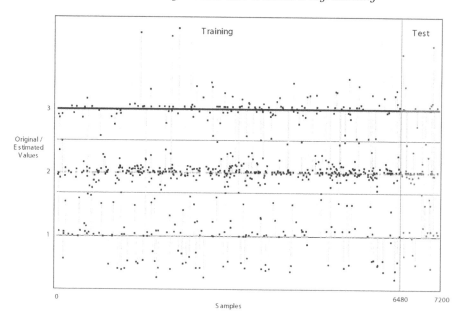

FIGURE 11.14: Graphical representation of the best result we obtained for the *Thyroid* data set, CV-partition 9: Comparison of original and estimated class values.

for the *Thyroid* data set, produced by GP; optimal thresholds are given as well as resulting confusion matrices for each data partition.

11.2.5.7 Classification Methods Applied in Detailed Test Series

For comparing GP-based classification with other machine learning methods, the following techniques for training classifiers were examined: Genetic programming (enhanced approach using extended parents and offspring selection), linear modeling, neural networks, the k-nearest-neighbor method, and support vector machines.

GP-Based Training of Classifiers

We have used the following parameter settings for our GP test series:

- Single population approach; population size: 500 – 1000

- Mutation rate: 10%

- Maximum formula tree height: 8

- Parent selection: Gender specific (random and roulette)

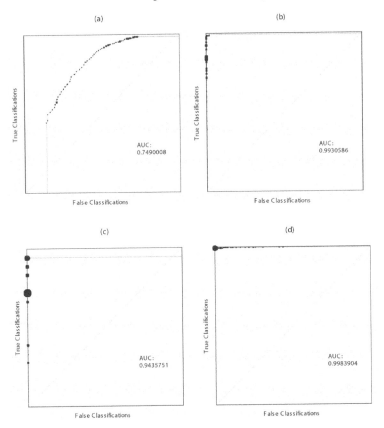

FIGURE 11.15: ROC curves and their area under the curve (AUC) values for classification models generated for *Thyroid* data, CV-set 9.

- Offspring selection: Strict offspring selection (success ratio as well as comparison factor set to 1.0)

- 1-elitism

- Termination criteria:

  - Maximum number of generations: 1000; not reached, all executions were terminated via the

  - Maximum selection pressure: 100

- Function set: All functions as described in Table 11.9.

- Fitness functions:

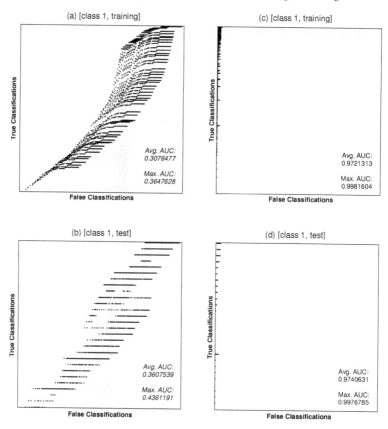

FIGURE 11.16: MROC charts and their maximum and average area under the curve (AUC) values for classification models generated for *Thyroid* data, CV-set 9.

- In order to keep the computational effort low, the mean squared errors function with early abortion was used as fitness function for the GP training process.

- The eventual selection of models is done by choosing those models that perform best on validation data (or, if no validation samples are specified, then the models' performance on training data is considered). For this selection we have used the classification specific evaluation function described in Section 11.2: The mean squared error is considered as well as class ranges, thresholds qualities, and AUC values; all other possible contributions have been neglected in the test series reported and discussed here. Thus, $c_1 = 4.0$, $c_k = 1.0$ for $k \in \{6, 7, 8\}$, and $c_k = 0.0$ for $k \in \{2, 3, 4, 5\}$.

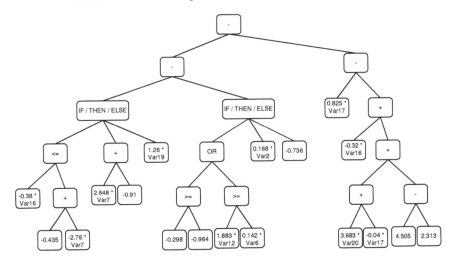

FIGURE 11.17: Graphical representation of a classification model (formula), produced for 10-fold cross validation partition 3 of the *Thyroid* data set.

In addition to splitting the given data into training and test data, extended GP-based training is implemented in such a way that a part of the given training data is not used for training models and serves as validation set; in the end, when it comes to returning classifiers, the algorithm returns those models that perform best on validation data. This approach has been chosen because it is assumed to help to cope with overfitting; it is also applied in other GP-based machine learning algorithms as for example described in [BL04]. In fact, this was also done in our standard GP tests for the *Melanoma* data set.

Linear Modeling

Given a data collection including m input features storing the information about N samples, a linear model is defined by the vector of coefficients $\theta_{1...m}$. For calculating the vector of modeled values e using the given input values matrix $u_{1...m}$, these input values are multiplied with the corresponding coefficients and added:

$$e = u_{1...m} * \theta \tag{11.54}$$

The vector of coefficients can be computed by simply applying matrix division. For conducting the test series documented here we have used the matrix division function provided by MATLAB®:

```
theta = InputValues \ TargetValues;
```

If a constant additive factor is to be included into the model (i.e., the coefficients vector), this command has to be extended:

```
r = size(InputValues,1);
theta = [InputValues ones(r,1)] \ TargetValues;
```

Theoretical background of this approach can be found in [Lju99].

Neural Networks

For training artificial neural network (ANN) models, three-layer feed-forward neural networks with one output neuron were created using the back-propagation as well as the Levenberg-Marquardt training method. Theoretical background and details can be found in [Nel01] (Chapter 11, "Neural Networks"), [Mar63], [Lev44], or [GMW82].

The following two approaches have been applied for training neural networks:

- On the one hand we have trained networks with 5 neurons in the hidden layer (referred to as "NN1" in the test series documentation in Section 11.2.5.8) as well as networks with 10 hidden neurons (referred to as "NN2" in the test series documentation); the number of iterations of the training process was set to 100 (in the first variant, "NN1") and 300 (in the second variant, "NN2"). In the context of analyzing the benchmark problems used here, higher numbers of nodes or iterations are likely to lead to overfitting (i.e., a better fit on the training data, but worse test results).
  The ANN training framework used to collect the results reported in this book is the NNSYSID20 package, a neural network toolbox implementing the Levenberg-Marquardt training method for MATLAB®; it has been implemented by Magnus Nørgaard at the Technical University of Denmark [Nør00].

- On the other hand, the multilayer perceptron training algorithm available in WEKA [WF05] has also been used for training classifiers. In this case the number of hidden nodes was set to $(a + c)/2$, where a is the number of attributes (features) and c the number of classes. The number of iterations was not pre-defined, but 10% of the training data were designated to be used as validation data; in order to combat the danger of overfitting, the training algorithm was terminated as soon as the error on validation data got worse in 20 iterations consecutively. This training method, which applies backpropagation learning, is in the following referred to as the "NN3" method.

kNN Classification

Unlike other data-based modeling methods based on linear models, neural networks or GP, k-nearest-neighbor classification works without creating any explicit models. During the training phase, the data are simply collected; when it comes to classifying a new, unknown sample x_{new}, the sample-wise distance between x_{new} and all other training samples x_{train} is calculated and the classification is done on the basis of those k training samples (x_{NN}) showing the smallest distances from x_{new}.

The distance between two samples is calculated as follows: First, all features are normalized by subtracting the respective mean values and dividing the remaining samples by the respective variables' standard deviation. Given a data matrix x including m features storing the information about N samples, the normalized values x_{norm} are calculated as

$$\forall(i \in [1, m])\forall(j \in [1, N]) : x_{norm}(i, j) = \frac{x(i, j) - \frac{1}{N}\sum_{k=1}^{N} x(i, k)}{\sigma(x(i, 1 \ldots N))} \quad (11.55)$$

where the standard deviation σ of a given variable x storing N values is calculated as

$$\sigma(x) = \sqrt{\frac{1}{N-1}\sum_{i=1}^{N}(x_i - \bar{x})^2} \quad (11.56)$$

with $\bar{x}$ denoting the mean value of x.

Then, on the basis of the normalized data, the distance between two samples a and b, $d(a, b)$, is calculated as the mean squared variable-wise distance:

$$d(a, b) = \frac{1}{n}\sum_{i=1}^{n}(a_{norm}(i) - b_{norm}(i))^2 \quad (11.57)$$

where n again is the number of features stored for each sample.

In the context of classification, the numbers of instances (of the k nearest neighbors) are counted for each given class and the algorithm automatically predicts that class that is represented by the highest number of instances. In the test series documented in this book we have applied weighting to kNN classification: The distance between x_{new} and any sample x_z is relevant for the classification statement, and the weight of "nearer" samples is higher than that of samples that are "further" away from x_{new}.

There is a lot of literature that can be found for kNN classification; very good explanations and compact overviews of kNN classification (including several possible variants and applications) are for example given in [DHS00] and [RN03].

Support Vector Machines

Support vector machines (SVMs) are a widely used approach in machine learning based on statistical learning theory [Vap98]; an example of the application of SVMs in the medical domain has been reported in [MIB⁺00], for example.

The most important aspect of SVMs is that it is possible to give bounds on the generalization error of the models produced, and to select the respectively best model from a set of models following the principle of structural risk minimization [Vap98]. SVM are designed to calculate hyperplanes that separate the data from each other and maximize the margin between sets of data points. While the basic training algorithm is only able to construct linear separators, so-called kernel functions can be used to calculate scalar products in higher-dimensional spaces; if the kernel functions used are nonlinear, then the separating boundaries will be nonlinear, too.

In this work we have used the SVM implementation described in [Pla99] and [KSBM01]; we have used the implementation of this algorithm which is available for the WEKA machine learning framework [WF05]. Polynomial kernels have been used as well as Gaussian radial basis function kernels with the γ parameter (defining the inverse variance) set to 0.01 and the complexity parameter c set to 10,000.

11.2.5.8 Detailed Test Series Results

The results summarized in this section have been partially published in [WAW06b], [WAW06e], and [WAW07].

Since the *Wisconsin* and the *Thyroid* data sets are publicly available, the results produced by GP are compared to those that have been published previously for various machine learning methods; the *Melanoma* is not openly available, therefore we have used all machine learning approaches mentioned for training classifiers for this data set.

All three data sets were investigated via 10-fold cross-validation (CV). For each data collection, each of the resulting 10 pairs of training and test data partitions has been used in 5 independent GP test runs; for the *Melanoma* data set, all machine learning algorithms mentioned previously have also been applied to all pairs of training and test data, the stochastic algorithms again applied 5 times independently.

Results for the *Wisconsin* Data Set

Table 11.12 summarizes the results for the 10-fold cross validation produced by GP with offspring selection; these figures boil down to the fact that extended GP has in this case been able to produce classifiers that on average correctly classify 97.91% of training samples and 97.53% of test samples.

Table 11.12: Summary of training and test results for the *Wisconsin* data set: Correct classification rates (average values and standard deviation values) for 10-fold CV partitions, produced by GP with offspring selection.

Partition	Training		Test	
	Avg.	*Std.Dev.*	*Avg.*	*Std.Dev.*
0	97.69%	0.27	97.06%	1.04
1	97.69%	0.85	97.65%	2.23
2	98.40%	0.72	97.94%	1.32
3	98.37%	0.56	98.24%	1.23
4	97.52%	0.78	97.06%	2.08
5	97.95%	0.77	97.94%	1.32
6	98.05%	0.43	97.05%	1.47
7	98.05%	0.47	97.65%	1.68
8	97.75%	0.62	97.65%	1.32
9	97.62%	0.74	97.06%	1.47
Avg.	**97.91%**	**0.62**	**97.53%**	**1.51**

In order to compare the quality of these results to those reported in the literature, Table 11.13 summarizes test accuracies that have been obtained using 10-fold cross validation. For each method listed we give the references to the respective articles in which these results have been reported[9]. Obviously the results summarized in Table 11.12 have to be considered surprisingly good as they outperform all other algorithms reported in the literature listed here. In [LC05], for example, recent results for several classification benchmark problems are documented; the *Wisconsin* data set was there analyzed using standard GP as well as three other GP-based classification variants (POPE-GP, DecMO-GP, and DecMOP-GP), and the respective results are also listed in Table 11.13.

Of course, for the sake of honesty we have to admit that the effort of GP to produce these classifiers is higher than the runtime or memory consumed by most other machine learning algorithms; in our GP tests using the *Wisconsin* data set and populations with 500 individuals the average number of generations executed was 51.6 and the average number of solutions evaluated ~1,296,742.

Results for the *Melanoma* Data Set

For the *Melanoma* data set no results are available in the literature; therefore we have tested all machine learning algorithms mentioned previously for getting an objective evaluation of our GP methods.

---

[9] An even more detailed listing of test results for this data set can be found in [JHC04].

Table 11.13: Comparison of machine learning methods: Average test accuracy of classifiers for the *Wisconsin* data set.

Algorithm	Test Accuracy
GP with OS	97.53%
Probit [WHMS03]	97.20%
RLP [BU95]	97.07%
SVM [WHMS03]	96.70%
C4.5 (decision tree) [HSC96]	96.0%
ANN [TG97]	95.61%
DecMOP-GP [LC05]	95.60%
DecMO-GP [LC05]	95.19%
POPE-GP [LC05]	95.08%
StandardGP [LC05]	93.82%

First, in Table 11.14 we summarize original vs. estimated classifications obtained by applying the classifiers produced by GP with offspring selection; in total, 97.17% of the training and 95.42% of the test samples are classified correctly (with standard deviations 0.87 and 2.13, respectively). These GP tests using the *Melanoma* data set were done with populations containing 1,000 individuals; the average number of generations executed was 54.4 and the average number of solutions evaluated ∼2,372,629.

Table 11.14: Confusion matrices for average classification results produced by GP with OS for the *Melanoma* data set.

Training		*Original Classification*	
		[0] (Benign)	*[1] (Malign)*
Estimated	*[0]*	1,043.21 (88.41%)	9.09 (0.77%)
Classification	*[1]*	24.28 (2.06%)	103.42 (8.76%)
Test		*Original Classification*	
		[0] (Benign)	*[1] (Malign)*
Estimated	*[0]*	115.18 (87.92%)	2.67 (2.04%)
Classification	*[1]*	3.33 (2.54%)	9.82 (7.50%)

Test results obtained using other machine learning algorithms are collected in Table 11.15. Support vector machine based training was done with radial as well as with polynomial kernel functions. Furthermore we used γ values 0.001 and 0.01. In standard GP (SGP) tests we used tournament parent selection ($k = 3$), 8% mutation, single point crossover and the same structural limitations as in GP with OS; in order to get a fair comparison, the population size was set to 1,000 and the number of generations to 2,500 yielding 2,500,000 evaluations per test run.

As we can see in Table 11.15, our GP implementation performs approximately as well as the support vector machines and neural nets applying those settings that are optimal in this test case: GP with OS was able to classify 95.42% of the test cases correctly, SVMs correctly classified 94.89% – 95.47% and neural nets (with validation set based stopping) 95.27% of the test cases evaluated. Standard GP as well as kNN, linear regression, and standard ANNs perform worse.

Even though it is nice to see that the average accuracy recorded for models produced by GP with OS is quite fine, the relatively high standard deviation of this method's performance (2.13, compared to 0.41 recorded for optimal SVMs) has to be seen as a negative aspect of these results.

Table 11.15: Comparison of machine learning methods: Average test accuracy of classifiers for the *Melanoma* data set.

Algorithm	Test Accuracy	
	Avg.	*Std.Dev.*
SVM (radial, $\gamma = 0.01$)	95.47%	0.41
GP with OS	*95.42%*	*2.13*
SVM (polynomial, $\gamma = 0.01$)	95.40%	0.56
SVM (radial, $\gamma = 0.001$)	95.27%	0.74
NN3	95.27%	1.91
SVM (polynomial, $\gamma = 0.001$)	94.89%	0.83
NN1	94.35%	2.39
kNN ($k = 3$)	93.59%	1.03
SGP	93.52%	3.72
NN2	92.90%	2.59
kNN ($k = 5$)	92.85%	0.94
Lin	92.45%	2.90

Results for the *Thyroid* Data Set

Finally, the results achieved for the *Thyroid* data set are to be reported here. Table 11.16 summarizes the results for the 10-fold cross validation produced by GP with offspring selection. For each class we characterize the classification accuracy on training and test data, giving average as well as standard deviation values for each partition. These figures boil down to the fact that extended GP has in this case been able to produce classifiers that on average correctly classify 99.10% of training samples and 98.76% of test samples, the total standard deviation values being 0.73 and 0.92, respectively.

Table 11.16: Summary of training and test results for the *Thyroid* data set: Correct classification rates (average values and standard deviation values) for 10-fold CV partitions, produced by GP with offspring selection.

Partition		Training			Test		
		Class 1	*Class 2*	*Class 3*	*Class 1*	*Class 2*	*Class 3*
0	*avg.*	94.67%	97.64%	99.63%	90.00%	95.68%	99.19%
	std.dev.	1.70	2.65	0.52	7.13	4.10	0.64
1	*avg.*	94.93%	98.67%	99.01%	88.75%	96.76%	99.43%
	std.dev.	3.58	4.23	0.46	5.23	5.86	0.44
2	*avg.*	96.67%	98.49%	99.49%	91.25%	96.22%	97.90%
	std.dev.	1.89	2.00	0.55	3.42	6.51	2.02
3	*avg.*	96.00%	98.19%	99.15%	90.00%	95.68%	99.46%
	std.dev.	2.87	1.56	0.42	5.59	4.10	0.31
4	*avg.*	95.33%	97.04%	99.19%	88.75%	96.22%	99.61%
	std.dev.	2.45	5.38	0.35	11.18	3.63	0.27
5	*avg.*	95.07%	96.62%	99.22%	95.00%	94.59%	99.37%
	std.dev.	1.92	5.29	0.40	5.23	4.27	0.29
6	*avg.*	93.47%	97.76%	99.16%	87.50%	94.59%	98.56%
	std.dev.	2.18	7.64	0.49	7.65	4.27	0.91
7	*avg.*	98.80%	98.97%	99.16%	87.50%	92.97%	99.40%
	std.dev.	2.18	5.92	0.49	7.65	4.52	0.30
8	*avg.*	94.40%	98.01%	99.23%	96.25%	94.05%	99.34%
	std.dev.	5.11	4.99	0.64	5.23	3.52	0.57
9	*avg.*	97.73%	96.62%	99.31%	91.25%	92.43%	99.55%
	std.dev.	2.69	2.65	0.52	3.42	3.52	0.15
Avg.	*avg.*	95.71%	97.80%	99.26%	90.63%	94.92%	99.18%
	std.dev.	2.66	4.23	0.48	6.17	4.43	0.59

In order to compare the quality of these results to those reported in the literature, Table 11.17 summarizes a selection of test accuracies that have been obtained using 10-fold cross validation; again, for each method listed we give the references to the respective articles in which these results have been reported. Obviously, the results summarized in Table 11.16 have to be considered quite fine, but not perfect as they are outperformed by results reported in [WK90] and [DAG01].

GP has also been repeatedly applied for solving the *Thyroid* problem; some of the results published are the following ones:

In [LH06] (Table 8), results produced by a pareto-coevolutionary GP classifier system for the *Thyroid* problem are reported, and here in Table 11.17 these results are stated as the "PGPC" results; in fact, these results are not the mean accuracy values but rather the median value, which is why these results are not totally comparable to other results stated here. Loveard and Ciesielski

[LC01] reported that classifiers for the *Thyroid* problem could be produced using GP with test accuracies ranging from 94.9% to 98.2% (depending on the range selection strategy used).

According to Banzhaf and Lasarczyk [BL04], GP-evolved programs consisting of register machine instructions turned out to eventually misclassify on average 2.29% of the given test samples, and that optimal classifiers are able to correctly classify 98.64% of the test data.

Furthermore, Gathercole and Ross [GR94] report classification errors between 1.6% and 0.73% as best result using tree-based GP, and that a classification error of 1.52% for neural networks is reported in [SJW92]. In fact, Gathercole and Ross reformulated the *Thyroid* problem to classifying cases as "class 3" or "not class 3"; as is stated in [GR94], it turned out to be relatively straight-forward for their GP implementation (DSS-GP) to produce function tree expressions which could distinguish between classes "1" and "2" completely correctly on both the training and test sets. "To be fair, in splitting up the problem into two phases (class *3* or not, then class *1* or *2*) the GP has been presented with an easier problem [...]. This could be taken in different ways: Splitting up the problem is mildly cheating, or demonstrating the flexibility of the GP approach." (Taken from [GR94].)

Table 11.17: Comparison of machine learning methods: Average test accuracy of classifiers for the *Thyroid* data set.

Algorithm	Accuracy	
	Training	**Test**
CART [WK90]	99.80%	99.36%
PVM [WK90]	99.80%	99.33%
Logical Rules [DAG01]	–	99.30%
GP [GR94]	–	$98.4\% - 99.27\%$
GP with OS	*99.10%*	*98.76%*
GP [BL04]	–	$97.71\% - 98.64\%$
GP [LC01]	–	$94.9\% - 98.2\%$
BP + local adapt. rates [SJW93]	99.6%	98.5%
ANN [SJW92]	–	98.48%
BP + genetic opt. [SJW93]	99.4%	98.4%
Quickprop [SJW93]	99.6%	98.3%
RPROP [SJW93]	99.6%	98.0%
PGPC [LH06]	–	97.44%

GP with strict offspring selection was here applied with populations of 1000 individuals; on average, the number of generations executed in our GP tests for the *Thyroid* test studies was 73.9, and on average 2,463,635.1 models were evaluated in each GP test run.

11.2.5.9   Conclusion

We have here described an enhanced genetic programming method that was successfully used for investigating machine learning problems in the context of medical classification. The approach works with hybrid formula structures combining logical expressions (as used for example in decision trees) and classical mathematical functions; the enhanced selection scheme originally successfully applied for solving combinatorial optimization problems using genetic algorithms was also applied yielding high quality results.

We have intensively investigated GP in the context of learning classifiers for three medical data collections, namely the *Wisconsin* and the *Thyroid* data sets taken from the UCI machine learning repository and the *Melanoma* data set, a collection that represents medical measurements which were recorded while investigating patients potentially suffering from skin cancer. The results presented in this section are indeed satisfying and make the authors believe that an application in a real-world framework in the context of medical data analysis using the techniques presented here is recommended. As documented in the test results summary, our GP-based classification approach is able to produce results that are – in terms of classification accuracy – at least comparable to or even better than the classifiers produced by classical machine learning algorithms frequently used for solving classification problems, namely linear regression, neural networks, neighborhood-based classification, or support vector machines as well as other GP implementations that have been used on the data sets investigated in our test studies.

11.3   Genetic Propagation

11.3.1   Test Setup

When speaking of analysis of genetic propagation as described in Section 6.1, we analyze how well which parts of the population succeed in propagating their genetic material to the next generation, i.e., to produce offspring that will be included in the next generation's population. In this section we shall report on tests in this area; major parts have been published in our article on offspring selection and its effects on genetic propagation in GP-based system identification [WAW08] as well as in [Win08].

We have here used the NO_x data set II already presented and described in Section 11.1.2.3. Originally, this data set includes 10 variables, each storing approximately 36,000 samples; the first 10,000 samples are neglected in the tests reported on here, approximately 18,000 samples are training, and 4,000 samples are validation (which is in this case equivalent to test) data. The last ~4,000 samples are again neglected.

In principle, we are using conventional GP (with tournament and proportional selection) as well as extended GP (with gender specific selection as well as offspring selection). The details of the test strategies used are given in Table 11.18.

Table 11.18: GP test strategies.

Strategy	Properties
(I) *Conventional GP*	|Pop| = 1000; Tournament parent selection ($k = 3$) nr. of rounds: 1000
(II) *Conventional GP*	|Pop| = 1000; Proportional parent selection; nr. of rounds: 1000

(III)
*Extended
GP*
	|Pop| = 500;
	Gender specific parent selection
	  (proportional, random);
	Offspring selection
	  (SuccessRatio = 1, MaxSelPres = 100)

In all three test strategies we used subtree exchange crossover, the time series analysis specific evaluation function (with early abortion as described in Section 11.1.1) for evaluating solutions, and applied 1-elitism as well as 15% mutation rate.

11.3.2 Test Results

We have executed independent test series with 5 executions for each test strategy; the results are to be summarized and analyzed here.

With respect to solution quality and effort[10], the extended GP algorithm clearly outperforms the conventional GP variants (as summarized in Table 11.19).

Table 11.19: Test results.

		I	**II**	**III**
Best	min.	1,390.21	3,022.12	1,201.23
Quality	avg.	1,513.84	5,014.96	1,481.69
(Training)	max.	2,431.54	10,013.12	2,012.27
Best	min.	8,231.76	12,312.83	4,531.56
Quality	avg.	10,351.96	15,747.69	8,912.61
(Test)	max.	13,945.23	21,315.23	16,123.34
Generations		500		64.31
Effort		1,000,000		898,332.23

Regarding parent analysis, in all test runs we documented the propagation count for each individual and sum these over all generations. So we get

$$pc_{total}(i) = \sum_{i \in [1;gen]} pc(i) \tag{11.58}$$

for each individual index i and assume that gen is the number of generations executed. Additionally, we form equally sized partitions of the population indices and sum up the pc_{total} values for each partition.

In Table 11.20 we give the average pc_{total} values for percentiles of the populations of test series I, II, and III; for test series I and II we collected the pc_{total} of 100 indices for forming a partition, and for test series III we collected 50 indices for each partition. The Figures 11.18 and 11.19 show pc_{total} values of exemplary test runs of the series I and II summed up for partitions of 10 solution indices each. Figure 11.20 shows pc_{total} values of exemplary test runs of series III summed up for partitions of 5 solution indices each.

---

[10]The number of solutions evaluated is here interpreted as the algorithm's total effort.

Table 11.20: Average overall genetic propagation of population partitions.

Population Percentile	Test Strategy		
	I	II	III
0	27.88%	10.31%	13.54%
1	21.29%	10.35%	11.20%
2	16.65%	10.31%	11.67%
3	12.64%	10.26%	10.91%
4	9.06%	10.25%	10.63%
5	6.08%	10.28%	9.85%
6	3.71%	10.24%	9.39%
7	1.88%	10.16%	8.83%
8	0.72%	10.10%	7.92%
9	0.10%	7.74%	6.07%

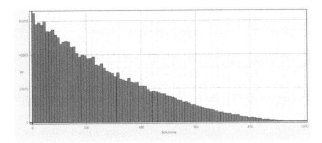

FIGURE 11.18: pc_{total} values for an exemplary run of series I.

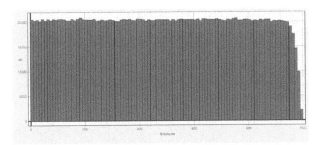

FIGURE 11.19: pc_{total} values for an exemplary run of series II.

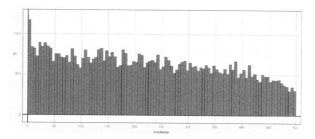

FIGURE 11.20: pc_{total} values for an exemplary run of series III.

As we see from the results given in Tables 11.19 and 11.20 and Figure 11.18, there is a rather high selection pressure when using tournament selection; the results are rather good and (as expected) less fit individuals are by far not able to contribute to the population as well as fitter ones, leading to a quick and drastic reduction of genetic diversity.

The results for test series II, as given in Tables 11.19 and 11.20 and Figure 11.19, are significantly different: The results are a lot worse (especially on training data) than those of algorithm variant I, and obviously there is no strong selection pressure as almost all individuals (or, rather the individuals at the respective indices) are able to contribute almost to the same extent. Only the worst ones are not able to propagate their genetic material to the next generations as well as better ones. This is due to the fact that in the presence of very bad individuals roulette wheel selection selects the best individuals approximately as often as those that perform middlingly well. Especially in data-based modeling there are often individuals that score extremely badly (due to divisions by very small values, for example), and in comparison to those all other ones are approximately equally fit.

Finally, test series III obviously produced the best results with respect to training as well as validation data (see also Table 11.19). Even more, the results that are given in Table 11.20, column III, and displayed in Figure 11.20, show that the combination of random and roulette parent selection and offspring selection results in a very moderate distribution of the pc_{total} values: Fitter individuals contribute more than less fit ones, but even the worst ones are still able to contribute to a significant extent. Thus, genetic diversity is increased which also contributes positively to the genetic programming process.

11.3.3 Summary

Thus, in order to sum up this section, offspring selection in GP-based system identification significantly influences the algorithm's ability to create high quality results as well as the genetic propagation dynamics: Not only fitter individuals are able to propagate their genetic make-up, but also less fit ones are able to contribute to the next population. This is also somehow the case when using proportional selection, but in the presence of individuals with very bad fitness values the selection pressure is almost lost which leads to solutions of rather bad quality. When using offspring selection, extremely bad individuals are eliminated immediately; when using OS in combination with gender specific parent selection (applying random and proportional selection mechanisms), GP is able to produce significantly better results than when using standard techniques. Parents diversification and thus increased genetic diversity in GP populations is considered one of the most influential aspects in this context.

11.3.4   Additional Tests Using Random Parent Selection

In addition to the tests reported on in the previous parts of this section we have also tested conventional as well as extended GP using random parent selection. Thus, we have two more test cases to be analyzed.

Table 11.21: Additional test strategies for genetic propagation tests.

Strategy	Properties
(IV) *Conventional GP*	$\|Pop\| = 2000$; Random parent selection nr. of rounds: 500
(V) *Extended GP*	$\|Pop\| = 500$; Random parent selection Offspring selection (SuccessRatio $= 1$, MaxSelPres $= 100$)

As we had expected, the test results obtained for standard GP with random parent selection were very bad; obviously, no suitable models were found. When using OS, on the contrary, the test results for random parent selection were not that bad at all: The models are (on training data) not quite as good as those obtained using random/roulette and OS or conventional GP with tournament parent selection, but still they perform (surprisingly) well on test data[11]. In Table 11.22 we summarize the respective result qualities.

---

[11]Of course, these remarks are only valid for the tests reported on here - here we do not give any general statement regarding result quality using random parent selection and OS.

Table 11.22: Test results in additional genetic propagation tests (using random parent selection).

		IV	V
Best	min.	>50,000.00	5,041.22
Quality	avg.	>50,000.00	7,726.11
(Training)	max.	>50,000.00	8,843.73
Best	min.	>50,000.00	7,129.31
Quality	avg.	>50,000.00	8,412.31
(Test)	max.	>50,000.00	12,653.98
Generations		500	102.86
Effort		1,000,000	1,324,302

In Table 11.23 we give the average pc_{total} values for percentiles of the populations of test series IV and V (collecting the pc_{total} values of 200 indices for forming a partition for series IV and 50 indices for each partition for series V). Obviously (and exactly as we had expected) random parent selection leads to all individuals having the approximately same success in propagating their genetic make-up. When using OS, the result is (even a little bit surprisingly) significantly different: Better individuals have a much higher chance to produce successful offspring than worse ones; the probability of the best 10%, for example, to produce successful children is almost twice as high as the probability of the worst 10% to do so.

Table 11.23: Average overall genetic propagation of population partitions for random parent selection tests.

Population Percentile	Test Strategy	
	IV	V
0	10.03 %	13.16 %
1	10.02 %	12.07 %
2	9.98 %	11.41 %
3	9.99 %	10.66 %
4	10.02 %	10.30 %
5	9.99 %	9.33 %
6	10.00 %	8.96 %
7	9.98 %	8.62 %
8	9.99 %	7.81 %
9	10.00 %	7.68 %

Obviously, random parent selection leads to an increased number of generations that have to be executed until a given selection pressure limit is reached. This is graphically shown in Figure 11.21, which gives the selection pressure progress for two exemplary test runs of the test series including OS, i.e., III and V. In the standard case using random / roulette parent selection and offspring selection, III, the selection pressure obviously rises faster than when using random parent selection in combination with strict offspring selection. Still, even though it takes longer when using random parent selection, the characteristics are very similar, i.e., it rises steadily with some notable fluctuations.

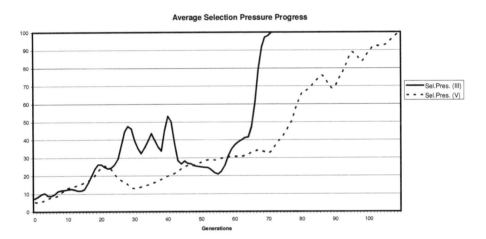

FIGURE 11.21: Selection pressure progress in two exemplary runs of test series III and V (extended GP with gender specific parent selection and strict offspring selection).

11.4 Single Population Diversity Analysis

11.4.1 GP Test Strategies

Within our first series of empirical tests regarding solutions similarity and diversity we analyzed the diversity of populations of single population GP processes. For testing the population diversity analysis method described in Section 6.2 and illustrating graphical representations of the results of these tests we have used the following two data sets:

- The NO_x data set contains the measurements taken from a 2 liter 4 cylinder BMW diesel engine at a dynamical test bench (simulated vehicle: BMW 320d Sedan); this data set has already been described in Section 11.1 as NO_x data set III.

- The *Thyroid* data set is a widely used machine learning benchmark data set containing the results of medical measurements which were recorded while investigating patients potentially suffering from hypothyroidism; further details regarding this data set can be found in Chapter 11.2.

Both data collections have been split into training and validation / test data partitions taking the first 80% of each data set as training samples available to the identification algorithm; the rest of the data is considered as validation data.

We have used various GP selection strategies for analyzing the NO_x and the *Thyroid* data sets:

- On the one hand, we have used standard GP with proportional as well as tournament selection (tournament size $k = 3$).

- On the other hand we have also intensively tested GP using offspring selection and gender specific parent selection (proportional and random selection).

In general, we have tested GP with populations of 1,000 solution candidates (with a maximum tree size of 50 and a maximum tree height of 5), standard subtree exchange crossover, structural as well as parametric node mutation and total 15% mutation rate; the mean squared errors function was used for evaluating the solutions on training as well as on validation (test) data. Other essential parameters vary depending on the test strategies; these are summarized in Table 11.24.

Table 11.24: GP test strategies.

Strategy	Properties
(A) *Standard GP*	Tournament parent selection (tournament size $k = 3$); Number of generations: 4000
(B) *Standard GP*	Proportional parent selection; Number of generations: 4000
(C) *GP with OS*	Gender specific parent selection; (Random & proportional) Success ratio: 0.8 Comparison factor: 0.8 (Maximum selection pressure: 50 (not reached) Number of generations: 4000
(D) *GP with OS*	Gender specific parent selection; (Random & proportional) Success ratio: 1.0 Comparison factor: 1.0 Maximum selection pressure: 100

11.4.2   Test Results

In Table 11.25 we summarize the quality of the best models produced using the GP test strategies (A) – (D); for the NO_x data set the quality is given as the mean squared error; for the *Thyroid* data set we give the classification accuracy, i.e., the ratio of samples that are classified correctly. The models are evaluated on training as well as on validation data; as each test strategy was executed 5 times independently, we here state mean average and standard deviation values.

Obviously, the test series (A) and (D) perform best; the results produced using offspring selection are better than those using standard GP. The classification results for the *Thyroid* data set are not quite as good as those reported in [WAW06e] and Section 11.2; this is due to the fact that we here used smaller models and concentrated on the comparison of GP strategies with respect to population diversity.

Solution quality analysis is of course important and interesting, but here we are more interested in a comparison of population diversity during the execution of the GP processes. We have calculated the similarity among the GP populations during the execution of the GP test series described in Table 11.24: The multiplicative similarity approach (as defined in Equations 9.63 – 9.66) has been chosen; all coefficients $c_1 \ldots c_{10}$ were set to 0.2, only the coefficient c_1 weighting the level difference contribution d_1 was set to 0.8.

Table 11.25: Test results: Solution qualities.

Results for NO_x test series				
	GP Strategy			
	(A)	**(B)**	**(C)**	**(D)**
Training (mse)	2.518	5.027	2.674	1.923
Training ($std(mse)$)	1.283	2.142	2.412	0.912
Validation (mse)	3.012	5.021	2.924	2.124
Validation ($std(mse)$)	1.431	3.439	2.103	1.042
Evaluated solutions, avg.	$4 \cdot 10^6$	$4 \cdot 10^6$	$10.2 \cdot 10^6$	$3.91 \cdot 10^6$
Generations (avg.)	4,000	4,000	4,000	98.2
Results for $Thyroid$ test series				
	GP Strategy			
	(A)	**(B)**	**(C)**	**(D)**
Training (cl. acc., avg.)	0.9794	0.9758	0.9781	0.9812
Training (cl. acc., std)	0.0032	0.0017	0.0035	0.0012
Validation (cl. acc., avg.)	0.9764	0.9675	0.9767	0.9804
Validation (cl. acc., std)	0.0029	0.0064	0.0069	0.0013
Evaluated solutions, avg.	$4 \cdot 10^6$	$4 \cdot 10^6$	$12.2 \cdot 10^6$	$5.1 \cdot 10^6$
Generations (avg.)	4,000	4,000	4,000	167.8

In Table 11.26 we give the average population similarity values calculated using Equation 6.7; again, as each test series was executed several times, we give the average and standard deviation values (written in italic letters). As we see in the first row, the average similarity values are approximately in the interval $[0.2; 0.25]$ at the beginning of the GP runs, i.e., after the initialization of the GP populations. In standard GP, as can be seen in the first column, the average similarity reaches values above 0.7 after 400 generations and stays at approximately this level until the end of the execution of the GP process; in the end, the average similarity was ∼0.87 in the NO_x tests and ∼0.81 in the *Thyroid* test series. Analyzing the second and the third column we notice that this is not the case in test series (B) and (C): The similarity values do in test series (B) by far not rise as high as in series (A) (especially when working on the *Thyroid* data set), and also in test series (C) we have measured significantly lower similarities than in series (A) (i.e., the population diversity was higher during the whole GP process). Obviously, the use of offspring selection with rather soft parameter settings (i.e., success ratio and comparison factor set to values below 1.0) does not have the same effects on the GP process as strict ones. The by far highest similarity values are documented for test series (D) using maximally strict offspring selection (which has produced the best quality models, as documented in Table 11.25): As is summarized in the far right column, during the whole evolutionary process the mutual similarity among the models increases steadily, while also the selection pressure increases. In the end, when the selection pressure reaches a high level (in these cases, the predefined limit was set to 100) and the algorithm stops, we see a very high similarity among the solution candidates, i.e., the population has converged and evolution is likely to have gotten stuck. This is in fact consistent with the impression already stated in [WAW06a] or [WAW06e], e.g.; here we see that this in fact really happens.

Table 11.26: Test results: Population diversity (average similarity values; avg., std.).

Gen.	GP Strategy (A)	(B)	(C)	Gen.	GP Strategy (D)
	NO$_x$ tests				
0	0.247	0.250	0.270	0	0.197
	0.041	*0.031*	*0.037*		*0.039*
100	0.723	0.491	0.517	10	0.397
	0.073	*0.051*	*0.038*		*0.039*
400	0.813	0.497	0.564	20	0.603
	0.035	*0.058*	*0.059*		*0.049*
1000	0.859	0.510	0.520	40	0.810
	0.021	*0.055*	*0.052*		*0.039*
4000	0.871	0.518	0.526	End of	0.985
(End of run)	*0.019*	*0.059*	*0.053*	run	*0.032*
	Thyroid tests				
Gen.	GP Strategy (A)	(B)	(C)	Gen.	GP Strategy (D)
0	0.206	0.205	0.208	0	0.197
	0.041	*0.040*	*0.036*		*0.040*
100	0.581	0.241	0.444	10	0.397
	0.047	*0.043*	*0.035*		*0.039*
400	0.737	0.321	0,610	20	0.602
	0.032	*0.058*	*0.026*		*0.049*
1000	0.808	0.341	0.692	40	0.810
	0.029	*0.049*	*0.031*		*0.041*
4000	0.812	0.343	0.701	End of	0.975
(End of run)	*0.038*	*0.056*	*0.030*	run	*0.019*

In Table 11.27 we summarize the maximum population diversity values calculated using Equation 6.8; again we give the average and standard deviation values (written in italic letters). As we see in the first (left) column, in standard GP with tournament selection the average maximum similarity reaches

Table 11.27: Test results: Population diversity (maximum similarity values; avg., std.).

Gen.	GP Strategy (A)	(B)	(C)	Gen.	GP Strategy (D)
NOₓ tests					
0	0.919	0.934	0.904	0	0.936
	0.116	*0.095*	*0.123*		*0.109*
100	0.995	0.825	0.944	10	0.961
	0.014	*0.074*	*0.059*		*0.049*
400	0.998	0.809	0.978	20	0.971
	0.006	*0.075*	*0.037*		*0.033*
1000	0.999	0.811	0.965	40	0.995
	0.005	*0.059*	*0.044*		*0.012*
4000	0.999	0.819	0.969	End of	0.996
(End of run)	*0.003*	*0.066*	*0.035*	run	*0.009*
Thyroid tests					
Gen.	GP Strategy (A)	(B)	(C)	Gen.	GP Strategy (D)
0	0.823	0.771	0.766	0	0.777
	0.127	*0.145*	*0.145*		*0.157*
100	0,958	0.749	0.840	10	0.873
	0.028	*0.123*	*0.094*		*0.101*
400	0.973	0.752	0.883	20	0.934
	0.032	*0.125*	*0.067*		*0.049*
1000	0.977	0.744	0.913	40	0.976
	0.022	*0.117*	*0.061*		*0.022*
4000	0.977	0.754	0.909	End of	0.999
(End of run)	*0.021*	*0.111*	*0.058*	run	*0.004*

values above 0.95 rather fast, i.e., for all models in the population rather similar solutions can be found. This is not the case when using proportional selection. When using offspring selection the same effect as in standard GP with tournament selection can be seen, especially in the NO_x test series.

The Figures 11.22 – 11.25 exemplarily show the average population diversity by giving the distribution of similarities among all individuals. The Figures 11.22 and 11.23 show the similarity distributions of an exemplary test run of series (A) at generation 200 and 4000; obviously, most similarity calculations returned similarity values between 0.7 and 1.0, and the distribution at generation 200 is comparable to the distribution at the end of the test run. For the GP runs incorporating offspring selection this is not the case, as we exemplarily see in Figures 11.24 and 11.25: After 20 generations most similarity values almost fit Gaussian distribution with mean value 0.8, and at the end of the run all models are very similar to each other (i.e., the population has converged, the selection pressure reaches the given limit and the algorithm

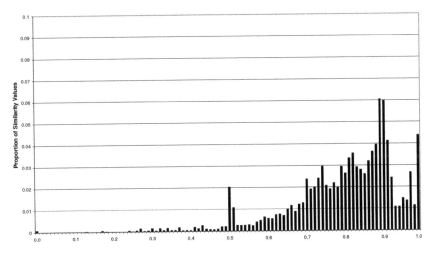

FIGURE 11.22: Distribution of similarity values in an exemplary run of NO_x test series A, generation 200.

stops).

Finally, Figure 11.26 shows the average similarity values for each model (calculated using Equation 6.5) for exemplary test runs of the *Thyroid* test series (A)[12] and (D). Obviously, the average similarity in standard GP reaches values in the range [0.7;0.8] very early and then stays at this level during the rest of the GP execution. When using gender specific selection and offspring selection, otherwise, the average similarity steadily increases during the GP process and almost reaches 1.0 at the end of the run, when the maximum selection pressure is reached.

11.4.3   Conclusion

Structural similarity estimation has been used for measuring the genetic diversity among GP populations: Several variations of genetic programming using different types of selection schemata have been tested using fine-grained similarity estimation, and two machine learning data sets have been used for these empirical tests. The test results presented show that population diversity differs a lot in the test runs depending on the selection schemata used.

---

[12]In fact, for the test run of series (A) we here only show the progress over the first 2000 generations.

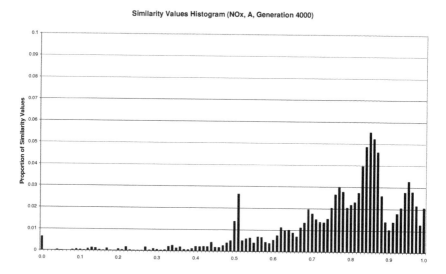

FIGURE 11.23: Distribution of similarity values in an exemplary run of NO_x test series A, generation 4000.

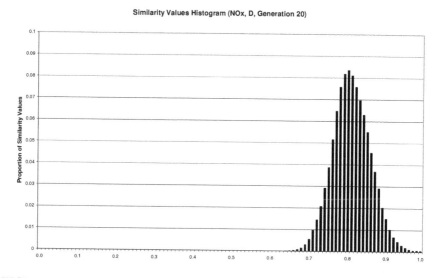

FIGURE 11.24: Distribution of similarity values in an exemplary run of NO_x test series (D), generation 20.

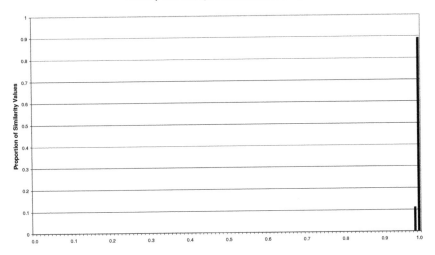

FIGURE 11.25: Distribution of similarity values in an exemplary run of NO_x test series (D), generation 95.

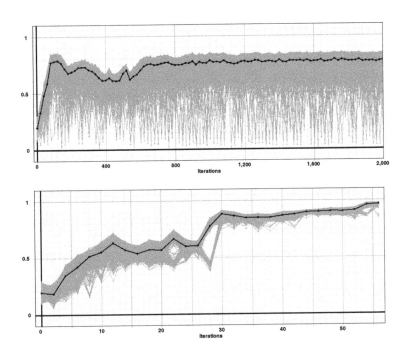

FIGURE 11.26: Population diversity progress in exemplary *Thyroid* test runs of series (A) and (D) (shown in the upper and lower graph, respectively).

11.5   Multi-Population Diversity Analysis

Our second series of empirical tests regarding solutions similarity and diversity was dedicated to the diversity of populations of multi-population GP processes; for testing the multi-population diversity analysis method described in Section 6.2 and illustrating graphical representations of the results of these tests we have again used the following two data sets: The NO_x data set III described in Section 11.1 as well as the *Thyroid* data set.

Both data collections have been split into training and validation / test data partitions: In the case of the NO_x data set the first 50% of the data set were used as training samples; in the case of the *Thyroid* data set the first 80% were considered by the training algorithms.

11.5.1   GP Test Strategies

In general, 4 different strategies for parallel genetic programming have been applied:

- Parallel island GP without interaction between the populations; i.e., all populations evolve independently.

- Parallel island GP with occasional migration after every 100th generation in standard GP and every 5th generation in GP with offspring selection: The worst 1% of each population p_i is replaced by copies of the best 1% of solutions in population p_{i-1}; the best solutions of the last population (in the case of n population that is p_n) replace the worst ones of the first population (p_1). The unidirectional ring migration topology has been used.

- Parallel island GP with migration after every 50th generation in standard GP and every 5th generation in GP with offspring selection: The worst 5% of each population p_i is replaced by copies of the best 5% of solutions in population p_{i-1}. Again, the unidirectional ring migration topology has been used.

- Finally, the SASEGASA algorithm as described in Chapter 5 has been used as well.

In all cases the algorithms have been initialized with 5 populations, each containing 200 solutions (in our case representing formulas, of course). Additionally, each of the first 3 strategies has been tested with standard GP settings as well as offspring selection; Table 11.28 summarizes the 7 test strategies that have been applied and whose results shall be discussed here.

11.5.2 Test Results

All test strategies summarized in Table 11.28 have been executed 5 times using the NO_x as well as the *Thyroid* data set. Multi-population diversity was measured using the equations given in Section 6.2.2: For each solution we calculate the average as well as the maximum similarities with solutions of all other populations of the respective algorithms (in the following, these values are denoted as $MPdiv$ values). Additionally, we have also collected all solutions of the algorithms' populations into temporary total populations and calculate the average as well as the maximum similarities of all solutions compared to all other ones (hereafter denoted as $SPdiv$ values).

Again, the multiplicative structural similarity approach (as defined in Equations 9.63 – 9.66) has been used for estimating the similarity of model structures; all coefficients $c_1 \ldots c_{10}$ were set to 0.2, only the coefficient c_1 weighting the level difference contribution d_1 was set to 0.8.

In the following we summarize these values for all test runs by stating the average values as well as standard deviations: Table 11.29 summarizes the results of the test runs using the *Thyroid* data set, 11.30 those of the test runs using the NO_x data set.

Figure 11.27 exemplarily illustrates the multi-population diversity in a test run of series F at iteration 50: The value represented in row i of column j in bar k gives the average similarity of model i of population k with all formulas stored in population j. Low multi-population similarity values are indicated by light cells; dark cells represent high similarity values.

Table 11.28: GP test strategies.

Strategy	Properties
(A) *Parallel standard GP*	Tournament parent selection (tournament size $k = 3$); Number of generations: 2000
(B) *Parallel GP with OS*	Random & roulette parent selection Strict Offspring selection (success ratio: 1.0, comparison factor: 1.0, maximum selection pressure: 200)
(C) *Parallel standard GP, 1% migration*	Tournament parent selection (tournament size $k = 3$); Number of generations: 2000 1% best / worst replacement after every 100th generation
(D) *Parallel GP with OS, 1% migration*	Random & roulette parent selection Strict Offspring selection (success ratio: 1.0, comparison factor: 1.0, maximum selection pressure: 200) 1% best / worst replacement after every 5th generation
(E) *Parallel standard GP, 5% migration*	Tournament parent selection (tournament size $k = 3$); Number of generations: 2000 5% best / worst replacement after every 50th generation
(F) *Parallel GP with OS, 5% migration*	Random & roulette parent selection Strict Offspring selection (success ratio: 1.0, comparison factor: 1.0, maximum selection pressure: 200) 5% best / worst replacement after every 5th generation
(G) *SASEGASA*	Random & roulette parent selection Strict Offspring selection (success ratio: 1.0, comparison factor: 1.0, maximum selection pressure: 200)

Table 11.29: Multi-population diversity test results of the GP test runs using the *Thyroid* data set.

Test Series	Iteration		MPdiv (avg)	MPdiv (max)	SPdiv (avg)	SPdiv (max)
			Results for the *Thyroid* data set			
A	300	avg	0.2433	0.3301	0.2048	0.8973
		std	0.0514	0.0496	0.0612	0.0291
	2000	avg	0.3592	0.3925	0.3628	0.9027
		std	0.0613	0.0610	0.0593	0.0351
B	20	avg	0.1698	0.2356	0.2130	0.9182
		std	0.0497	0.0317	0.0317	0.0852
	End of Run	avg	0.3915	0.4037	0.3592	0.9850
		std	0.0599	0.0769	0.0820	0.0202
C	300	avg	0.1778	0.2788	0.1836	0.6543
		std	0.0587	0.0549	0.0296	0.0971
	2000	avg	0.4145	0.4885	0.3834	0.9236
		std	0.0551	0.0762	0.0665	0.0417
D	20	avg	0.3276	0.4269	0.3394	0.9312
		std	0.0486	0.1094	0.0175	0.0459
	End of Run	avg	0.4412	0.5822	0.3866	0.9736
		std	0.0734	0.0635	0.0772	0.0283
E	300	avg	0.3395	0.6271	0.2715	0.6116
		std	0.0441	0.0975	0.0139	0.0811
	2000	avg	0.5329	0.8710	0.3991	0.9129
		std	0.0833	0.0509	0.0921	0.0821
F	20	avg	0.3721	0.5024	0.2711	0.5192
		std	0.0629	0.0822	0.0981	0.0601
	End of Run	avg	0.5915	0.8802	0.4576	0.9828
		std	0.1034	0.0996	0.0514	0.0437
G	20	avg	0.4839	0.5473	0.3173	0.5237
		std	0.0823	0.0419	0.0581	0.0623
	50	avg	0.4325	0.5512	0.3228	0.5828
		std	0.0518	0.0920	0.0672	0.0660
	100	avg	0.5102	0.7168	0.3783	0.7296
		std	0.0730	0.0724	0.0861	0.0740
	200	avg	0.8762	0.9314	0.4206	0.9512
		std	0.0505	0.0458	0.0792	0.0249
	End of Run	avg	–	–	0.9792	0.9934
		std	–	–	0.0256	0.0162

11.5.3 Discussion

As we see in Tables 11.29 and 11.30, the average diversity among populations in parallel island GP without interaction (i.e., in test series (A) and (B)) rises up to values between 0.35 and 0.4, no matter whether or not OS is applied; the maximum values eventually reach values between 0.45 and 0.5. Considering all solutions collected in temporary total populations, as expected the average similarities reach values below 0.4, the maximum similarities almost reach 1.0.

The similarity values monitored in test series (C) and (D) are, in comparison to those of series (A) and (B), slightly higher, but not dramatically. This does not hold for the next pair of test series (with 5% migration): The similarity values calculated for test series (E) and (F) are significantly higher than those of test series $(A) - (D)$; in other words, the exchange of only 5% of the populations' models can lead to a significant decrease of population diversity among populations of multi-population GP.

When using the SASEGASA, the diversity among populations is high in the beginning and then steadily decreases as the algorithm is executed. This

Table 11.30: Multi-population diversity test results of the GP test runs using the NO_x data set III.

Test Series	Iteration		Results for the NO_x data set			
			MPdiv (avg)	MPdiv (max)	SPdiv (avg)	SPdiv (max)
A	300	avg	0.3187	0.3991	0.2773	0.8613
		std	0.0124	0.0685	0.0726	0.0799
	2000	avg	0.3689	0.4627	0.3300	0.9887
		std	0.0288	0.0390	0.0390	0.0434
B	20	avg	0.1997	0.1498	0.2902	0.8992
		std	0.0698	0.0912	0.0604	0.0634
	End of	avg	0.3723	0.4811	0.3440	0.9743
	Run	std	0.0233	0.0244	0.0254	0.0482
C	300	avg	0.2515	0.3323	0.1935	0.8293
		std	0.0968	0.0685	0.0607	0.1062
	2000	avg	0.3329	0.4741	0.2821	0.9311
		std	0.0365	0.0323	0.0402	0.0441
D	20	avg	0.2985	0.3922	0.3791	0.8862
		std	0.0870	0.0825	0.0487	0.0829
	End of	avg	0.5544	0.6839	0.4208	0.9661
	Run	std	0.0542	0.1039	0.0280	0.0332
E	300	avg	0.5002	0.6697	0.3111	0.6037
		std	0.0588	0.0696	0.0474	0.0453
	2000	avg	0.6002	0.8523	0.4745	0.9763
		std	0.0538	0.0263	0.0728	0.0910
F	20	avg	0.3597	0.5248	0.3901	0.5839
		std	0.0743	0.0769	0.0662	0.0775
	End of	avg	0.5607	0.9080	0.4877	0.9906
	Run	std	0.0931	0.0799	0.0249	0.0181
G	20	avg	0.4471	0.5303	0.2694	0.4670
		std	0.0619	0.0897	0.0802	0.0522
	50	avg	0.4923	0.6102	0.3025	0.6120
		std	0.0854	0.0749	0.0550	0.0902
	100	avg	0.5889	0.6939	0.3923	0.7972
		std	0.1184	0.0835	0.0812	0.0805
	200	avg	0.9047	0.9148	0.5741	0.9128
		std	0.0387	0.0258	0.1253	0.0401
	End of	avg	–	–	0.9683	0.9932
	Run	std	–	–	0.0412	0.0319

is of course due to the reunification of populations as soon as the maximum selection pressure is reached.

By executing these test series and analyzing the results as given in this section we have demonstrated how multi-population diversity can be monitored using similarity measures as those described in Section 9.4.2. Reference values are given by parallel GP without migration; of course, the higher the migration rates become, the more migration affects the diversity among GP populations. When using the SASEGASA, rather high multi-population specific diversity is given in the early stages of the parallel GP process, and due to the merging of population the diversity decreases and in the end reaches diversity values comparable to those of single population GP with offspring selection.

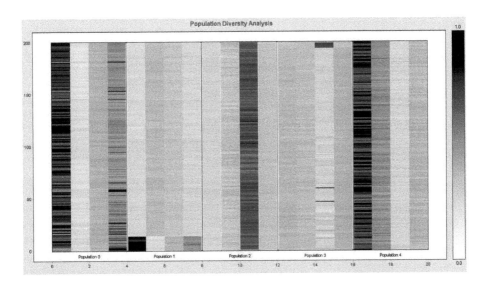

FIGURE 11.27: Exemplary multi-population diversity of a test run of *Thyroid* series F at iteration 50, grayscale representation.

11.6   Code Bloat, Pruning, and Population Diversity

11.6.1   Introduction

In Chapter 2.6 we have described one of the major problems of genetic programming, namely permanent code growth, often also referred to as bloat; evolution is also seen as "survival of the fattest," and, as Langdon and Poli expressed it, fitness-based selection leads to the fact that "fitness causes bloat" [LP97]. There are several approaches for combating this unwanted unlimited growth of chromosome size, some of them being

- limiting the size and / or the height of the program trees,

- pruning programs, and

- punishing complex programs by decreasing their quality depending on their respective tree representations' size and / or height.

Of course, there is no optimal strategy for fixing formula size parameters, population size, or pruning strategies a priori (see also remarks in Chapter 2). Still, some code prevention strategies are surely more recommendable than others; we here report on an exemplary test series for characterizing some of the possible approaches.

In all other test series executed and reported on in other sections in this book we have used fixed complexity limits (limiting size and height of program trees); we shall here report on our tests regarding code growth in GP-based structure identification applying the pruning strategies presented in Section 9.3.2 as well as structure tree size dependent fitness manipulation and fixed size limits (partially with additional pruning). All these approaches have been tested using standard GP as well as extended GP including gender specific selection and offspring selection. As an example, we have tested these GP variants on the NO_x data set II presented and described in Section 11.1.2.3; population diversity, formula complexity parameters as well as additional pruning effort (only in case of applying pruning, of course) have been monitored and shall be reported on here.

We have again used 50% of the given data for training models (namely samples 10,000 – 28,000), and 10% as validation data (samples 28,001 – 32,000 used by pruning strategies) and ~7.5% as test data (samples 32,001 – 35,000). As we are also aware of the problem of overfitting, we have systematically collected each GP run's best models with respect to best fit on training as well as on validation data (using the *mse* function for estimating the formulas' qualities). The algorithm is designed to optimize formulas with respect to training data; validation data are only used for pruning strategies (if used at all). At the end of each test run, the models with best fit on training as well as on validation data are analyzed, and in order to fight overfitting we select the best model on validation data as the result returned by the algorithm. Test data, which are not available to the algorithm, are used for demonstrating that this strategy is a reasonable one: Analyzing the evaluation of the best models on test data we see that those that are best on validation data perform better on test data than those that were optimally fit to training data.

During the GP process, the standard mean squared error function was used; the time series specific fitness function considering plain values as well as differential and integral values was used for selecting those models that perform best on training and validation data. All three components (i.e., plain values, differentials, and integral values) have been weighted using equal weighting factors. When comparing the quality of the results documented in the following sections we again state the fitness values calculated using the mean squared errors function.

11.6.2 Test Strategies

In detail, the following test strategies have been applied: On the one hand the parameters for standard and extended GP are summarized in Table 11.31, and the code growth prevention parameters are summarized in Table 11.32. In all tests the initial population was created using a size limit of 50 nodes and a maximum height of 6 levels for each structure tree.

Table 11.31: GP parameters used for code growth and bloat prevention tests.

Variant	Parameters
1 *(Standard GP, SGP)*	Population size: 1000 2000 generations Single point crossover; structural and parametric node mutation Parent selection: Tournament selection $(k = 3)$

2
*(Extended GP,
EGP)*
	Population size: 1000
	Single point crossover; structural and
	  parametric node mutation
	Parent selection: Gender specific selection
	  (random & proportional)
	Strict offspring selection
	  (maximum selection pressure: 100)

In the following table and in the explanations given afterwards, md is the maximum deterioration limit and mc the maximum coefficient of deterioration and structure complexity reduction as described in Section 9.3.2. For ES-based pruning, mr denotes the maximum number of rounds, and mur the maximum number of unsuccessful rounds.

In those tests including increased pruning (as applied in test series (h) and (i)) the initial pruning ratio is set to 0.3, i.e., in the beginning 30% of the population are pruned. Then, during the process execution, the pruning rate steadily increases and finally reaches 0.8; in standard GP runs, the rate is increased linearly, and in extended GP including offspring selection we compute the actual pruning ratio in relation to the actual selection pressure (so that in the end, when the selection pressure has reached its maximum value, the pruning rate has also reached its maximum, namely 0.8).

Furthermore, fs stands for the formula's size (i.e., the number of nodes in the corresponding structure tree), and pf is the fitness punishment factor: If structure complexity based punishment is applied, then the fitness f of a model is modified as $f' = f * (1 + pf)$ (if $pf > 0$).

Table 11.32: Summary of the code growth prevention strategies applied in these test series.

Variant	Characteristics
a	No code growth prevention strategy
b	20% systematic pruning: $md = 0$, $mc = 1$
c	20% ES-based pruning: $md = 0$, $mc = 1$, $\lambda = 5$, $mr = 5$, $mur = 1$
d	50% ES-based pruning: $md = 0.5$, $mc = 1$, $\lambda = 10$, $mr = 10$, $mur = 2$
e	100% ES-based pruning: $md = 2$, $mc = 1.5$, $\lambda = 20$, $mr = 10$, $mur = 2$
f	Increasing ES-based pruning: $md = 1$, $mc = 1.5$, $\lambda = 10$, $mr = 10$, $mur = 2$
g	Quality punishment: $pf = (fs - 50)/50$
h	Fixed limits: Maximum tree height 6, maximum tree size 50
i	Fixed limits: Maximum tree height 6, maximum tree size 50 combined with occasional ES-based pruning standard GP: every 5th, extended GP: every 2nd generation $md = 1$, $mc = 1$, $\lambda = 10$, $mr = 5$, $mur = 2$

Please note that in strategies (b) and (c) pruning is done after each generation step, whereas in (d) – (g) it is done after each creation of a new model by crossover and / or mutation. In standard GP this does not make any difference, but when using offspring selection the decision whether to prune after each creation or after each generation has major effects on the algorithmic process.

The mean squared errors function (with early stopping, see Section 9.2.5.5) was used here since we mainly concentrate on pruning and population dynamics relevant aspects. Furthermore, all variables (including the target variable) were linearly scaled to the interval [-100; +100].

11.6.3   Test Results

Once again, all test strategies have been executed 5 times independently; formula complexity has been monitored (and protocolled after each generation step) as well as structural population diversity which was protocolled after every 10th generation: The multiplicative similarity approach (as defined in Equations 9.63 – 9.66) has again been chosen; all coefficients $c_1 \ldots c_{10}$ were set to 0.2, only the coefficient c_1 weighting the level difference contribution d_1 was set to 0.8. The similarity of models was calculated symmetrically (as described in Equation 6.4).

11.6.3.1   No Formula Size Limitation

Exactly as we had expected, extreme code growth also occurs in GP-based structure identification; Figure 11.28 illustrates the progress of formula complexity in terms of formula size in exemplary test runs of series 1a and 2a: The average formula size is given as well as minimum and maximum values and the progress of the best individual's size.

As we see here, formulas tend to grow very big rather quickly; when using offspring selection, this effect is even a bit more obvious: On average, in standard GP the formula size has reached 212.84 after 30 iterations; when using OS the average formula size was even higher after 30 generations (namely 276.35).

11.6.3.2   Light Pruning

The results of test series (b) and (c) can be summarized in the following way: Without any further mechanisms that limit the structural complexity of formula trees, light pruning as described in strategies (b) and (c) is not an appropriate way to prevent GP from growing enormous formula structures. After 100 generations, the average formula size in standard GP has grown to 471.34 in test series (1b) and 333.65 in test runs of series (1c) (average standard deviation: 204.29 and 238.27, respectively); in extended GP the average formula size at generation 30 on average reached 293.26 and 276.12 in test runs (2b) and (2c), the respective standard deviations being 157.23 and 124.80.

Systematically analyzing the results of the pruning phases performed in test runs (b) and (c) we can compare the performances of ES-based and systematic

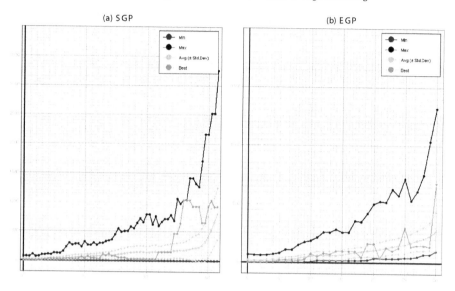

FIGURE 11.28: Code growth in GP without applying size limits or complexity punishment strategies (left: standard GP, right: extended GP).

pruning. For this purpose we have collected the pruning performance statistics for the tests (b) and (c) and summarize them in Table 11.33:

Table 11.33: Performance of systematic and ES-based pruning strategies.

Parameter	Systematic pruning	ES-based pruning
Solutions evaluated for pruning one solution	161.02	54.56
Runtime consumed (per iteration)	31.27 sec	12.23 sec
Average coefficient of deterioration and reduction of structural complexity	0.2495	0.4053

Obviously, both pruning methods performed approximately equally well and were able to reduce the complexity of the formulas that were supposed to be pruned. Additionally, we also see that especially for bigger model structures the runtime consumption is a lot higher when using systematic pruning; in the course of a GP process it is not considered necessary or even beneficial to reduce models as much as possible. Therefore we shall in the following test runs concentrate on ES-based pruning phases. Thus, we suggest using systematic pruning as a preparation step for results analysis, but not during the execution of GP-based training processes.

11.6.3.3 Medium Pruning

Medium pruning, as applied in test series (d), is in fact able to reduce the size of the formulas stored in the GP populations significantly.

Table 11.34: Formula size progress in test series (d).

Test series	Iteration	Formula size	
		avg	std
(1d)	20	21.83	32.12
	50	74.24	111.84
	100	123.67	144.78
	500	167.51	156.89
	2000	168.23	147.56
(2d)	10	10.77	13.27
	20	90.43	52.79
	50	228.02	112.51
	End of run	283.98	172.33

Table 11.35: Quality of results produced in test series (d).

Test series	Evaluation data	Best model selection basis			
		Training data		Validation data	
		avg	std	avg	std
(1d)	Training data	1,178.13	205.20	8,231.38	1,041.87
	Validation data	17,962.78	762.97	15,850.49	1,309.10
	Test data	7,162.48	690.10	5,996.27	927.09
(2d)	Training data	1,823.43	823.56	6,005.74	729.47
	Validation data	14,590.83	1,476.25	10,506.30	981.35
	Test data	6,341.28	770.42	4,439.27	918.72

The best results obtained in the (d) test series are summarized in Table 11.35: For each test run we have collected the models with best fit on training data as well as those that perform best on validation data; average values are given as well as standard deviations. Obviously, rather strong overfitting has happened here; as we had expected, the production of very large formulas leads to over-fit formulas that are not able to perform well on samples that were not used during the training phase.

11.6.3.4 Strong Pruning

Rather strong pruning was applied in test series (e), and as we see in Table 11.36, the formulas produced by GP are significantly smaller than those produced in the previous test series. Still, we observed the fact that genetic diversity is lost very quickly: Already in early stages of the evolutionary processes, the average structural similarity of solutions reaches a very high level (which is documented in the two most right columns of Table 11.36).

The quality of the best models produced is very bad (above 5,000), which is why we do here not state any further details about the evaluation of these models on the given data partitions. We suppose that this low results quality is connected to the loss of population diversity (and of course also the fact that the pruning operations applied were allowed to decrease the models' quality).

Table 11.36: Formula size and population diversity progress in test series (e).

Test series	Iteration	Formula size		Solutions similarity	
		avg	*std*	*avg*	*std*
(1e)	50	12.82	15.76	0.8912	0.0912
	100	18.27	18.15	0.9371	0.0289
	500	19.75	23.52	0.9685	0.0187
	2000	21.39	20.87	0.9891	0.0095
(2e)	10	15.77	9.23	0.9574	0.0318
	20	19.86	10.83	0.9825	0.0247
	50	21.64	16.34	0.9921	0.0082
	End of run	20.03	18.27	0.9943	0.0093

11.6.3.5 Increased Pruning

As light, medium, and strong pruning did not lead to the desired results, we have also tried increasing pruning as defined in test strategy (f). As we see in Table 11.37, this strategy performs rather well: The size of the formulas produced by GP rises especially in early stages of the GP process, but then decreases and on average finally reaches values between 80 and 100.

In addition to this, the population diversity stays higher in the beginning than in GP tests including constantly strong pruning, but eventually decreases and the solutions finally show higher similarities due to the increased pruning in later algorithmic stages.

Table 11.37: Formula size and population diversity progress in test series (f).

Test series	Iteration	Formula size		Solutions similarity	
		avg	std	avg	std
(1f)	50	62.72	95.76	0.3674	0.0943
	100	91.27	130.77	0.3897	0.1059
	500	92.43	107.41	0.6820	0.1124
	2000	87.02	90.68	0.8035	0.0861
(2f)	10	40.78	31.47	0.5235	0.0612
	20	63.59	59.34	0.7052	0.0803
	50	80.26	40.99	0.9450	0.0588
	End of run	79.45	47.67	0.9967	0.0156

The quality values of the results produced in this test series are summarized in Table 11.38. Obviously, less overfitting has happened than in the tests with light or medium pruning.

Table 11.38: Quality of results produced in test series (f).

Test series	Evaluation data	Best model selection basis			
		Training data		Validation data	
		avg	std	avg	std
(1f)	Training data	2,597.35	542.04	7,781.28	827.83
	Validation data	8,904.91	611.02	5,981.52	974.31
	Test data	3,786.51	800.38	2,830.78	427.08
(2f)	Training data	2,275.24	649.11	3,814.93	850.89
	Validation data	9,712.98	767.56	5,862.62	518.53
	Test data	4,912.38	1,198.58	2,275.03	931.62

11.6.3.6 Complexity Dependent Quality Punishment

In fact, our GP test runs including complexity dependent quality punishment, i.e., those of test strategy (g), were also able to produce acceptable results for the NO_x data set investigated here. As we see in Table 11.39, in standard GP the formula sizes are rather high in the beginning and then decrease steadily, whereas in GP with offspring selection the models on average include between 50 and 60 nodes during the whole execution of the GP processes. Population diversity values are comparable to those reported for GP tests without pruning or quality dependent punishment as summarized for example in Section 11.4.

Figure 11.29 illustrates the formula complexity progress of an exemplary GP run of test series (2g). The qualities of the models with best fit on training and validation are summarized in Table 11.40.

Table 11.39: Formula size and population diversity progress in test series (g).

Test series	Iteration	Formula size		Solutions similarity	
		avg	std	avg	std
(1g)	50	140.76	90.75	0.3824	0.0534
	100	92.62	71.23	0.3916	0.0620
	500	73.73	64.99	0.6381	0.0825
	2000	79.07	47.61	0.7202	0.0696
(2g)	10	50.24	64.67	0.4873	0.0836
	20	60.71	59.01	0.5412	0.0741
	50	65.34	48.33	0.8904	0.0852
	End of run	58.82	41.87	0.9315	0.0423

Table 11.40: Quality of results produced in test series (g).

Test series	Evaluation data	Best model selection basis			
		Training data		Validation data	
		avg	std	avg	std
(1g)	Training data	1,837.84	526.10	4,729.42	480.36
	Validation data	12,902.67	767.35	4,531.73	588.30
	Test data	2,597.73	835.41	2,708.36	825.64
(2g)	Training data	1,402.19	593.84	3,121.86	773.91
	Validation data	9,345.87	738.60	3,949.64	962.03
	Test data	2,853.62	812.51	2,618.94	664.07

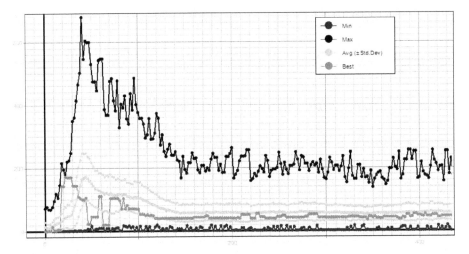

FIGURE 11.29: Progress of formula complexity in one of the test runs of series (1g), shown for the first ~400 iterations.

11.6.3.7 Fixed Size Limits

In the case of fixed size limits the crossover and mutation operators have to consider limits for the complexity of models. Model size and population diversity statistics for test series (h) are summarized in Table 11.41; in GP with offspring selection all formulas eventually are maximally big, and the solutions similarity values show results comparable to those reported in Section 11.4. Table 11.42 summarizes the quality of the results produced, again evaluated on training, validation, and test data. Figure 11.30 illustrates the formula complexity progress of exemplary GP test runs of series (1h) and (2h).

Table 11.41: Formula size and population diversity progress in test series (h).

Test series	Iteration	Formula size		Solutions similarity	
		avg	std	avg	std
(1h)	50	37.4182	6.3174	0.4151	0.0935
	100	40.2866	5.8133	0.7231	0.0729
	500	41.7823	4.3973	0.8175	0.0326
	2000	44.2108	5.0450	0.8629	0.0271
(2h)	10	22.4965	8.3763	0.3973	0.0386
	20	27.6203	4.2514	0.6022	0.0493
	50	44.9120	6.4871	0.8907	0.0371
	End of run	50.0000	0.0000	0.9751	0.0189

Table 11.42: Quality of results produced in test series (h).

Test series	Evaluation data	Best model selection basis			
		Training data		Validation data	
		avg	std	avg	std
(1h)	Training data	1,774.94	300.51	4,168.30	1,186.62
	Validation data	10,801.77	923.04	4,248.37	858.02
	Test data	5,791.25	1,266.51	2,610.64	930.44
(2h)	Training data	1,568.12	382.04	3,083.64	502.75
	Validation data	9,641.89	833.71	3,738.13	504.89
	Test data	4,802.30	1,371.22	1,374.61	704.73

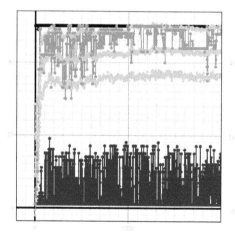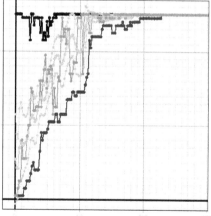

FIGURE 11.30: Progress of formula complexity in one of the test runs of series (1h) (shown left) and one of series (2h) (shown right).

In addition to total statistics we shall also discuss two selected models returned by one of the test runs of series (2h): Model b_t is the model that performs best on training data (shown in Figure 11.31), b_v the one that performs best on validation data (shown in Figure 11.32). The error distributions on training, validation, and test data partitions are illustrated in Figure 11.33.

Table 11.43 characterizes the performance of b_t and b_v by means of mean squared errors as well as the integral values. For this we have calculated the sum of the target values on training, validation, and test data and compared these integral values to those calculated using the models under investigation. Obviously, b_t shows a better integral fit on training (and also validation) data, but when it comes to test data, the model that performed best on validation data (b_v) produces much more satisfying results (with an integral error of only 2.354% on test data).

Table 11.43: Comparison of best models on training and validation data (b_t and b_v, respectively).

	b_t	b_v
Training quality (MSE)	1,434.65	2,253.62
Validation quality (MSE)	9,187.53	3,748.61
Test quality (MSE)	2,936,40	1,461.95
Target training values integral	$6.010 * 10^6$	
Estimated training values integral	$6.037 * 10^6$	$6.084 * 10^6$
	(-0.452%)	(+1.220%)
Target validation values integral	$4.660 * 10^5$	
Estimated validation values integral	$4.620 * 10^5$	$4.517 * 10^6$
	(+0.872%)	(+3.173%)
Target test values integral	$3.978 * 10^5$	
Estimated test values integral	$3.198 * 10^5$	$3.886 * 10^6$
	(+24.395%)	(+2.354%)

11.6.3.8 Fixed Size Limits and Occasional Pruning

Finally, test series with fixed size limits and occasional pruning have also been executed and analyzed; the results regarding formula complexity, population diversity, and results qualities are summarized in Tables 11.44 and 11.45.

Obviously, the results produced are (with respect to evaluation quality) comparable to those produced in the previous series. Still, of course the formula sizes are a bit smaller (due to pruning), and also overfitting seems

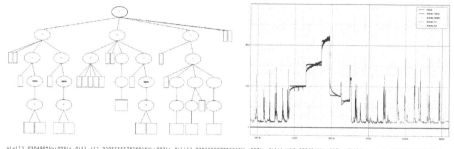

+(+([1,030496*Var003(t-0)]|-([1,02955555783691*Var007(t-8)]|[1,02955555783691*Var007(t-9)])|/([0,859604*Var007(t-7)]|Si gnun[+(-19,6774250843158
|[0,925277*Var007(t-0)])))]/(([0,796922*Var008(t-10)]|Si gnun[+(-13,4653753764306|[1,009622*Var007(t-1)])))))]+(+([1,064909*Var006(t-6)]
|[0,245936*Var001(t-5)]|-8,11313346378627]|-([1,0246687670851 2*Var007(t-10)]|[1,02466876708512*Var007(t-11)])|-([1,02993590500011*Var007(t-10)]
|[1,02993590500011*Var007(t-11)])))|[1,062877*Var006(t-6)]|[0,792079*Var008(t-10)]|Exp[Sqrt([1,008566*Var007(t-1)])))/([0,796922*Var008(t-10)]
|Si gnun[+(-19,6774250843158]|[1,009622*Var007(t-1)])))))*([1,030496*Var003(t-0)]|+([1,070487*Var007(t-0)]|Si gnun[Si n([1,0158164*Var004(t-6)])])|-
([1, gnun[0,925277*Var007(t-0)])|[1,028206*Var003(t-7)])|Exp[Si n([0,859604*Var007(t-7)])))]-([1,02015787437991*Var007(t-9)]
|[1,02015787437991*Var007(t-10)])|-([1,02955555783691*Var007(t-8)]|[1,02955555783691*Var007(t-9)]))

FIGURE 11.31: Model with best fit on training data: Model structure and full evaluation.

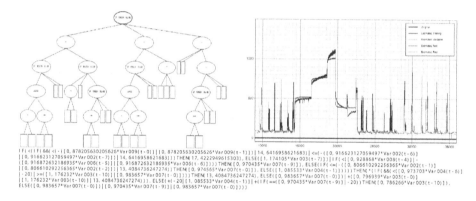

IF(<(IF(&&(<(-([0,878205630205626*Var009(t-0)]|[0,878205630205626*Var009(t-1)])|14,6416958621683)|<=(-([0,91662317059497*Var002(t-6)]
|[0,916623127059497*Var002(t-7)])|14,6416958621683))|THEN[17,4222949615303),ELSE([1,174105*Var003(t-7)]|IF(<([0,928868*Var008(t-4)])|-
([0,916872632186935*Var006(t-5)]|[0,916872632186935*Var006(t-6)])))THEN[0,970435*Var007(t-9)],ELSE(IF(<=(-([0,806610292256365*Var002(t-1)]
|[0,806610292256365*Var002(t-2)])|13,4084736247274])THEN[0,974565*Var007(t-0)],ELSE([1,085533*Var004(t-1)])))THEN[*(IF(&&(<([0,973703*Var004(t-6)]
|-20)|>=([1,176232*Var003(t-10)]|[0,983657*Var007(t-0)])))THEN[13,4084736247274),ELSE([1,085533*Var004(t-1)]|+(IF([0,796939*Var003(t-0)]
|[1,176232*Var003(t-10)]|13,4084736247274)),ELSE(+(-20|[1,085533*Var004(t-1)]|+(IF(==([0,970435*Var007(t-9)]|-20))THEN[0,786266*Var003(t-10)]),
ELSE([0,983657*Var007(t-0)]|[0,970435*Var007(t-9)]|[0,983657*Var007(t-0)])))

FIGURE 11.32: Model with best fit on validation data: Model structure and full evaluation.

to have decreased: Even though the fit on training data is not as good as on previous test series, the quality on test data is still very good and comparable to the test performance reached in test series (g) and (h).

11.6.4 Conclusion

In this section we have demonstrated the effects of code bloat and selected prevention strategies for GP. As expected and known from literature, without any limitations or size reducing strategies GP tends to produce bigger and bigger models that fit the given training data, but of course this also increases the probability of producing over-fit models. Pruning strategies have been analyzed, and the test results show that only strong pruning is able to prevent GP from producing bigger and bigger models, which again decreases

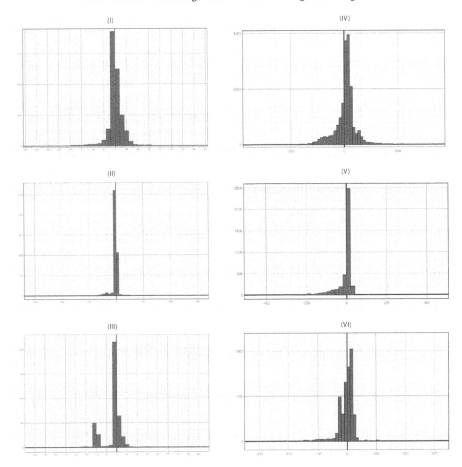

FIGURE 11.33: Errors distributions of best models: Charts I, II, and III show the errors distributions of the model with best fit on training data evaluated on training, validation, and test data, respectively; charts IV, V, and VI show the errors distributions of the model with best fit on validation data evaluated on training, validation, and test data, respectively.

population diversity and leads to results which are not optimal. Complexity dependent fitness punishment as well as fixed size limits enable GP to produce quite good results; occasional pruning in combination with fixed size limits can help to decrease overfitting.

Table 11.44: Formula size and population diversity progress in test series (i).

Test series	Iteration	Formula size		Solutions similarity	
		avg	std	avg	std
(1i)	50	34.8365	6.1534	0.4682	0.0852
	100	37.1863	4.9901	0.7413	0.0711
	500	39.2217	5.2673	0.8388	0.0450
	2000	40.1260	4.9724	0.8992	0.0251
(2i)	10	18.5330	6.6114	0.4307	0.0518
	20	21.5286	5.3083	0.7202	0.0772
	50	38.5143	5.6305	0.9248	0.0403
	End of run	48.2051	4.6228	0.9859	0.0178

Table 11.45: Quality of results produced in test series (i).

Test series	Evaluation data	Best model selection basis			
		Training data		Validation data	
		avg	std	avg	std
(1i)	Training data	2,258.22	561.27	5,869.40	1.233.09
	Validation data	6,608.26	1,463.49	4,819.26	730.51
	Test data	2,238.61	983.57	1,811.05	834.83
(2i)	Training data	1,723.07	623.11	4,209.57	499.89
	Validation data	6,361.46	921.26	3,607.13	736.05
	Test data	3,289.33	945.79	1,434.63	739.22

Conclusion and Outlook

In this book we have discussed basic principles as well as algorithmic improvements in the context of genetic algorithms (GAs) and genetic programming (GP); new problem independent theoretical concepts have been described which are used in order to substantially increase achievable solution qualities. The application of these concepts to significant combinatorial optimization problems as well as structure identification in time series analysis and classification has also been described.

We have presented enhanced concepts for GAs, which enable a self-adaptive interplay of selection and solution manipulation operators. By using these concepts we want to avoid the disappearance and support the combination of alleles from the gene pool that represent solution properties of highly fit individuals (introduced as *relevant genetic information*). As we have shown in several test series, relevant genetic information is often lost in conventional implementations of GAs and GP; if this happens, it can only be reintroduced into the population's gene pool by mutation. This dependence on mutation can be reduced by using generic selection principles such as offspring selection (which is also used in the SASEGASA) or self-adaptive population size adjustment (as used by the RAPGA). The survival of essential genetic information by supporting the *survival of relevant alleles rather than the survival of above-average chromosomes* is the main goal of both these approaches.

In the empirical part of this book we have documented and discussed our experiences in applying these new algorithmic concepts to benchmark as well as real world problems. Concretely, we have used traveling salesman problems as well as vehicle routing problems as representatives of combinatorial optimization problems; time series and classification analysis problems have been used as application areas of data-based structure identification with genetic programming. We have compared the results achievable with standard implementations of GAs and GP to the results achieved using extended algorithmic concepts that do not depend on a concrete problem representation and its operators; the influences of the new concepts on population dynamics in GA and GP populations have also been analyzed.

Nevertheless, there is still a lot of work to be done in the context of the research areas we have dealt with in this book. Furthermore, there are a lot of potential synergies which have to be considered and should be explored.

- The most important aspect is the following one: As the enhanced algorithmic concepts discussed in this book are problem independent, they can be applied to any kind of optimization problem which can be tackled by a GA or GP. Of course, there are numerous kinds of optimization problems beside traveling salesman and vehicle routing problems which can be solved successfully by genetic algorithms; regarding GP we have up to now more or less only gained experience in using offspring selection in the context of data-based modeling, but there is a huge variety of other problems which should also be tried to be solved using the approaches discussed in this book.

- HeuristicLab (HL) is our environment for developing and testing optimization methods, tuning parameters, and solving a multitude of problems. The development of HL was started in 2002 and has meanwhile led to a stable and productive optimization platform; it is continuously enhanced and a topic of several publications ([WA04c], [WA04a], [WA04b], [WA05a], and [WWB+07]). On the respective website[13] the interested reader can find information about the design of HeuristicLab, its development over the years, installable software packages, documentation, and publications in the context of HeuristicLab and the research group HEAL[14].

One of the most beneficial aspects of HeuristicLab is its plug-in based architecture. In software engineering in general, plug-in based software systems have become very popular; by not only splitting the source code into different modules, but compiling these modules into enclosed ready-to-use software building blocks, the development of a whole application or complex software system is reduced to the task of selecting, combining, and distributing the appropriate modules. Due to the support of dynamic location and loading techniques offered in modern application frameworks as for example Java or .NET, the modules do not need to be statically linked during compilation, but can be dynamically loaded at runtime. Thus, the core application can be enriched by adding these building blocks, which are therefore called "plug-ins" as they are additionally plugged into the program.

Several problem representations, solution encodings, and numerous algorithmic concepts have so far been developed for HeuristicLab, realizing a large number of heuristic and evolutionary algorithms (genetic algorithms, genetic programming, evolution strategies, tabu search, etc.) for

---

[13]http://www.heuristiclab.com.
[14]Heuristic and Evolutionary Algorithms Laboratory, Linz / Hagenberg, Austria.

a wide range of problem classes including the traveling salesman problem, the vehicle routing problem, real-valued test functions in different dimensions, and, last, but not least, also data-based modeling.

Still, not only the software platform itself is flexible and extensible, also the algorithms provided in HL are (since version 2.0) not fixed and hard-coded, but can be parameterized and even designed by the user. This is possible by realizing all solution generating and processing procedures as operators working on single solutions or sets of solutions.

By providing a set of plug-ins, each realizing a specific solution representation or operation, the process of developing new heuristic algorithms is revolutionized. Algorithms do not need to be programmed anymore, but can be created by combining operators of different plug-ins. This approach has a huge advantage: By providing a graphical user interface for selecting plug-ins and combining operators, no programming or software engineering skills are necessary for this process. As a consequence, algorithms can be modified, tuned, or developed by experts of different fields with little or even no knowledge in the field of software development.

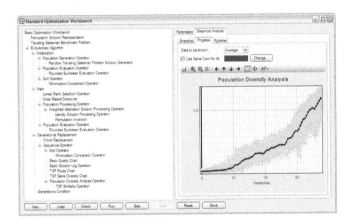

FIGURE 11.34: A simple workbench in HeuristicLab 2.0.

In Figure 11.34 we show a screenshot of a simple HeuristicLab workbench (version HL 2.0): A structure identification problem is solved by a GP algorithm using offspring selection. All relevant parts of the algorithm (as for example population initialization, crossover, generational replacement, and offspring selection) can be seen in the left part of the workbench GUI; these parts can be easily rearranged or replaced by users who are not necessarily experts in heuristic optimization or even computer science. Thus, we want to transfer algorithm development competence from experts in heuristic optimization to users working on concrete applications; users, who work in domains other than heuristic optimization, will thus no longer have to use heuristics as black box techniques (as it is frequently done nowadays), but can use them as algorithms which can be modified and easily tuned to specific problem situations.

- One of our current research interests is to combine agent-based software development techniques with heuristic optimization methods. Here again Genetic Programming is one of the fields that would on the one hand profit from the intrinsic parallelization of software agents as well as improve the quality and expressiveness of found models. Agents could be programmed to identify different variables in the given data sets and examine a broader range of correlations. Each of these agents represents a GP process evolving a population of formulas (models); at given synchronization points, these agents exchange information among each other. Unlike other parallel GP approaches, in which parts of populations are exchanged which in principle all have the same goal (namely to solve a given identification task), we here want to establish an information exchange mechanism by which partial information about relationships in the data is passed and shared among the identification agents.

The probably most important goal of such a parallel GP approach is to develop an automated mechanism that can identify not only singular relationships in data, but rather whole information networks that describe lots of relationships that can be found. This incorporates the use of GP agents that aim to identify models for different target variables. So it should become possible to identify classes of equivalent models that differ only in the way certain input variables are described; these results will hopefully help to find answers to one of the most important questions in system identification, namely which of the potential models are best suited for further theoretical analyses.

We hope that one of the results of this book will be an increased interest in population dynamics analysis as well as generic algorithmic developments as for example enhanced selection methods for evolutionary algorithms. By showing the general applicability of enhanced selection concepts in GAs applied to combinatorial problems as well as in GP, we hope that we have been able to inspire readers to apply these concepts to other problems as well as to include them in other variants of evolutionary algorithms.

Symbols and Abbreviations

Symbol Description

ANN	Artificial neural network
AUC	Area under a ROC curve
CGA	Canonical genetic algorithm
(C)VRP(TW)	(Capacitated) vehicle routing problem (with time windows)
CX	Cyclic crossover for the TSP
EA	Evolutionary algorithm
ERX	Edge recombination crossover for the TSP
ES	Evolution strategy
GA	Genetic algorithm
GP	Genetic programming
HL	HeuristicLab
kNN	k-nearest neighbor algorithm
NO_x	Nitric oxides
OS	Offspring selection
OX	Order crossover for the TSP
PMX	Partially matched crossover for the TSP
RAPGA	Relevant alleles preserving genetic algorithm
RBX	Route-based crossover for the CVRP
ROC	Receiver operating characteristic
SASEGASA	Self-adaptive segregative genetic algorithm including aspects of simulated annealing
SBX	Sequence-based crossover for the CVRP
SGA	Standard genetic algorithm
TS	Tabu search
TSP	Traveling salesman problem

References

[AA05]   W. Ashlock and D. Ashlock. Single parent genetic programming. In D. Corne et al., editors, *Proceedings of the 2005 IEEE Congress on Evolutionary Computation*, volume 2, pages 1172–1179, Edinburgh, UK, 2-5 September 2005. IEEE Press.

[AD95]   J. Antes and U. Derigs. A new parallel tour algorithm for the vehicle routing problem with time windows. Technical report, Institut für Wirtschaftsinformatik und Operations Research, Universität Köln, 1995.

[AD04]   E. Alba and B. Dorronsoro. Solving the vehicle routing problem by using cellular genetic algorithms. In J. Gottlieb and G. R. Raidl, editors, *Evolutionary Computation in Combinatorial Optimization*, volume 3004 of *Lecture Notes in Computer Science*, pages 11–20, Coimbra, Portugal, 2004. Springer.

[AD06]   E. Alba and B. Dorronsoro. Computing nine new best-so-far solutions for capacitated VRP with a cellular genetic algorithm. *Information Processing Letters*, 98(6):225–230, June 2006.

[AdRWL05]   D. Alberer, L. del Re, S. Winkler, and P. Langthaler. Virtual sensor design of particulate and nitric oxide emissions in a DI diesel engine. In *Proceedings of the 7th International Conference on Engines for Automobile ICE 2005*, number 2005-24-063, 2005.

[Aff01a]   M. Affenzeller. A new approach to evolutionary computation: Segregative genetic algorithms (SEGA). In J. Mira and A. Prieto, editors, *Connectionist Models of Neurons, Learning Processes, and Artificial Intelligence*, volume 2084 of *Lecture Notes in Computer Science*, pages 594–601. Springer, 2001.

[Aff01b]   M. Affenzeller. Segregative genetic algorithms (SEGA): A hybrid superstructure upwards compatible to genetic algorithms for retarding premature convergence. *International Journal of Computers, Systems and Signals (IJCSS)*, 2(1):18–32, 2001.

[Aff01c]   M. Affenzeller. Transferring the concept of selective pressure from evolutionary strategies to genetic algorithms. In Z. Bubnicki and A. Grzech, editors, *Proceedings of the 14th International Conference on Systems Science*, volume 2, pages 346–353. Oficyna Wydawnicza Politechniki Wroclawskiej, 2001.

[Aff02] M. Affenzeller. New variants of genetic algorithms applied to problems of combinatorial optimization. In R. Trappl, editor, *Cybernetics and Systems 2002*, volume 1, pages 75–80. Austrian Society for Cybernetic Studies, 2002.

[Aff03] M. Affenzeller. *New Hybrid Variants of Genetic Algorithms: Theoretical and Practical Aspects*. Schriften der Johannes Kepler Universität Linz. Universitätsverlag Rudolf Trauner, 2003.

[Aff05] M. Affenzeller. *Population Genetics and Evolutionary Computation: Theoretical and Practical Aspects*. Trauner Verlag, 2005.

[AK89] E. Aarts and J. Korst. *Simulated Annealing and Boltzman Machines: A Stochastic Approach to Combinatorial Optimization and Neural Computing*. John Wiley and Sons, 1989.

[Alb05] E. Alba. *Parallel Metaheuristics: A New Class of Algorithms*. Wiley Interscience, 2005.

[Alt94a] L. Altenberg. Emergent phenomena in genetic programming. In A. Sebald and L. Fogel, editors, *Evolutionary Programming — Proceedings of the Third Annual Conference*, pages 233–241, San Diego, CA, USA, 24-26 February 1994. World Scientific Publishing.

[Alt94b] L. Altenberg. The Schema Theorem and Price's Theorem. In D. Whitley and M. Vose, editors, *Foundations of Genetic Algorithms 3*, pages 23–49, Estes Park, Colorado, USA, 31 July–2 August 1994. Morgan Kaufmann. Published 1995.

[And71] T. W. Anderson. *The Statistical Analysis of Time Series*. Wiley, 1971.

[And76] O. D. Anderson. *Time Series Analysis and Forecasting: the Box-Jenkins Approach*. Butterworth, 1976.

[Ang93] P. J. Angeline. *Evolutionary Algorithms and Emergent Intelligence*. PhD thesis, Ohio State University, 1993.

[Ang94] P. J. Angeline. Genetic programming and emergent intelligence. In K. E. Kinnear, Jr., editor, *Advances in Genetic Programming*, chapter 4, pages 75–98. MIT Press, 1994.

[Ang98] P. J. Angeline. Subtree crossover causes bloat. In J. R. Koza et al., editors, *Genetic Programming 1998: Proceedings of the*

Third Annual Conference, pages 745–752, University of Wisconsin, Madison, Wisconsin, USA, 22-25 July 1998. Morgan Kaufmann.

[AT99] E. Alba and J. M. Troya. A survey of parallel distributed genetic algorithms. *Complexity (USA)*, 4(4):31–52, 1999.

[AW03] M. Affenzeller and S. Wagner. SASEGASA: An evolutionary algorithm for retarding premature convergence by self-adaptive selection pressure steering. In J. Mira and J. R. Alvarez, editors, *Computational Methods in Neural Modeling*, volume 2686 of *Lecture Notes in Computer Science*, pages 438–445. Springer, 2003.

[AW04a] M. Affenzeller and S. Wagner. Reconsidering the selection concept of genetic algorithms from a population genetics inspired point of view. In R. Trappl, editor, *Cybernetics and Systems 2004*, volume 2, pages 701–706. Austrian Society for Cybernetic Studies, 2004.

[AW04b] M. Affenzeller and S. Wagner. SASEGASA: A new generic parallel evolutionary algorithm for achieving highest quality results. *Journal of Heuristics - Special Issue on New Advances on Parallel Meta-Heuristics for Complex Problems*, 10:239–263, 2004.

[AW05] M. Affenzeller and S. Wagner. Offspring selection: A new self-adaptive selection scheme for genetic algorithms. In B. Ribeiro, R. F. Albrecht, A. Dobnikar, D. W. Pearson, and N. C. Steele, editors, *Adaptive and Natural Computing Algorithms*, Springer Computer Science, pages 218–221. Springer, 2005.

[AWW07] M. Affenzeller, S. Wagner, and S. Winkler. Self-adaptive population size adjustment for genetic algorithms. In *Proceedings of Computer Aided Systems Theory: EuroCAST 2007*, Lecture Notes in Computer Science, pages 820–828. Springer, 2007.

[AWW08] M. Affenzeller, S. Winkler, and S. Wagner. *Advances in Evolutionary Algorithms*, chapter Evolutionary Systems Identification: New Algorithmic Concepts and Applications. Artificial Intelligence. I-tech, 2008.

[B+05] H.-G. Beyer et al., editors. *GECCO 2005: Proceedings of the 2005 Conference on Genetic and Evolutionary Computation*, Washington DC, USA, 25-29 June 2005. ACM Press.

[Bal93] N. Balakrishnan. Simple heuristics for the vehicle routing problem with soft time windows. *Journal of the Operational Research Society*, 44(3):279–287, 1993.

[BB94] M. Bohanec and I. Bratko. Trading accuracy for simplicity in decision trees. *Machine Learning*, 15:223 – 250, 1994.

[BD91] P. J. Brockwell and R. A. Davis. *Time Series: Theory and Methods*. Springer, 1991.

[BD96] P. J. Brockwell and R. A. Davis. *A First Course in Time Series Analysis*. Springer, 1996.

[Ber98] M. Bergmann. *Tourenplanung mit Zeitfenstern - ein Überblick*. Shaker Verlag, 1998.

[BES01] E. Bradley, M. Easley, and R. Stolle. Reasoning about nonlinear system identification. *Artificial Intelligence*, 133:139–188, December 2001.

[Bey01] H. G. Beyer. *The Theory of Evolution Strategies*. Springer, 2001.

[BG05a] O. Bräysy and M. Gendreau. Vehicle Routing Problem with Time Windows, Part I: Route Construction and Local Search Algorithms. *Transportation Science*, 39(1):104–118, 2005.

[BG05b] O. Bräysy and M. Gendreau. Vehicle routing problem with time windows, Part II: Metaheuristics. *Transportation Science*, 39(1):119–139, 2005.

[BGK04] E. K. Burke, S. Gustafson, and G. Kendall. Diversity in genetic programming: An analysis of measures and correlation with fitness. *IEEE Transactions on Evolutionary Computation*, 8(1):47–62, 2004.

[BH74] M. Bellmore and S. Hong. Transformation of multisalesman problem to the standard traveling salesman problem. *Journal of the Association of Computer Machinery*, 21:500–504, 1974.

[BJ76] G. E. P. Box and G. M. Jenkins. *Time Series Analysis: Forecasting and Control*. Holden-Day, 1976.

[BJN+98] C. Bernhart, E. L. Johnson, G. L. Nemhauser, M. W. P. Savelsbergh, and P. H. Vance. Branch-and-price: Column generation for solving huge integer programs. *Mathematical Programming: State of the Art*, 46(3):316–329, 1998.

[BL04] W. Banzhaf and C. W. G. Lasarczyk. Genetic programming of an algorithmic chemistry. In U.-M. O'Reilly, T. Yu, R. L. Riolo, and B. Worzel, editors, *Genetic Programming Theory and Practice II*, chapter 11, pages 175–190. Springer, Ann Arbor, 13-15 May 2004.

[BMR93] L. Bianco, A. Mingozzi, and S. Ricciardelli. The traveling salesman problem with cumulative costs. *NETWORKS: Networks: An International Journal*, 23:81–91, 1993.

[BNKF98] W. Banzhaf, P. Nordin, R. E. Keller, and F. D. Francone. *Genetic Programming – An Introduction; On the Automatic Evolution of Computer Programs and its Applications.* Morgan Kaufmann, San Francisco, CA, USA, January 1998.

[Boc58] F. Bock. An algorithm for solving traveling-salesman and related network optimization problems. *14th National Meeting of the ORSA*, 1958.

[BR03] C. Blum and A. Roli. Metaheuristics in combinatorial optimization: Overview and conceptual comparison. *ACM Computing Surveys*, 35(3):268 – 308, 2003.

[Bra97] A. Bradley. The use of the area under the ROC curve in the evaluation of machine learning algorithms. *Pattern Recognition*, 30:1145–1159, 1997.

[BU95] C. E. Brodley and P. E. Utgoff. Multivariate decision trees. *Machine Learning*, 19(1):45–77, 1995.

[Car94] H. Cartwright. Getting the timing right - the use of genetic algorithms in scheduling. *Proceedings of Adaptive Computing and Information Processing Conference*, pages 393–411, 1994.

[Cav75] D. Cavicchio. *Adaptive Search Using Simulated Evolution.* PhD thesis, University of Michigan, 1975.

[CE69] N. Christofides and S. Eilon. An algorithm for the vehicle dispatching problem. *Operational Research Quarterly*, 20:309–318, 1969.

[CGT99] R. Cheng, M. Gen, and Y. Tsujimura. A tutorial survey of job-shop scheduling problems using genetic algorithms. Part II: Hybrid genetic search strategies. *Computers & Industrial Engineering*, 37(1-2):51–55, 1999.

[Cha01] C. Chatfield, editor. *Time Series and Forecasting.* Chapman and Hall, 2001.

[CJ91a] R. J. Collins and D. R. Jefferson. Antfarm: Towards simulated evolution. In C. G. Langton, C. Taylor, J. Doyne Farmer, and S. Rasmussen, editors, *Artificial Life II*, pages 579–601. Addison-Wesley, Redwood City, CA, 1991.

[CJ91b] R. J. Collins and D. R. Jefferson. Representations for artificial organisms. In Jean-Arcady Meyer and Stewart W. Wilson, editors, *Proceedings of the First International Conference on*

Simulation of Adaptive Behavior: From Animals to Animats, pages 382–390. MIT Press, 1991.

[CL98] T. G. Crainic and G. Laporte. *Fleet Management and Logistics*. Khuwer, 1998.

[CMT81] N. Christofides, A. Mingozzi, and P. Toth. Exact algorithms for solving the vehicle routing problem based on spanning trees and shortest path relaxations. *Mathematical Programming*, 20(3):255–282, 1981.

[CO07] S. Christensen and F. Oppacher. Solving the artificial ant on the Santa Fe trail problem in 20,696 fitness evaluations. In D. Thierens et al., editors, *GECCO '07: Proceedings of the 9th Annual Conference on Genetic and Evolutionary Computation*, volume 2, pages 1574–1579, London, 7-11 July 2007. ACM Press.

[CP80] H. P. Crowder and M. Padberg. Solving large-scale symmetric travelling salesman problems to optimality. *Management Science*, 26:495–509, 1980.

[CP94] W. Sam Chung and R. A. Perez. The schema theorem considered insufficient. *Proceedings of the Sixth IEEE International Conference on Tools with Artificial Intelligence*, pages 748–751, 1994.

[CP97] E. Cantú-Paz. A survey of parallel genetic algorithms. Technical Report IlliGAL 97003, University of Illinois at Urbana-Champaign, 1997.

[CP01] E. Cantú-Paz. *Efficient and Accurate Parallel Genetic Algorithms*. Kluwer Academic Publishers, 2001.

[CP⁺03a] E. Cantú-Paz et al., editors. *Genetic and Evolutionary Computation – GECCO 2003, Part I*, volume 2723 of *Lecture Notes in Computer Science*, Chicago, IL, USA, 12-16 July 2003. Springer.

[CP⁺03b] E. Cantú-Paz et al., editors. *Genetic and Evolutionary Computation – GECCO 2003, Part II*, volume 2724 of *Lecture Notes in Computer Science*. Springer, 12-16 July 2003.

[CPG99] E. Cantu-Paz and D. E. Goldberg. On the scalability of parallel genetic algorithms. *Evolutionary Computation*, 7(4):429–449, 1999.

[Cra85] N. L. Cramer. A representation for the adapative generation of simple sequential programs. *International Conference on Genetic Algorithms and Their Applications (ICGA85)*, pages 183–187, 1985.

[Cro58] G. Croes. A method for solving travelling-salesman problems. *Operations Research*, 6:791–812, 1958.

[CTE$^+$06] P. Collet, M. Tomassini, M. Ebner, S. Gustafson, and A. Ekárt, editors. *Proceedings of the 9th European Conference on Genetic Programming*, volume 3905 of *Lecture Notes in Computer Science*, Budapest, Hungary, 10 - 12 April 2006. Springer.

[CW64] G. Clarke and J. Wright. Scheduling of vehicles from a central depot to a number of delivery points. *Operations Research*, 12:568–581, 1964.

[D$^+$04a] K. Deb et al., editors. *Genetic and Evolutionary Computation – GECCO-2004, Part I*, volume 3102 of *Lecture Notes in Computer Science*, Seattle, WA, USA, 26-30 June 2004. Springer-Verlag.

[D$^+$04b] K. Deb et al., editors. *Genetic and Evolutionary Computation – GECCO-2004, Part II*, volume 3103 of *Lecture Notes in Computer Science*, Seattle, WA, USA, 26-30 June 2004. Springer-Verlag.

[DAG01] W. Duch, R. Adamczak, and K. Grabczewski. A new methodology of extraction, optimization and application of crisp and fuzzy logical rules. *IEEE Transactions on Neural Networks*, 12:277–306, 2001.

[Dar98] C. Darwin. *The Origin of Species*. Wordsworth Classics of World Literature. Wordsworth Editions Limited, 1998.

[Dav85] L. Davis. Applying adaptive algorithms to epistatic domains. In *Proceedings of the International Joint Conference on Artificial Intelligence*, 1985.

[Dei04] M. Deistler. System identification and time series analysis: Past, present, and future. In *Stochastic Theory and Control: Proceedings of a Workshop held in Lawrence, Kansas*, Lecture Notes in Control and Information Sciences, pages 97–110. Springer Berlin / Heidelberg, 2004.

[DeJ75] K. A. DeJong. *An Analysis of the Behavior of a Class of Genetic Adaptive Systems*. PhD thesis, University of Michigan, 1975.

[DG89] K. Deb and D. E. Goldberg. An investigation of niche and species formation in genetic function optimization. In *Proceedings of the Third International Conference on Genetic Algorithms*, pages 42–50. Morgan Kaufmann, 1989.

[DH02]   J. E. Devaney and J. G. Hagedorn. The role of genetic pro-
          gramming in describing the microscopic structure of hydrat-
          ing plaster. In E. Cantú-Paz et al., editors, *Late Breaking
          Papers at the Genetic and Evolutionary Computation Confer-
          ence (GECCO-2002)*, pages 91–98, New York, NY, July 2002.
          AAAI.

[DHS00]   R. O. Duda, P. E. Hart, and D. G. Stork. *Pattern Classifica-
          tion*. Wiley Interscience, 2nd edition, 2000.

[DLJD00] D. Dumitrescu, B. Lazzerini, L. C. Jain, and A. Dumitrescu.
          Evolutionary Computation. The CRC Press International Se-
          ries on Computational Intelligence. CRC Press, 2000.

[dN06]   L. M. de Menezes and N. Y. Nikolaev. Forecasting with genet-
          ically programmed polynomial neural networks. *International
          Journal of Forecasting*, 22(2):249–265, April-June 2006.

[Dom90]   W. Domschke. *Logistik: Rundreisen und Touren*. Oldenburg
          Verlag München Wien, 1990.

[DOMK+01] S. Dreiseitl, L. Ohno-Machado, H. Kittler, S. Vinterbo, H. Bill-
          hardt, and M. Binder. A comparison of machine learning
          methods for the diagnosis of pigmented skin lesions. *Journal
          of Biomedical Informatics*, 34:28–36, 2001.

[DR59]   G. B. Dantzig and R. H. Ramser. The Truck Dispatching
          Problem. *Management Science*, 6:80–91, 1959.

[dRLF+05] L. del Re, P. Langthaler, C. Furtmüller, S. Winkler, and M. Af-
          fenzeller. NO_x virtual sensor based on structure identification
          and global optimization. In *Proceedings of the SAE World
          Congress 2005*, number 2005-01-0050, 2005.

[Dro98]   S. Droste. Genetic programming with guaranteed quality. In
          J. R. Koza et al., editors, *Genetic Programming 1998: Pro-
          ceedings of the Third Annual Conference*, pages 54–59, Uni-
          versity of Wisconsin, Madison, Wisconsin, USA, 22-25 July
          1998. Morgan Kaufmann.

[E+07]   M. Ebner et al., editors. *Proceedings of the 10th European
          Conference on Genetic Programming*, volume 4445 of *Lecture
          Notes in Computer Science*, Valencia, Spain, 11 - 13 April
          2007. Springer.

[Eic07]   C. F. Eick. *Evolutionary Programming: Genetic Program-
          ming* (http://www2.cs.uh.edu/~ceick/6367/eiben6.ppt). De-
          partment of Computer Science, University of Houston, Texas,
          2007.

[EKK04] J. Eggermont, J. N. Kok, and W. A. Kosters. Detecting and pruning introns for faster decision tree evolution. In X. Yao et al., editors, *Parallel Problem Solving from Nature - PPSN VIII*, volume 3242 of *LNCS*, pages 1071–1080, Birmingham, UK, 18-22 September 2004. Springer-Verlag.

[EN00] A. Ekart and S. Z. Nemeth. A metric for genetic programs and fitness sharing. In R. Poli et al., editors, *Genetic Programming, Proceedings of EuroGP'2000*, volume 1802 of *LNCS*, pages 259–270, Edinburgh, 15-16 April 2000. Springer-Verlag.

[EN01] A. Ekart and S. Z. Nemeth. Selection based on the pareto nondomination criterion for controlling code growth in genetic programming. *Genetic Programming and Evolvable Machines*, 2(1):61–73, March 2001.

[ES03] A. E. Eiben and J. E. Smith. *Introduction to Evolutionary Computing*. Springer, 2003.

[FBF+03] P. Flach, H. Blockeel, C. Ferri, J. Hernández-Orallo, and J. Struyf. Decision support for data mining: Introduction to ROC analysis and its applications. *Data mining and decision support: Integration and collaboration*, 2003.

[FE05] J. E. Fieldsend and R. M. Everson. Formulation and comparison of multi-class ROC surfaces. *Proceedings of the ICML 2005 Workshop on ROC Analysis in Machine Learning*, pages 41–48, 2005.

[FG97] D. B. Fogel and A. Ghozeil. Schema processing under proportional selection in the presence of random effects. *IEEE Transactions on Evolutionary Computation*, 1(4):290–293, 1997.

[FG98] D. B. Fogel and A. Ghozeil. The schema theorem and the misallocation of trials in the presence of stochastic effects. *Proceedings of the 7th International Conference on Evolutionary Programming VI*, 1447:313–321, 1998.

[FJM97] M. L. Fisher, K. Jörnsteen, and O. B. G. Madsen. Vehicle routing with time windows: Two optimization algorithms. *Operations Research*, 45(3):488–492, 1997.

[FM91] B. R. Fox and M. B. McMahon. Genetic operators for sequencing problems. In Gregory J. E. Rawlins, editor, *Foundations of Genetic Algorithms*, pages 284–300. Morgan Kaufmann Publishers, 1991.

[Fog93] D. B. Fogel. Applying evolutionary programming to selected travelling salesman problems. *Cybernetics and Systems*, 24:27–36, 1993.

[Fog94] D. B. Fogel. An introduction to simulated evolutionary optimization. *IEEE Transactions on Neural Networks*, 5(1):3–14, 1994.

[For81] R. Forsyth. BEAGLE – A Darwinian approach to pattern recognition. *Kybernetes*, 10:159–166, 1981.

[FP93] C. Foisy and J. Potvin. Implementing an insertion heuristc on parallel hardware. *Computers and Operations Research*, 20(7):737–745, 1993.

[FP98] P. Funes and J. Pollack. Evolutionary body building: Adaptive physical designs for robots. *Artificial Life*, 4(4):337–357, Fall 1998.

[FPS06] G. Folino, C. Pizzuti, and G. Spezzano. Improving cooperative GP ensemble with clustering and pruning for pattern classification. In M. Keijzer et al., editors, *GECCO 2006: Proceedings of the 8th Annual Conference on Genetic and Evolutionary Computation*, volume 1, pages 791–798, Seattle, Washington, USA, 8-12 July 2006. ACM Press.

[FPSS96] U. M. Fayyad, G. Piatetsky-Shapiro, and P. Smyth. From data mining to knowledge discovery: An overview. *Advances in Knowledge Discovery and Data Mining*, 1996.

[GAMRRP07] M. Garcia-Arnau, D. Manrique, J. Rios, and A. Rodriguez-Paton. Initialization method for grammar-guided genetic programming. *Knowledge-Based Systems*, 20(2):127–133, March 2007. AI 2006, The 26th SGAI International Conference on Innovative Techniques and Applications of Artificial Intelligence.

[Gao03] Y. Gao. Population size and sampling complexity in genetic algorithms. In *Proceedings of the Genetic and Evolutionary Computation Conference (GECCO) 2003*, 2003.

[Gas67] T. Gaskell. Bases for vehicle fleet scheduling. *Operational Research Quarterly*, 18:281–295, 1967.

[GAT06] A. L. Garcia-Almanza and E. P. K. Tsang. Simplifying decision trees learned by genetic programming. In *Proceedings of the 2006 IEEE Congress on Evolutionary Computation*, pages 7906–7912, Vancouver, 6-21 July 2006. IEEE Press.

[GB89] J. J. Grefenstette and J. Baker. How genetic algorithms work: A critical look at implicit parallelism. In J. D. Schaffer, editor, *Proceedings of the Third International Conference on Genetic Algorithms*. Morgan Kaufmann Publishers, 1989.

[GBD80] B. L. Golden, L. Bodin, and T. Doyle. Approximate traveling salesman algorithm. *Operations Research*, 28:694–711, 1980.

[GGRG85] J. J. Grefenstette, R. Gopal, B. Rosmaita, and D. Van Gucht. Genetic algorithms for the traveling salesperson problem. *International Conference on Genetic Algorithms and Their Applications*, pages 160–168, 1985.

[GL85] D. Goldberg and R. Lingle. Alleles, loci, and the traveling salesman problem. *International Conference on Genetic Algorithms*, 1985.

[GL97] F. Glover and F. Laguna. *Tabu Search*. Kluwer Academic Publishers, 1997.

[Glo86] F. Glover. Future paths for integer programming and links to artificial intelligence. *Computers & Operations Research*, 13:533–549, 1986.

[GM74] B. Gillett and L. Miller. A heuristic for the vehicle–dispatch problem. *Operations Research*, 22:340–349, 1974.

[GMW82] P. Gill, W. Murray, and M. Wright. *Practical Optimization*. Academic Press, 1982.

[Gol84] B. L. Golden. Introduction to and recent advances in vehicle routing methods. *Transportation Planning Models*, pages 383–418, 1984.

[Gol89] D. E. Goldberg. *Genetic Algorithms in Search, Optimization and Machine Learning*. Addison Wesley Longman, 1989.

[Gom63] R. E. Gomory. An algorithm for integer solutions to linear programs. In R. L. Graves and P. Wolfe, editors, *Recent Advances in Mathematical Programming*, pages 269–302. McGraw-Hill, New York, 1963.

[GR94] C. Gathercole and P. Ross. Dynamic training subset selection for supervised learning in genetic programming. In Y. Davidor, H.-P. Schwefel, and R. Männer, editors, *Parallel Problem Solving from Nature III*, volume 866 of *LNCS*, pages 312–321, Jerusalem, 9-14 October 1994. Springer-Verlag.

[Grö77] M. Grötschel. *Polyedrische Charakterisierung kombinatorischer Optimierungsprobleme*. PhD thesis, University of Bonn, 1977.

[Gru94] F. Gruau. *Neural Network Synthesis using Cellular Encoding and the Genetic Algorithm*. PhD thesis, Laboratoire de l'Informatique du Parallilisme, Ecole Normale Supirieure de Lyon, France, 1994.

[GS90] M. Gorges-Schleuter. *Genetic Algorithms and Population Structures — A Massively Parallel Algorithm.* PhD thesis, University of Dortmund, 1990.

[Ham58] C. L. Hamblin. Computer languages. *The Australian Journal of Science*, 20:135–139, 1958.

[Ham62] C. L. Hamblin. Translation to and from Polish notation. *Computer Journal*, 5:210–213, 1962.

[Ham94] J. D. Hamilton. *Time Series Analysis.* Princeton University Press, 1994.

[HC89] D. L. Hartl and A. G. Clark. *Principles of Population Genetics.* Sinauer Associates Inc., 2nd edition, 1989.

[Hel00] K. Helsgaun. An effective implementation of the Lin-Kernighan traveling salesman heuristic. *European Journal of Operational Research*, 126(1):106–130, 2000.

[Hey88] J. B. Heywood. *Internal Combustion Engine Fundamentals.* McGraw-Hill, 1988.

[HGL93] A. Homaifar, S. Guan, and G. E. Liepins. A new approach on the traveling salesman problem by genetic algorithms. In *Proceedings of the 5th International Conference on Genetic Algorithms*, pages 460–466. Morgan Kaufmann Publishers Inc., 1993.

[HHM04] H. Tuan Hao, N. Xuan Hoai, and R. I. McKay. Does it matter where you start? A comparison of two initialisation strategies for grammar guided genetic programming. In R. I. Mckay and S.-B. Cho, editors, *Proceedings of The Second Asian-Pacific Workshop on Genetic Programming*, Cairns, Australia, 6-7 December 2004.

[HM82] J. Hanley and B. McNeil. The meaning and use of the area under a receiver operating characteristic (ROC) curve. *Radiology*, 143:29–36, 1982.

[HOFLe04] J. Hernández-Orallo, C. Ferri, C. Lachiche, and P. Flach (editors). *ROC Analysis in Artificial Intelligence, 1st International Workshop ROCAI-2004.* 2004.

[Hol75] J. H. Holland. *Adaption in Natural and Artifical Systems.* University of Michigan Press, 1975.

[HRv07] K. Holladay, K. Robbins, and J. von Ronne. FIFTH: A stack based GP language for vector processing. In M. Ebner et al., editors, *Proceedings of the 10th European Conference on Genetic Programming*, volume 4445 of *Lecture Notes in Computer*

Science, pages 102–113, Valencia, Spain, 11 - 13 April 2007. Springer.

[HS95] D. P. Helmbold and R. E. Schapire. Predicting nearly as well as the best pruning of a decision tree. *Proceedings of the Eighth Annual Conference on Computational Learning Theory*, pages 61–68, 1995.

[HSC96] H. J. Hamilton, N. Shan, and N. Cercone. RIAC: A rule induction algorithm based on approximate classification. Technical Report CS 96-06, Regina University, 1996.

[IIS98] T. Ito, H. Iba, and S. Sato. Non-destructive depth-dependent crossover for genetic programming. In W. Banzhaf et al., editors, *Proceedings of the First European Workshop on Genetic Programming*, volume 1391 of *LNCS*, pages 71–82, Paris, 14-15 April 1998. Springer-Verlag.

[Jac94] D. Jacquette. *Philosophy of Mind*. Prentice Hall, 1994.

[Jac99] C. Jacob. Lindenmayer systems and growth program evolution. In T. S. Hussain, editor, *Advanced Grammar Techniques Within Genetic Programming and Evolutionary Computation*, pages 76–79, Orlando, Florida, USA, 13 July 1999.

[JCC⁺92] D. Jefferson, R. Collins, C. Cooper, M. Dyer, M. Flowers, R. Korf, C. Taylor, and A. Wang. Evolution as a theme in artificial life: The genesys/tracker system. *Artificial Life II*, pages 417–434, 1992.

[Jes42] R. J. Jessen. Statistical investigation of a sample survey for obtaining farm facts. Research Bulletin 304, Iowa State College of Agriculture, 1942.

[JHC04] I. Jonyer, L. B. Holder, and D. J. Cook. Attribute-value selection based on minimum description length. In *Proceedings of the International Conference on Artificial Intelligence*, pages 1154–1159, 2004.

[JM05] X. Jiang and Y. Motai. Incremental on-line PCA for automatic motion learning of eigen behavior. *Proceedings of the 1st International Workshop on Automatic Learning and Real-Time ALaRT '05*, pages 153–164, 2005.

[K⁺06] M. Keijzer et al., editors. *GECCO 2006: Proceedings of the 8th Annual Conference on Genetic and Evolutionary Computation*, Seattle, Washington, USA, 8-12 July 2006. ACM Press.

[Kar77] R. M. Karp. Probabilistic analysis of partitioning algorithms of the traveling salesman problem in the plane. *Mathematics of Operations Research*, 2:209–224, 1977.

[Kar79] R. M. Karp. A patching algorithm for the nonsymmetric traveling salesman problem. *SIAM Journal of Computing*, 8:561–573, 1979.

[KBAK99] J. R. Koza, F.H. Bennett III, D. Andre, and M. A. Keane. The design of analog circuits by means of genetic programming. In P. Bentley, editor, *Evolutionary Design by Computers*, chapter 16, pages 365–385. Morgan Kaufmann, San Francisco, USA, 1999.

[Kei96] M. Keijzer. Efficiently representing populations in genetic programming. In P. J. Angeline and K. E. Kinnear, Jr., editors, *Advances in Genetic Programming 2*, chapter 13, pages 259–278. MIT Press, Cambridge, MA, USA, 1996.

[Kei02] M. Keijzer. *Scientific Discovery using Genetic Programming*. PhD thesis, Danish Technical University, Lyngby, Denmark, March 2002.

[Ken73] M. G. Kendall. *Time Series*. Griffin, 1973.

[KGV83] S. Kirkpatrick, C. D. Gelatt, and M. P. Vecchi. Optimization by simulated annealing. *Science*, 220:671–680, 1983.

[KIAK99] J. R. Koza, F. H. Bennett III, D. Andre, and M. A. Keane. *Genetic Programming III: Darvinian Invention and Problem Solving*. Morgan Kaufmann Publishers, 1999.

[Kin93] K. E. Kinnear, Jr. Generality and difficulty in genetic programming: Evolving a sort. In S. Forrest, editor, *Proceedings of the 5th International Conference on Genetic Algorithms, ICGA-93*, pages 287–294, University of Illinois at Urbana-Champaign, 17-21 July 1993. Morgan Kaufmann.

[KKS+03a] J. R. Koza, M. A. Keane, M. J. Streeter, W. Mydlowec, J. Yu, and G. Lanza. *Genetic Programming IV: Routine Human-Competitive Machine Intelligence*. Kluwer Academic Publishers, 2003.

[KKS+03b] J. R. Koza, M. A. Keane, M. J. Streeter, W. Mydlowec, J. Yu, and G. Lanza. *Genetic Programming IV: Routine Human-Competitive Machine Learning*. Kluwer Academic Publishers, 2003.

[KM97] N. Kohl and O.B.G. Madsen. An optimization algorithm for the vehicle routing problem with time windows based on lagrangean relaxation. *Operations Research*, 45(3):395–406, 1997.

[KO90] M. G. Kendall and J. K. Ord. *Time Series*. Edward Arnold, 1990.

[KOL+04] M. Keijzer, U.-M. O'Reilly, S. M. Lucas, E. Costa, and T. Soule, editors. *Genetic Programming 7th European Conference, EuroGP 2004, Proceedings*, volume 3003 of *LNCS*, Coimbra, Portugal, 5-7 April 2004. Springer-Verlag.

[Koz89] J. R. Koza. Hierarchical genetic algorithms operating on populations of computer programs. In N. S. Sridharan, editor, *Proceedings of the Eleventh International Joint Conference on Artificial Intelligence IJCAI-89*, volume 1, pages 768–774. Morgan Kaufmann, 20-25 August 1989.

[Koz92a] J. R. Koza. A genetic approach to the truck backer upper problem and the inter-twined spiral problem. In *Proceedings of IJCNN International Joint Conference on Neural Networks*, volume IV, pages 310–318. IEEE Press, 1992.

[Koz92b] J. R. Koza. *Genetic Programming: On the Programming of Computers by Means of Natural Selection*. The MIT Press, 1992.

[Koz94] J. R. Koza. *Genetic Programming II: Automatic Discovery of Reusable Programs*. The MIT Press, 1994.

[KP98] R. Kohavi and F. Provost. Glossary of terms. *Machine Learning, Special Issue on Applications of Machine Learning and the Knowledge Discovery Process*, 30:271–274, 1998.

[KRKT87] A. Kolen, A. Rinnooy-Kan, and H. Trienekens. Vehicle routing with time windows. *Operations Research*, 35(2):266–274, 1987.

[KSBM01] S. S. Keerthi, S. K. Shevade, C. Bhattacharyya, and K. R. K. Murthy. Improvements to platt's SMO algorithm for SVM classifier design. *Neural Computation*, 13(3):637–649, 2001.

[KTC+05] M. Keijzer, A. Tettamanzi, P. Collet, J. I. van Hemert, and M. Tomassini, editors. *Proceedings of the 8th European Conference on Genetic Programming*, volume 3447 of *Lecture Notes in Computer Science*, Lausanne, Switzerland, 30 March - 1 April 2005. Springer.

[Kus98] I. Kuscu. Evolving a generalised behavior: Artificial ant problem revisited. In V. William Porto, N. Saravanan, D. Waagen, and A. E. Eiben, editors, *Seventh Annual Conference on Evolutionary Programming*, volume 1447 of *LNCS*, pages 799–808, Mission Valley Marriott, San Diego, California, USA, 25-27 March 1998. Springer-Verlag.

[Lan95] W. B. Langdon. Evolving data structures using genetic programming. In L. Eshelman, editor, *Genetic Algorithms: Proceedings of the Sixth International Conference (ICGA95)*, pages 295–302, Pittsburgh, PA, USA, 15-19 July 1995. Morgan Kaufmann.

[Lan98]   W. B. Langdon.   *Genetic Programming and Data Structures: Genetic Programming + Data Structures = Automatic Programming!*, volume 1 of *Genetic Programming*. Kluwer, Boston, 24 April 1998.

[Lan99]   W. B. Langdon.   Size fair tree crossovers.   In Eric Postma and Marc Gyssen, editors, *Proceedings of the Eleventh Belgium/Netherlands Conference on Artificial Intelligence (BNAIC'99)*, pages 255–256, Kasteel Vaeshartelt, Maastricht, Holland, 3-4 November 1999.

[Lan00]   W. B. Langdon.   Size fair and homologous tree genetic programming crossovers.   *Genetic Programming and Evolvable Machines*, 1(1/2):95–119, April 2000.

[Lap92]   G. Laporte.   The vehicle routing problem: An overview of exact and approximate algorithms. *European Journal of Operational Research*, 59:345–358, 1992.

[Lar99]   J. Larsen. *Parallelization of the Vehicle Routing Problem with Time Windows*. PhD thesis, Department of Computer Science, University of Copenhagen, 1999.

[LAWR05]   P. Langthaler, D. Alberer, S. Winkler, and L. Del Re. Design eines virtuellen Sensors für Partikelmessung am Dieselmotor. In M. Horn, M. Hofbauer, and N. Dourdoumas, editors, *Proceedings of the 14th Styrian Seminar on Control Engineering and Process Automation (14. Steirisches Seminar über Regelungstechnik und Prozessautomatisierung)*, pages 71–87, 2005.

[LC01]   T. Loveard and V. Ciesielski. Representing classification problems in genetic programming. In *Proceedings of the Congress on Evolutionary Computation*, volume 2, pages 1070–1077, COEX, World Trade Center, 159 Samseong-dong, Gangnam-gu, Seoul, Korea, 2001. IEEE Press.

[LC05]   D. P. X. Li and V. Ciesielski. Multi-objective techniques in genetic programming for evolving classifiers. *Proceedings of the 2005 Congress on Evolutionary Computation (CEC '05)*, pages 183–190, 2005.

[Lev44]   K. Levenberg. A method for the solution of certain non-linear problems in least squares. *The Quarterly of Applied Mathematics*, 2:164–168, 1944.

[Lev66] V. I. Levenshtein. Binary codes capable of correcting deletions, insertions, and reversals. *Soviet Physics Doklady*, 10(8):707–710, 1966.

[LGX97] Y. Leung, Y. Gao, and Z. B. Xu. Degree of population diversity - a perspective on premature convergence in genetic algorithms and its Markov chain analysis. *IEEE Transactions on Neural Networks*, 8(5):1165–1176, 1997.

[LH06] P. Lichodzijewski and M. I. Heywood. Pareto-coevolutionary genetic programming for problem decomposition in multi-class classification. *Proceedings of the Genetic and Evolutionary Computation Conference GECCO'07*, pages 464–471, 2006.

[Lin65] S. Lin. Computer solutions of the traveling salesman problem. *Systems Technical Journal*, 44:2245–2269, 1965.

[Lju99] L. Ljung. *System Identification – Theory For the User, 2nd edition*. PTR Prentice Hall, Upper Saddle River, N.J., 1999.

[LK73] S. Lin and B. W. Kernighan. An effective heuristic algorithm for the traveling-salesman problem. *Operations Research*, 21:498–516, 1973.

[LKM+99] P. Larranaga, C. M. H. Kuijpers, R. H. Murga, I. Inza, and D. Dizdarevic. Genetic algorithms for the travelling salesman problem: A review of representations and operators. *Artificial Intelligence Review*, 13:129–170, 1999.

[LLRKS85] E. L. Lawler, J. K. Lenstra, A. Rinnooy-Kan, and D. B. Shmoys. *The Travelling Salesman Problem*. Wiley, New York, 1985.

[LN00] W. B. Langdon and J. P. Nordin. Seeding GP populations. In Riccardo Poli, Wolfgang Banzhaf, William B. Langdon, Julian F. Miller, Peter Nordin, and Terence C. Fogarty, editors, *Genetic Programming, Proceedings of EuroGP'2000*, volume 1802 of *LNCS*, pages 304–315, Edinburgh, 15-16 April 2000. Springer-Verlag.

[LP97] W. B. Langdon and R. Poli. Fitness causes bloat. In P. K. Chawdhry, R. Roy, and R. K. Pant, editors, *Soft Computing in Engineering Design and Manufacturing*, pages 13–22. Springer-Verlag London, 23-27 June 1997.

[LP98] W. B. Langdon and R. Poli. Why ants are hard. In J. R. Koza, W. Banzhaf, K. Chellapilla, K. Deb, M. Dorigo, D. B. Fogel, M. H. Garzon, D. E. Goldberg, H. Iba, and R. Riolo, editors, *Genetic Programming 1998: Proceedings of the Third Annual*

Conference, pages 193–201, University of Wisconsin, Madison, Wisconsin, USA, 22-25 July 1998. Morgan Kaufmann.

[LP02] W. B. Langdon and R. Poli. *Foundations of Genetic Programming*. Springer Verlag, Berlin Heidelberg New York, 2002.

[LS97] S. Luke and L. Spector. A comparison of crossover and mutation in genetic programming. In J. R. Koza et al., editors, *Genetic Programming 1997: Proceedings of the Second Annual Conference*, pages 240–248, Stanford University, CA, USA, 13-16 July 1997. Morgan Kaufmann.

[LW95] J. Y. B. Lee and P. C. Wong. The effect of function noise on GP efficiency. In X. Yao, editor, *Progress in Evolutionary Computation*, volume 956 of *Lecture Notes in Artificial Intelligence*, pages 1–16. Springer-Verlag, Heidelberg, Germany, 1995.

[Mah40] P. Mahalanobis. A sample survey of the acreage under jute in Bengal. *Sankhyu*, 4:511–530, 1940.

[Man97] Y. Mansour. Pessimistic decision tree pruning based on tree size. *Proceedings of the Fourteenth International Conference on Machine Learning*, pages 195–201, 1997.

[Mar63] D. W. Marquardt. An algorithm for least-squares estimation of nonlinear parameters. *SIAM Journal on Applied Mathematics*, 11:431–441, 1963.

[McC60] J. L. McCarthy. Recursive functions of symbolic expressions and their computation by machine, part I. *Communications of the ACM*, 3(4):184–195, 1960.

[McK00] R. I. McKay. Fitness sharing in genetic programming. In D. Whitley et al., editors, *Proceedings of the Genetic and Evolutionary Computation Conference (GECCO-2000)*, pages 435–442, Las Vegas, Nevada, USA, 10-12 July 2000. Morgan Kaufmann.

[Men27] K. Menger. Zur allgemeinen Kurventheorie. *Fundamenta Mathematicae*, 10:96–115, 1927.

[Müh89] H. Mühlenbein. Parallel genetic algorithms, population genetics and combinatorial optimization. *Proceedings of the 3rd International Conference on Genetic Algorithms*, pages 416–421, 1989.

[MH99] N. F. McPhee and N. J. Hopper. Analysis of genetic diversity through population history. In W. Banzhaf et al., editors, *Proceedings of the Genetic and Evolutionary Computation Con-*

ference, volume 2, pages 1112–1120, Orlando, Florida, USA, 13-17 July 1999. Morgan Kaufmann.

[MIB+00] K. Morik, M. Imhoff, P. Brockhausen, T. Joachims, and U. Gather. Knowledge discovery and knowledge validation in intensive care. *Artificial Intelligence in Medicine*, 19:225–249, 2000.

[Mic92] Z. Michalewicz. *Genetic Algorithms + Data Structures = Evolution Programs*. Springer, 1992.

[Min89] J. Mingers. An empirical comparison of pruning methods for decision tree induction. *Machine Learning*, 4:227 – 243, 1989.

[Mit96] M. Mitchell. *An Introduction to Genetic Algorithms*. The MIT Press, 1996.

[Mit00] T. M. Mitchell. *Machine Learning*. McGraw-Hill, New York, 2000.

[MJK07] D. C. Montgomery, C. L. Jennings, and M. Kulahci. *Introduction to Time Series Analysis and Forecasting*. Wiley & Sons, 2007.

[MK00] Y. Maeda and S. Kawaguchi. Redundant node pruning and adaptive search method for genetic programming. In D. Whitley et al., editors, *Proceedings of the Genetic and Evolutionary Computation Conference (GECCO-2000)*, page 535, Las Vegas, Nevada, USA, 10-12 July 2000. Morgan Kaufmann.

[Mor91] F. Morrison. *The Art of Modeling Dynamic Systems: Forecasting for Chaos, Randomness, and Determinism*. John Wiley & Sons, Inc, 1991.

[MP43] W. S. McCulloch and W. H. Pitts. A logical calculus of the ideas imminent in nervous activity. In *Bulletin of Mathematical Biophysics*, volume 5, pages 115–137, 1943.

[NB95] P. Nordin and W. Banzhaf. Complexity compression and evolution. In L. Eshelman, editor, *Genetic Algorithms: Proceedings of the Sixth International Conference (ICGA95)*, pages 310–317, Pittsburgh, PA, USA, 15-19 July 1995. Morgan Kaufmann.

[Nel01] O. Nelles. *Nonlinear System Identification*. Springer Verlag, Berlin Heidelberg New York, 2001.

[Nol97] D. Nolan. Quantitative parsimony. *British Journal for the Philosophy of Science*, 48(3):329–343, 1997.

[Nor97] P. Nordin. *Evolutionary Program Induction of Binary Machine Code and its Applications.* PhD thesis, Universität Dortmund, Fachbereich Informatik, 1997.

[Nør00] M. Nørgaard. Neural network based system identification toolbox. Technical Report 00-E-891, Technical University of Denmark, 2000.

[NV92] A. E. Nix and M. D. Vose. Modeling genetic algorithms with markov chains. *Annals of Mathematics and Artificial Intelligence*, 5(1):79–88, 1992.

[OO94] U.-M. O'Reilly and F. Oppacher. The troubling aspects of a building block hypothesis for genetic programming. In L. D. Whitley and M. D. Vose, editors, *Foundations of Genetic Algorithms 3*, pages 73–88, Estes Park, Colorado, USA, 31 July–2 August 1994. Morgan Kaufmann. Published 1995.

[O'R95] U.-M. O'Reilly. *An Analysis of Genetic Programming.* PhD thesis, Carleton University, Ottawa-Carleton Institute for Computer Science, Ottawa, Ontario, Canada, 22 September 1995.

[O'R97] U.-M. O'Reilly. Using a distance metric on genetic programs to understand genetic operators. In *IEEE International Conference on Systems, Man, and Cybernetics, Computational Cybernetics and Simulation*, volume 5, pages 4092–4097, Orlando, Florida, USA, 12-15 October 1997.

[OSH87] I. M. Oliver, D. J. Smith, and J. R. C. Holland. A study of permutation crossover operators on the travelling salesman problem. In J. J. Grefenstette, editor, *Genetic algorithms and their applications: Proceedings of the Second International Conference on Genetic Algorithms*, pages 224–230, Hillsdale, NJ, 1987. Lawrence Erlbaum Assoc.

[Osm93] I. H. Osman. Metastrategy simulated annealing and tabu search algorithms for the vehicle routing problem. *Annals of Operations Research*, 41(1–4):421–451, 1993.

[Pan83] A. Pankratz. *Forecasting With Univariate Box-Jenkins Models: Concepts and Cases.* Wiley, 1983.

[Pan91] A. Pankratz. *Forecasting With Dynamic Regression Models.* Wiley, 1991.

[PB96] J.-Y. Potvin and S. Bengio. The Vehicle Routing Problem with Time Windows - Part II: Genetic Search. *INFORMS Journal on Computing*, 8(2):165–172, 1996.

[Per94] T. Perkis. Stack-based genetic programming. In *Proceedings of the 1994 IEEE World Congress on Computational Intelligence*, volume 1, pages 148–153, Orlando, Florida, USA, 27-29 June 1994. IEEE Press.

[PL97a] R. Poli and W. B. Langdon. An experimental analysis of schema creation, propagation and disruption in genetic programming. In T. Back, editor, *Genetic Algorithms: Proceedings of the Seventh International Conference*, pages 18–25, Michigan State University, East Lansing, MI, USA, 19-23 July 1997. Morgan Kaufmann.

[PL97b] R. Poli and W. B. Langdon. Genetic programming with one-point crossover. In P. K. Chawdhry, R. Roy, and R. K. Pant, editors, *Soft Computing in Engineering Design and Manufacturing*, pages 180–189. Springer-Verlag London, 23-27 June 1997.

[PL97c] R. Poli and W. B. Langdon. A new schema theory for genetic programming with one-point crossover and point mutation. In J. R. Koza et al., editors, *Genetic Programming 1997: Proceedings of the Second Annual Conference*, pages 278–285, Stanford University, CA, USA, 13-16 July 1997. Morgan Kaufmann.

[Pla99] J. Platt. Fast training of support vector machines using sequential minimal optimization. In B. Schoelkopf, C. Burges, and A. Smola, editors, *Advances in Kernel Methods - Support Vector Learning*, pages 185–208. MIT Press, 1999.

[PM01a] R. Poli and N. F. McPhee. Exact GP schema theory for headless chicken crossover and subtree mutation. In *Proceedings of the 2001 Congress on Evolutionary Computation CEC2001*, pages 1062–1069, COEX, World Trade Center, 159 Samseong-dong, Gangnam-gu, Seoul, Korea, 27-30 May 2001. IEEE Press.

[PM01b] R. Poli and N. F. McPhee. Exact schema theorems for GP with one-point and standard crossover operating on linear structures and their application to the study of the evolution of size. In J. F. Miller, M. Tomassini, P. L. Lanzi, C. Ryan, A. G. B. Tettamanzi, and W. B. Langdon, editors, *Genetic Programming, Proceedings of EuroGP'2001*, volume 2038 of *LNCS*, pages 126–142, Lake Como, Italy, 18-20 April 2001. Springer-Verlag.

[PM01c] R. Poli and N. F. McPhee. Exact schema theory for GP and variable-length GAs with homologous crossover. In L. Spec-

348   *References*

tor et al., editors, *Proceedings of the Genetic and Evolutionary Computation Conference (GECCO-2001)*, pages 104–111,
San Francisco, California, USA, 7-11 July 2001. Morgan Kaufmann.

[PM03a] R. Poli and N. F. McPhee. General schema theory for genetic programming with subtree-swapping crossover: Part I.
Evolutionary Computation, 11(1):53–66, March 2003.

[PM03b] R. Poli and N. F. McPhee. General schema theory for genetic programming with subtree-swapping crossover: Part II.
Evolutionary Computation, 11(2):169–206, June 2003.

[PMR04] R. Poli, N. F. McPhee, and J. E. Rowe. Exact schema theory and markov chain models for genetic programming and
variable-length genetic algorithms with homologous crossover.
Genetic Programming and Evolvable Machines, 5(1):31–70,
March 2004.

[Pol97] R. Poli. Evolution of graph-like programs with parallel distributed genetic programming. In T. Back, editor, *Genetic
Algorithms: Proceedings of the Seventh International Conference*, pages 346–353, Michigan State University, East Lansing,
MI, USA, 19-23 July 1997. Morgan Kaufmann.

[Pol99a] R. Poli. New results in the schema theory for GP with one-
point crossover which account for schema creation, survival
and disruption. Technical Report CSRP-99-18, University of
Birmingham, School of Computer Science, December 1999.

[Pol99b] R. Poli. Parallel distributed genetic programming. In David
Corne, Marco Dorigo, and Fred Glover, editors, *New Ideas in
Optimization*, Advanced Topics in Computer Science, chap-
ter 27, pages 403–431. McGraw-Hill, Maidenhead, Berkshire,
England, 1999.

[Pol00a] R. Poli. Exact schema theorem and effective fitness for GP
with one-point crossover. In D. Whitley et al., editors, *Proceedings of the Genetic and Evolutionary Computation Conference (GECCO-2000)*, pages 469–476, Las Vegas, Nevada,
USA, 10-12 July 2000. Morgan Kaufmann.

[Pol00b] R. Poli. Hyperschema theory for GP with one-point crossover,
building blocks, and some new results in GA theory. In R. Poli,
W. Banzhaf, W. B. Langdon, J. F. Miller, P. Nordin, and
T. C. Fogarty, editors, *Genetic Programming, Proceedings of
EuroGP'2000*, volume 1802 of *LNCS*, pages 163–180, Edin-
burgh, 15-16 April 2000. Springer-Verlag.

[Pol00c] R. Poli. A macroscopic exact schema theorem and a redefinition of effective fitness for GP with one-point crossover. Technical Report CSRP-00-1, University of Birmingham, School of Computer Science, February 2000.

[Pol01] R. Poli. Exact schema theory for genetic programming and variable-length genetic algorithms with one-point crossover. *Genetic Programming and Evolvable Machines*, 2(2):123–163, June 2001.

[Pop92] K. Popper. *The Logic of Scientific Discovery*. Taylor & Francis, 1992.

[PP01] F. Previdi and T. Parisini. Model-free fault detection: a spectral estimation approach based on coherency functions. *International Journal of Control*, 74:1107–1117, 2001.

[PR93] J. Potvin and J. Rousseau. A parallel route building algorithm for the vehicle routing and scheduling problem with time windows. *European Journal of Operations Research*, 66:331–340, 1993.

[Pri04] C. Prins. A simple and effective evolutionary algorithm for the vehicle routing problem. *Computers & Operations Research*, 31(12):1985–2002, 2004.

[PRM01] R. Poli, J. E Rowe, and N. F. McPhee. Markov models for GP and variable-length GAs with homologous crossover. Technical Report CSRP-01-6, University of Birmingham, School of Computer Science, January 2001.

[PS82] C. Papadimitriou and K. Steiglitz. *Combinatorial Optimization: Algorithms and Complexity*. Prentice-Hall, 1982.

[PTMC02] F. B. Pereira, J. Tavares, P. Machado, and E. Costa. GVR: A New Genetic Representation for the Vehicle Routing Problem. In *AICS '02: Proceedings of the 13th Irish International Conference on Artificial Intelligence and Cognitive Science*, pages 95–102, London, UK, 2002. Springer-Verlag.

[PTT01] D. Peña, G. C. Tiao, and R. S. Tsay. *A Course in Time Series Analysis*. Wiley, 2001.

[Que03] Christian Queinnec. *LISP in Small Pieces*. Cambridge University Press, 2003.

[Raw91] G. J. E. Rawlins, editor. *Foundations of Genetic Algorithms*, volume 1. Morgan Kaufmann Publishers, 1991.

[RB96] J. P. Rosca and D. H. Ballard. Discovery of subroutines in genetic programming. In Peter J. Angeline and K. E. Kinnear,

Jr., editors, *Advances in Genetic Programming 2*, chapter 9, pages 177–202. MIT Press, Cambridge, MA, USA, 1996.

[Rec73] I. Rechenberg. *Evolutionsstrategie*. Friedrich Frommann Verlag, 1973.

[Ree95] C. Reeves. *Modern Heuristic Techniques for Combinatorial Optimization*. McGraw-Hill International Ltd., 1995.

[Rei91] G. Reinelt. TSPLIB - A traveling salesman problem library. *ORSA Journal on Computing*, 3:376–384, 1991.

[RF99] J. L. Rodríguez-Fernández. Ockham's razor. *Endeavour*, 23:121–125, 1999.

[RN03] S. J. Russell and P. Norvig. *Artificial Intelligence: A Modern Approach*. Prentice Hall, 2nd edition, 2003.

[Ros95a] J. Rosca. Towards automatic discovery of building blocks in genetic programming. In E. V. Siegel and J. R. Koza, editors, *Working Notes for the AAAI Symposium on Genetic Programming*, pages 78–85, MIT, Cambridge, MA, USA, 10–12 November 1995. AAAI.

[Ros95b] J. P. Rosca. Entropy-driven adaptive representation. In Justinian P. Rosca, editor, *Proceedings of the Workshop on Genetic Programming: From Theory to Real-World Applications*, pages 23–32, Tahoe City, California, USA, 9 July 1995.

[Ros97] J. P. Rosca. Analysis of complexity drift in genetic programming. In J. R. Koza et al., editors, *Genetic Programming 1997: Proceedings of the Second Annual Conference*, pages 286–294, Stanford University, CA, USA, 13-16 July 1997. Morgan Kaufmann.

[RS94] N. J. Radcliffe and P. D. Surry. Fitness variance of formae and performance prediction. In L. D. Whitley and M. D. Vose, editors, *Foundations of Genetic Algorithms*, volume 3, pages 51–72. Morgan Kaufmann Publishers, 1994.

[Rus95] R. A. Russell. Hybrid heuristics for the vehicle routing problem with time windows. *Transportation Science*, 29:156–166, 1995.

[Sam59] A. L. Samuel. Some studies in machine learning using the game of checkers. In *IBM Journal of Research and Development*, volume 3, pages 211 – 229, 1959.

[Sav85] M. W. P. Savelsbergh. Local search in routing problems with time windows. *Annals of Operations Research*, 4:285–305, 1985.

[Sch75] H.-P. Schwefel. *Evolutionsstrategie und numerische Opti-mierung*. PhD thesis, Technische Universität Berlin, 1975.

[Sch94] H.-P. Schwefel. *Numerische Optimierung von Computer-Modellen mittels der Evolutionsstrategie*. Birkhäuser Verlag, Basel, Switzerland, 1994.

[SD88] M. Solomon and J. Desrosiers. Time window constrained rout-ing and scheduling problems. *Transportation Science*, 22(1):1–13, 1988.

[SF98] T. Soule and J. A. Foster. Removal bias: a new cause of code growth in tree based evolutionary programming. In *1998 IEEE International Conference on Evolutionary Computation*, pages 781–186, Anchorage, Alaska, USA, 5-9 May 1998. IEEE Press.

[SFD96] T. Soule, J. A. Foster, and J. Dickinson. Code growth in genetic programming. In J. R. Koza et al., editors, *Genetic Programming 1996: Proceedings of the First Annual Confer-ence*, pages 215–223, Stanford University, CA, USA, 28–31 July 1996. MIT Press.

[SFP93] R. E. Smith, S. Forrest, and A. S. Perelson. Population di-versity in an immune systems model: Implications for genetic search. In *Foundations of Genetic Algorithms*, volume 2, pages 153–166. Morgan Kaufmann Publishers, 1993.

[SG99] K. Sterelny and P. E. Griffiths. *Sex and Death: An Introduc-tion to Philosophy of Biology*. University of Chicago Press, 1999.

[SH98] P. W. H. Smith and K. Harries. Code growth, explicitly de-fined introns, and alternative selection schemes. *Evolutionary Computation*, 6(4):339–360, Winter 1998.

[SHF94] E. Schöneburg, F. Heinzmann, and S. Feddersen. *Genetische Algorithmen und Evolutionsstrategien*. Addison-Wesley, 1994.

[Sig86] I. K. Sigal. Computational implementation of a combined branch and bound algorithm for the travelling salesman prob-lem. *Computational Mathematics and Mathematical Physics*, 26:14–19, 1986.

[SJW92] W. Schiffmann, M. Joost, and R. Werner. Optimization of the backpropagation algorithm for training multilayer percep-trons. Technical Report 15, University of Koblenz, Institute of Physics, 1992.

[SJW93] W. Schiffmann, M. Joost, and R. Werner. Comparison of op-timized backpropagation algorithms. *Proceedings of the Eu-*

ropean Symposium on Artificial Neural Networks ESANN '93, pages 97–104, 1993.

[Smi80] S. F. Smith. *A Learning System Based on Genetic Adaptive Algorithms*. PhD thesis, University of Pittsburgh, 1980.

[SMM⁺91] T. Starkweather, S. McDaniel, K. Mathias, D. Whitley, and C. Whitley. A comparison of genetic scheduling operators. *Proceedings of the Fourth International Conference on Genetic Algorithms*, pages 69–76, 1991.

[SOG04] K. Sastry, U.-M. O'Reilly, and D. E. Goldberg. Population sizing for genetic programming based on decision making. In U.-M. O'Reilly et al., editors, *Genetic Programming Theory and Practice II*, chapter 4, pages 49–65. Springer, Ann Arbor, 13-15 May 2004.

[Sol86] M. M. Solomon. On the worst-case performance of some heuristics for the vehicle routing and scheduling problem with time window constraints. *Networks*, 16:161–174, 1986.

[Sol87] M. M. Solomon. Algorithms for the Vehicle Routing and Scheduling Problem with Time Window Constraints. *Operations Research*, 35:254–265, 1987.

[SP94] M. Srinivas and L. Patnaik. Adaptive probabilities of crossover and mutation in genetic algorithms. In *IEEE Trans. on Systems, Man, and Cybernetics*, volume 24, pages 656–667, 1994.

[SPWR02] C. R. Stephens, R. Poli, A. H. Wright, and J. E. Rowe. Exact results from a coarse grained formulation of the dynamics of variable-length genetic algorithms. In W. B. Langdon et al., editors, *GECCO 2002: Proceedings of the Genetic and Evolutionary Computation Conference*, pages 578–585, New York, 9-13 July 2002. Morgan Kaufmann Publishers.

[Sri99] A. Srinivasan. Note on the location of optimal classifiers in n-dimensional ROC space (technical report PRG-TR-2-99). Technical report, Oxford University Computing Laboratory, 1999.

[SW97] C. R. Stephens and H. Waelbroeck. Effective degrees of freedom in genetic algorithms and the block hypothesis. *Proceedings of the Seventh International Conference on Genetic Algorithms (ICGA97)*, pages 34–40, 1997.

[SW99] C. R. Stephens and H. Waelbroeck. Schemata evolution and building blocks. *Evolutionary Computation*, 7(2):109–124, 1999.

[SWM91] T. Starkweather, D. Whitley, and K. Mathias. Optimization using distributed genetic algorithms. *Parallel Problem Solving from Nature*, pages 176–185, 1991.

[T+07] D. Thierens et al., editors. *GECCO 2007: Proceedings of the 9th Annual Conference on Genetic and Evolutionary Computation*, London, UK, 7-11 July 2007. ACM Press.

[Tai93] E. D. Taillard. Benchmarks for basic scheduling problems. *European Journal of Operational Research*, 64:278–285, 1993.

[TG97] I. Taha and J. Ghosh. Evaluation and ordering of rules extracted from feedforward networks. *Proceedings of the IEEE International Conference on Neural Networks*, pages 221–226, 1997.

[TH02] M. Terrio and M. I. Heywood. Directing crossover for reduction of bloat in GP. In W. Kinsner, A. Seback, and K. Ferens, editors, *IEEE CCECE 2003: IEEE Canadian Conference on Electrical and Computer Engineering*, pages 1111–1115. IEEE Press, 12-15 May 2002.

[Tha95] S. R. Thangiah. *Vehicle Routing with Time Windows using Genetic Algorithms*, chapter Chapter 11, pages 253–278. The Practical Handbook of Genetic Algorithms: New Frontiers. CRC Press, 1995.

[THL94] J. Ting-Ho-Lo. Synthetic approach to optimal filtering. *IEEE Transactions on Neural Networks*, 5:803–811, 1994.

[TMPC03] J. Tavares, P. Machado, F. B. Pereira, and E. Costa. On the influence of GVR in vehicle routing. In *SAC'03: Proceedings of the 2003 ACM Symposium on Applied Computing*, pages 753–758. ACM, 2003.

[Tom95] M. Tomassini. A survey of genetic algorithms. *Annual Reviews of Computational Physics*, 3:87–118, 1995.

[TOS94] S. Thangiah, I. Osman, and T. Sun. Hybrid genetic algorithm simulated annealing and tabu search methods for vehicle routing problem with time windows. Technical report, Computer Science Department, Slippery Rock University, 1994.

[TPS96] S. R. Thangiah, J.-Y. Potvin, and T. Sun. Heuristic approaches to vehicle routing with backhauls and time windows. *International Journal on Computers and Operations Research*, 23(11):1043–1057, 1996.

[Vap98] V. Vapnik. *Statistical Learning Theory*. Wiley, New York, 1998.

[vB95] A. v. Breedam. Vehicle routing: Bridging the gap between theory and practice. *Belgian Journal of Operations Research, Statistics and Computer Science*, 35(1):63–80, 1995.

[vBS04] R. van Basshuysen and F. Schäfer. *Internal Combustion Engine Handbook*. SAE International, 2004.

[VL91] M. D. Vose and G. E. Liepins. Punctuated equilibria in genetic search. *Complex Systems*, 5:31–44, 1991.

[Voi31] B. F. Voigt. *Der Handlungsreisende, wie er sein soll und was er zu thun hat, um Aufträge zu erhalten und eines glücklichen Erfolgs in seinen Geschäften gewiss zu sein*. Von einem alten Commis Voyageur, 1831.

[Vos99] M. D. Vose. *The Simple Genetic Algorithm: Foundations and Theory*. MIT Press, Cambridge, MA, 1999.

[W18] M. Thorburn W. The myth of occam's razor. *Mind*, 27:345–353, 1918.

[WA04a] S. Wagner and M. Affenzeller. HeuristicLab grid - a flexible and extensible environment for parallel heuristic optimization. In Z. Bubnicki and A. Grzech, editors, *Proceedings of the 15th International Conference on Systems Science*, volume 1, pages 289–296. Oficyna Wydawnicza Politechniki Wroclawskiej, 2004.

[WA04b] S. Wagner and M. Affenzeller. HeuristicLab grid - a flexible and extensible environment for parallel heuristic optimization. *Journal of Systems Science*, 30(4):103–110, 2004.

[WA04c] S. Wagner and M. Affenzeller. The heuristicLab optimization environment. Technical report, Institute of Formal Models and Verification, Johannes Kepler University, Linz, Austria, 2004.

[WA05a] S. Wagner and M. Affenzeller. HeuristicLab: A generic and extensible optimization environment. In B. Ribeiro, R. F. Albrecht, A. Dobnikar, D. W. Pearson, and N. C. Steele, editors, *Adaptive and Natural Computing Algorithms*, Springer Computer Science, pages 538–541. Springer, 2005.

[WA05b] S. Wagner and M. Affenzeller. SexualGA: Gender-specific selection for genetic algorithms. In N. Callaos, W. Lesso, and E. Hansen, editors, *Proceedings of the 9th World Multi-Conference on Systemics, Cybernetics and Informatics (WM-SCI) 2005*, volume 4, pages 76–81. International Institute of Informatics and Systemics, 2005.

[WAW04a] S. Winkler, M. Affenzeller, and S. Wagner. Identifying nonlinear model structures using genetic programming techniques. In R. Trappl, editor, *Cybernetics and Systems 2004*, volume 1, pages 689–694. Austrian Society for Cybernetic Studies, 2004.

[WAW04b] S. Winkler, M. Affenzeller, and S. Wagner. New methods for the identification of nonlinear model structures based upon genetic programming techniques. In Z. Bubnicki and A. Grzech, editors, *Proceedings of the 15th International Conference on Systems Science*, volume 1, pages 386–393. Oficyna Wydawnicza Politechniki Wroclawskiej, 2004.

[WAW05a] S. Winkler, M. Affenzeller, and S. Wagner. Genetic programming based model structure identification using on-line system data. In F. Barros, A. Bruzzone, C. Frydman, and N. Gambiasi, editors, *Proceedings of Conceptual Modeling and Simulation Conference CMS 2005*, pages 177–186. Frydman, LSIS, Université Paul Cézanne Aix Marseille III, 2005.

[WAW05b] S. Winkler, M. Affenzeller, and S. Wagner. New methods for the identification of nonlinear model structures based upon genetic programming techniques. *Journal of Systems Science*, 31(1):5–13, 2005.

[WAW06a] S. Winkler, M. Affenzeller, and S. Wagner. Advances in applying genetic programming to machine learning, focussing on classification problems. In *Proceedings of the 9th International Workshop on Nature Inspired Distributed Computing NIDISC '06, part of the Proceedings of the 20th IEEE International Parallel & Distributed Processing Symposium IPDPS 2006*. IEEE, 2006.

[WAW06b] S. Winkler, M. Affenzeller, and S. Wagner. Automatic data based patient classification using genetic programming. In R. Trappl, R. Brachman, R.A. Brooks, H. Kitano, D. Lenat, O. Stock, W. Wahlster, and M. Wooldridge, editors, *Cybernetics and Systems 2006*, volume 1, pages 251–256. Austrian Society for Cybernetic Studies, 2006.

[WAW06c] S. Winkler, M. Affenzeller, and S. Wagner. HeuristicModeler: A multi-purpose evolutionary machine learning algorithm and its applications in medical data analysis. In A. Bruzzone, A. Guasch, M. Piera, and J. Rozenblit, editors, *Proceedings of the International Mediterranean Modelling Multiconference I3M 2006*, pages 629–634. Piera, LogiSim, Barcelona, Spain, 2006.

[WAW06d] S. Winkler, M. Affenzeller, and S. Wagner. Sets of receiver operating characteristic curves and their use in the evaluation of multi-class classification. In *Proceedings of the Genetic and Evolutionary Computation Conference GECCO 2006*, volume 2, pages 1601–1602. Association for Computing Machinery (ACM), 2006.

[WAW06e] S. Winkler, M. Affenzeller, and S. Wagner. Using enhanced genetic programming techniques for evolving classifiers in the context of medical diagnosis - an empirical study. In *Proceedings of the GECCO 2006 Workshop on Medical Applications of Genetic and Evolutionary Computation (MedGEC 2006)*. Association for Computing Machinery (ACM), 2006.

[WAW07] S. Winkler, M. Affenzeller, and S. Wagner. Advanced genetic programming based machine learning. *Journal of Mathematical Modelling and Algorithms*, 6(3):455–480, 2007.

[WAW08] S. Winkler, M. Affenzeller, and S. Wagner. Offspring selection and its effects on genetic propagation in genetic programming based system identification. In Robert Trappl, editor, *Cybernetics and Systems 2008*, volume 2, pages 549–554. Austrian Society for Cybernetic Studies, 2008.

[WC99] P. A. Whigham and P. F. Crapper. Time series modelling using genetic programming: An application to rainfall-runoff models. In L. Spector et al., editors, *Advances in Genetic Programming 3*, chapter 5, pages 89–104. MIT Press, Cambridge, MA, USA, June 1999.

[WEA+06] S. Winkler, H. Efendic, M. Affenzeller, L. Del Re, and S. Wagner. On-line modeling based on genetic programming. *International Journal on Intelligent Systems Technologies and Applications*, 2(2/3):255–270, 2006.

[Wei06] W. S. Wei. *Time Series Analysis – Univariate and Multivariate Methods*. Addison-Wesley, 2006.

[Wen95] O. Wendt. *Tourenplanung durch Einsatz naturanaloger Verfahren*. Deutscher Universitätsverlag, 1995.

[WER06] S. Winkler, H. Efendic, and L. Del Re. Quality pre-assessment in steel industry using data based estimators. In S. Cierpisz, K. Miskiewicz, and A. Heyduk, editors, *Proceedings of the IFAC Workshop MMM'2006 on Automation in Mining, Mineral and Metal Industry*. International Federation for Automatic Control, 2006.

[WF05] I. H. Witten and E. Frank. *Data Mining: Practical Machine Learning Tools and Techniques*. Morgan Kaufmann, San Francisco, 2005.

[WH87] P. H. Winston and B. K. P. Horn. *LISP*. Addison Wesley, 1987.

[Whi93] D. Whitley, editor. *Foundations of Genetic Algorithms*, volume 2. Morgan Kaufmann Publishers, 1993.

[Whi95] P. A. Whigham. A schema theorem for context-free grammars. In *1995 IEEE Conference on Evolutionary Computation*, volume 1, pages 178–181, Perth, Australia, 29 November - 1 December 1995. IEEE Press.

[Whi96a] P. A. Whigham. *Grammatical Bias for Evolutionary Learning*. PhD thesis, School of Computer Science, University College, University of New South Wales, Australian Defence Force Academy, Canberra, Australia, 14 October 1996.

[Whi96b] P. A. Whigham. Search bias, language bias, and genetic programming. In John R. Koza, David E. Goldberg, David B. Fogel, and Rick L. Riolo, editors, *Genetic Programming 1996: Proceedings of the First Annual Conference*, pages 230–237, Stanford University, CA, USA, 28–31 July 1996. MIT Press.

[WHMS03] J. Wen-Hua, D. Madigan, and S. L. Scott. On bayesian learning of sparse classifiers. Technical Report 2003-08, Avaya Labs Research, 2003.

[Wig60] E. P. Wigner. The unreasonable effectiveness of mathematics in the natural sciences. In *Communications on Pure and Applied Mathmatics*, volume XIII, pages 1–14. John Wiley & Sons, Inc, New York, 1960.

[Win04] S. Winkler. Identifying nonlinear model structures using genetic programming. Master's thesis, Johannes Kepler University, Linz, Austria, 2004.

[Win08] S. Winkler. *Evolutionary System Identification - Modern Concepts and Practical Applications*. PhD thesis, Institute for Formal Models and Verification, Johannes Kepler University Linz, 2008.

[WK90] S. M. Weiss and I. Kapouleas. An empirical comparison of pattern recognition, neural nets, and machine learning classification methods. In J. W. Shavlik and T. G. Dietterich, editors, *Readings in Machine Learning*, pages 177–183. Kaufmann, San Mateo, CA, 1990.

[WL96] A. S. Wu and R. K. Lindsay. A survey of intron research in genetics. In H.-M. Voigt, W. Ebeling, I. Rechenberg, and H.-P. Schwefel, editors, *Parallel Problem Solving From Nature IV. Proceedings of the International Conference on Evolutionary*

Computation, volume 1141 of *LNCS*, pages 101–110, Berlin, Germany, 22-26 September 1996. Springer-Verlag.

[Wri43] S. Wright. Isolation by distance. *Genetics*, 28:114–138, 1943.

[WSF89] D. Whitley, T. Starkweather, and D. Fuguay. Scheduling problems and traveling salesman: The genetic edge recombination operator. *Proceedings of the Third International Conference on Genetic Algorithms and Their Applications*, pages 133–140, 1989.

[WWB+07] S. Wagner, S. Winkler, R. Braune, G. Kronberger, A. Beham, and M. Affenzeller. Benefits of plugin-based heuristic optimization software systems. 2007.

[YA94] Y. Yoshida and N. Adachi. A diploid genetic algorithm for preserving population diversity - pseudo-meiosis GA. In *Lecture Notes in Computer Science*, volume 866, pages 36–45. Springer, 1994.

[YN97] T. Yamada and R. Nakano. Genetic algorithms for job-shop scheduling problems, 1997.

[Zha97] B.-T. Zhang. A taxonomy of control schemes for genetic code growth. Position paper at the Workshop on Evolutionary Computation with Variable Size Representation at ICGA-97, 20 July 1997.

[Zha00] B.-T. Zhang. Bayesian methods for efficient genetic programming. *Genetic Programming and Evolvable Machines*, 1(3):217–242, July 2000.

[Zhu00] K. Q. Zhu. A new genetic algorithm for VRPTW. In *Proceedings of the International Conference on Artificial Intelligence*, 2000.

[ZM96] B.-T. Zhang and H. Mühlenbein. Adaptive fitness functions for dynamic growing/pruning of program trees. In P. J. Angeline and K. E. Kinnear, Jr., editors, *Advances in Genetic Programming 2*, chapter 12, pages 241–256. MIT Press, Cambridge, MA, USA, 1996.

[Zwe93] M. H. Zweig. Receiver-operating characteristic (ROC) plots: A fundamental evaluation tool in clinical medicine. *Clinical Chemistry*, 39:561–577, 1993.

Index

λ-interchange, 148, 152
k-change methods, *see* k-opt
k-opt, 127, 129
k-optimal, 129
I1 heuristic, 145, 146
2-change, *see* 2-opt
2-opt, 128, 138, 148, 153
2-opt*, 153
2-optimal, 128
3-opt, 129, 138
3-optimal, 129

actual population size, 71, 75, 76
actual selection pressure, 69, 70, 85
adaptive mutation schemes, 9
ADF, *see* genetic programming
adjacency representation, 12, **130**, 132
allele, 5, 10, 11
alternating edges crossover, 12, **131**, 132
artificial ant, 44
artificial evolution, 67
artificial neural network, 26, 47, 276
assignment problem, 125
asynchronous migration, 21

binary representation, 3, 4, 7, 10, **11**, 14
biological evolution, 3
bitwise mutation, 8
bloat, 60, 61, 306, 307, 309, 312, 313, 315, 317, 318
Box-Jenkins approach, 235
building block, 10, 14, 15

building block hypothesis, 10, 16, 76, 97, 102

canonical genetic algorithm, 3, 4, 10, 14
capacitated vehicle routing problem (CVRP), *see* vehicle routing problem
capacitated vehicle routing problem with time windows (VRPTW), *see* vehicle routing problem
capacity constraint, 142
cellular genetic algorithm, 148
chromosome, 5, 28
classification, 251
    classification rate, 252
    classifier evaluation, 265
    dynamic range selection, 266
    graphical classifier analysis, 270
    medical data analysis, 263, 272, 278
    using PG, 251
cluster first route second method, 147
coarse-grained parallel GA, 17, **18**, 21, 79, 214
coefficient of determination function, 182
column generation method, 143
combinatorial optimization, 13, 121, 122
communication topology, 20
comparison factor, 70
confusion matrix, 252
construction heuristic, 126, 143
continuous crossover, 13

crossover, **2**, 5, 11, 21, 31, 65, 76
  alternating edges crossover,
    12, **131**, 132
  continuous crossover, 13
  cyclic crossover, 105, **135**
  discrete crossover, 13
  edge recombination crossover,
    100, **136**, 215
  enhanced edge recombination
    crossover, 138
  heuristic crossover, 132
  maximal preservative
    crossover, 100, 215
  multiple point crossover, 7
  order crossover, 99, **134**, 214
  partially matched crossover,
    105, **133**, 151
  route-based crossover, **150**,
    224, 229
  sequence-based crossover,
    148, 224, 229
  single point crossover, 7
  structure tree crossover, 31
  subtour chunks crossover, 131
  uniform crossover, 7
crowding, 66
cut-inversion mutation, *see*
    inversion mutation
cutting-plane method, 125
cyclic crossover (CX), 105, **135**

data-based modeling, **157**, 167,
    188, 189, 191–194, 235,
    251
  a priori knowledge, 168
  artificial neural network, 276
  autoregressive modeling, 172
  classification, 251
    evaluation, 258, 259
    medical data analysis, 263,
      272, 278
    using GP, 251
  data preprocessing, 167
  functions basis, 173
  implementation in HL, 170

  k-nearest neighbor
    classification, 277
  linear modeling, 275
  overfitting, 61, **158**, 306, 312,
    313, 318
  parameter optimization, 188,
    189
  polynomial models, 159
  pruning, 191–194
  static vs. dynamic modeling,
    171
  support vector machine, 278
  time series analysis, 235
    evaluation, 236
  training, validation and test
    data, 171
  using genetic programming,
    169, 170
  virtual sensor, 236
defining length, 15
definite neighborhood, 127
delete-n replacement, 10
delete-n-last replacement, 10
deme, *see* subpopulation
diesel engine emissions, 237
diffusion model, 19
diploid chromosomes, 5
discrete crossover, 13
distance constrained vehicle
    routing problem
    (DVRP), 139
distance matrix, 122
distance measure, 217
divide-and-conquer, 17
dynamic programming, 143
dynamic range selection, 266

edge map, 137
edge recombination crossover
    (ERX), 100, **136**, 215
elitism, 9, 222
enhanced edge recombination
    crossover (EERX), 138
essential genetic information, **67**,
    68, 97, 98, 104, 207

Euclidean distance, 123, 222
evaluation, 30
evolution strategy, 1, 9, 189
evolutionary algorithms, 69
evolutionary computation, 1
exact algorithm, 142
exact GA schema theory, 16
exact GP schema theory, 54
exchange mutation, **138**, 148, 152
exhaust gas recirculation, 239

Federal Test Procedure, 239
fine-grained parallel GA, 17, **19**
fitness, 5
fitness function, 6, 67
fitness value, 6, 15, 70
fitness-sharing, 66
forma theory, 16
function, 28, 174
    arity, 177
    evaluation, 174
    neutral elements, 177
    parametrization, 177
    string representation, 175
functions basis, 173

generational replacement, **9**, 73
genetic algorithm, 1, **2**, 69, 72,
    130, 147
genetic diversity, 22, **66**, 73,
    90–92, 100, 207, 219,
    292, 293, 300, 306, 307,
    309, 312, 313, 315, 317,
    318
genetic drift, 79–82
genetic programming, 1, 25, **25**
    applications, 43
    automatically defined
        function (ADF), 35
    basic steps, 37
    bibliography, 62, 63
    bloat, 60, 61, 306, 307, 309,
        312, 313, 315, 317, 318
    challenges, 59

    chromosome representation,
        28, 178
    evaluation, 30
    genetic operators, 31
    history, 26
    hyperschema, 56, 58
    linear genetic programming,
        36, 37
    macroscopic schema theorem,
        58
    schema, 50–52, 54
    schema theories, 50–52, 54
    structure tree, 178
genetic propagation, 89, 285, 286,
    288–290
genome, 5
genotype, 5
genotypical identity, 75
giant tour heuristic, 147
global convergence, 84, 210
global parallel GA, 18
graphical genetic programming,
    37

Hamiltonian cycle, 122
haploid chromosomes, 5
heuristic crossover, 132
HeuristicLab, 170, 173
higher order representation, 8
hybrid parallel GA, 20
hyperschema, 56, 58

improvement heuristic, 126, 143
insertion mutation, **138**, 148
integer programming, 124
inversion mutation, **139**, 148
island, *see* island model
island GA, *see* coarse-grained
    parallel GA
island model, 17, 79, 82, 214

job shop scheduling problem, 13

k-nearest neighbor classification,
    277

Lagrange relaxation-based
method, 143
Levenshtein distance, 200
Lin-Kerninghan algorithm, 121
linear correlation, 239, 246, 248
linear genetic programming, 36
linear modeling, 47, 275
linear regression, 47, 275
linear-rank selection, 6
LISP, 28
local adaption, 188
local premature convergence, 21,
79, 80, 83–86
local search, 127, 155, 222
local search (LSM), 154
Łukasiewicz notation, 29

master–slave parallel GA, *see*
global parallel GA
MATLAB®, 165
maximal preservative crossover
(MPX), 100, 215
maximum effort, 75
maximum selection pressure, 71,
84, 210
mean squared error function, 49,
158, 182
melanoma data set, 263, 279
migration, 18, **20**, 21, 79, 82
multiple depots vehicle routing
problem (MDVRP), 139
multiple point crossover, 7
multiple traveling salesman
problem (MTSP), 130
multiplexer, 43
mutation, **2**, 8, 11, 21, 31, 101,
138
2-opt mutation, 128, 148, 153
2-opt* mutation, 153
3-opt mutation, 129
bitwise mutation, 8
exchange mutation, **138**, 148,
152
insertion mutation, **138**, 148
inversion mutation, **139**, 148

local search mutation, 154
one level exchange mutation,
154
or-opt mutation, 154
real-valued mutation, 13
relocate mutation, 148, 152
shift mutation, 9
simple inversion mutation,
138, 209, 211, 215
structure tree mutation, 31
two level exchange mutation,
154

natural evolution, *see* biological
evolution
nearest neighbor algorithm, 126
neighborhood, 65
neighborhood search, *see* local
search
neighborhood structure, 127
noise, 158

objective function, 6
Occam's razor, 60, 159
offspring selection, 14, 68, 69, **70**,
71–73, 77–79, 103, 186,
208
one level exchange (M1), **154**,
222, 231
one point crossover, *see* single
point crossover
or-opt, 154
order crossover (OX), 99, **134**, 214
order of schema, 15
ordinal representation, 132
overfitting, 61, **158**, 306, 312, 313,
318

panmictic population, 18, 80
parameter optimization, 188, 189
parent analysis, 89, 285, 286,
288–290
parent selection, 2, 6, 78, 228
partially matched crossover
(PMX), 105, **133**, 151

partitioning heuristic, 127
path representation, 12, **133**, 134, 148, 221
PDGP, 37
Polish notation, *see* Lukasiewicz notation
population, 25–27, 36, 38, 39
population diversity, *see* genetic diversity
population dynamics, 89
    genetic diversity, 90–92, 292, 293, 300, 306, 307, 309, 312, 313, 315, 317, 318
    genetic propagation, 89, 285, 286, 288–290
    parent analysis, 89, 285, 286, 288, 289
    parents analysis, 290
population size adjustment, 73, 113
population-based heuristic, 65
position-based mutation, *see* insertion mutation
postman problem, 123
pre-selection, 66
prefix notation, 29
premature convergence, 3, 9, 66, 67, **71**, 79, 83
premature stagnation, *see* premature convergence
problem representation, *see* representation
program component GP schema, 51
proportional selection, *see* roulette wheel selection
pruning, 191–194, 306, 307, 309, 312, 313, 315, 317, 318
push forward insertion heuristic, **145**, 146

quality, *see* fitness value

R^2 function, *see* coefficient of determination function

random selection, 74
RAPGA, 68, 69, **73**, 75, 77, 78, 113
real-valued benchmark test functions, 216
real-valued mutation, 13
real-valued representation, 13
receiver operating characteristic analysis, 254, 270
    AUC, 254
    MROC analysis, 256
    ROC curve, 254, 270
recombination, *see* crossover
relocate mutation, 148, 152
repair operator, 131, 147
replacement, 9
representation, 8, 11, 77, 147
reunification, 83, 84
rooted tree GP schema, 52
roulette wheel selection, **6**, 74, 222
route planning problem, 121
route-based crossover (RBX), **150**, 224, 229
runtime consumption, 185

Santa Fe trail, 45
SASEGASA, 17, 68, **79**, 83, 210
savings heuristic, 143, 145
scalable selection pressure, 74
schema, 14, 50–52, 54
schema theorem, 10, 16, 76, 102
schemata analysis, 132
SEGA, 82–84
selection, 2, 6, 78
    offspring selection, 14, 68, 69, **70**, 71–73, 77–79, 103, 186, 208
    parent selection, 2, 6, 78, 228
selection pressure, 7, 19, 21, 78, 79, 81, 100
sequence-based crossover (SBX), **148**, 224, 229
shift mutation, 9
similarity value, 217

simple inversion mutation, **138**, 209, 211, 215
simulated annealing, 11, 70
single machine scheduling, 140
single point crossover, 7
soft time window model, 139
Solomon heuristic, *see* push forward insertion heuristic
solution similarity, 92, 198
    structure tree solution similarity, 198, 200, 202
    TSP solution similarity, 92
split delivery vehicle routing problem (SDVRP), 139
standard genetic algorithm (SGA), 3, 6, 98
strict offspring selection, 106
structural identity, *see* genotypical identity
structure tree, **28**, 179, 194, 195, 198, 200, 202
    adjusted evaluation, 184
    combined evaluation, 183
    crossover, 180
    evaluation, 30, 181, 183–185, 236, 258, 259
      early stopping, 185
    genetic item, 202
    hybrid structure, 264
    initialization, 179
    mutation, 181
    pruning, 194
      ES-inspired pruning, 195
      exhaustive pruning, 195
    similarity, 198, 200, 202
      evaluation-based similarity, 198
      structural similarity, 200
structure tree crossover
    crossover, 31
    mutation, 31
subpopulation, 20, 79–82, 207, 212, 220
subtour chunks crossover, 131

success ratio, 70, 71
success rule, 70
successive method, 144
support vector machine, 278
swap mutation, *see* exchange mutation
sweep heuristic, 144, 146
symbolic expression, 29
symbolic regression, 46, 157
synchronous migration, 21
system identification, 157, 166, **166**, 167
    basic steps, 166
    parameter identification, 167
    structural identification, 166

tabu search, 11
terminal, 28, 173
    constant, 173
    differential, 173
    evaluation, 173
    parametrization, 176
    string representation, 175
    variable, 173
thyroid data set, 263, 281
time dependent traveling salesman problem (TDTSP), 124
time series analysis, 235, 237, 242
    Box-Jenkins approach, 235
total enumeration, 124
tournament replacement, 10
tournament selection, 7, 222
trajectory-based heuristic, 65, 224
transmission probability, 16
traveling salesman problem (TSP), 76, 97, **121**, 208
traveling salesman problem with time windows (TSPTW), 124
traveling salesman subtour problem (TSSP), 123
triangle inequality, 123, 130
trip delimiter representation, 147, 148

two level exchange (M2), **154**,
    222, 231

uniform crossover, 7

variables selection, 167
    backward search, 168
    exhaustive search, 167
    sequential search, 168
variance accounted for function,
    182
vehicle routing problem (VRP),
    13, 121, **139**, **141**, 147,
    221
village-town-city model, 80
virtual sensor, 236
    diesel engine emissions, 237,
    242
    NO_x, 238, 243, 246
    soot, 242

Wisconsin data set, 263, 278